Applied Innovation and Technology Management

Technology is not just limited to technology companies. Managing innovation and technology is no longer a luxury and needs to be understood by all sectors around the world and by both technical and non-technical managers. This book series explores existing and emerging technologies that address current challenges within innovation and technology managements. Each title is developed to provide a set of frameworks, tools and methods that can be adopted by researchers, managers and student in engineering, innovation and technology fields. Research, policy and practice-based books in the series cover topics such as roadmapping, portfolio management, technology forecasting, R&D management, health technologies, bio technologies, transportation management, smart cities, and open innovation, among many others.

Gülgün Kayakutlu • M. Özgür Kayalica
Editors

Decision Making Using AI in Energy and Sustainability

Methods and Models for Policy and Practice

Springer

Editors
Gülgün Kayakutlu
Energy Institute
Istanbul Technical University
Istanbul, Türkiye

M. Özgür Kayalica
Energy Institute
Istanbul Technical University
Istanbul, Türkiye

Faculty of Management
Istanbul Technical University
Istanbul, Türkiye

ISSN 2662-9402 ISSN 2662-9410 (electronic)
Applied Innovation and Technology Management
ISBN 978-3-031-38386-1 ISBN 978-3-031-38387-8 (eBook)
https://doi.org/10.1007/978-3-031-38387-8

This Springer imprint is published by the registered company Springer Nature Switzerland AG
The registered company address is: Gewerbestrasse 11, 6330 Cham, Switzerland

Paper in this product is recyclable.

Foreword

During recent years, we have seen access to energy resources become a major factor in determining the winners of wars, oil producers banding together to create new global alliances, and the cost of energy undermining the national security of many countries. Clearly, energy and geopolitics have always been interconnected. Decision makers find themselves in difficult situations where many energy problems are very difficult to overcome. In addition, the energy transition and the race for technological dominance create more challenges.

Artificial intelligence (AI) has a huge impact on science and technology. In particular, AI in energy and sustainability. For example, AI can predict the demand and supply of renewable energy, optimize efficiency in energy systems, and improve the management of natural energy resources. IFIP TG12 WG 11 "AI in Energy and Sustainability" contributes to this field with a new edited book.

The book focuses on innovative applications of AI that address the critical challenges in energy systems, climate change, energy efficiency and sustainability policies, sustainable transportation and smart cities, as well as other important pressing issues in decisions making using energy research.

The book is organized into six parts. The first part includes chapters that deal with climate change and sustainability policies. The second part deals with climate change mitigations. Sustainable energy markets are the focus of the third part. Research in this area is very timely since we are witnessing the rise of electricity, oil, and gas prices, in many parts of the world. The fourth part of the book deals with energy efficiency and the fifth part with smart cities. Any discussion on "smart cities" does not involve only technologies, but also the engagement of decision makers and citizens. The final part deals with different trends in the near future in selected topics such as applications in agriculture, microgrids, and electrical mobility.

The book is organized very carefully and the material covers a spectrum of issues. The book will be a valuable material for energy researchers and decision makers. The editors have done an excellent job in producing this book.

Distinguished Emeritus Professor
University of Florida
Gainesville, FL, USA

Panos M. Pardalos

Preface

This book is the result of the hard work IFIP TG12 WG 11 "AI in Energy and Sustainability" has performed. Both editors and the majority of the authors are members of this working group, who design and develop new models and algorithms for AI applications in the energy and sustainability fields of research. We have brought a wide range of work together focused on decision making in the field. The reader will find AI use for decision making in climate change, sustainability, energy efficiency, use of new technologies, electrical vehicles, organic agriculture implementations and more in this book. Most of the papers in this book have evolved since 2020 and the authors had the opportunity to update their work with fine-tuning until November 2022. We would like to thank all the contributors for sharing their precious research work with us. We are also grateful to have the Foreword of our book written by Professor Panos Pardalos, one of the globally known energy optimisation researchers. We strongly believe that decision makers in the energy and sustainability field will make great use of the applications and results shared in our book *Decision Making Using AI in Energy and Sustainability* that will be printed as part of the Springer Nature series.

Istanbul, Türkiye

Gülgün Kayakutlu
M. Özgür Kayalıca

Contents

Part I
Sustainability Policies

Chapter 1
Climate Change – Can AI Help Understanding and More Effective Facing of Various Interrelated Impacts?

Eunika Mercier-Laurent

Abstract Climate change is an extremely complex problem involving quite all disciplines. Today's main discussions are on reducing carbon emissions. Related actions are only "quick fixes" generating weak effects on a quickly evolving global process. Energy, water and transportation optimization are possible using adequate techniques of Artificial Intelligence (AI). AI can help reduce the multiple impacts on the condition of understanding this phenomenon as a whole and related dynamics. Inspired by the AI problem-solving approach, knowledge-based AI allows defining a set of factors affecting climate change and their interactions with each other. This deep understanding is vital for the selection of suited techniques to address them simultaneously using a system and holistic approach. Inspired by Strategic Knowledge Management, an assessment for evaluation of the current situation and measuring the progress is proposed and prototyped.

Keywords Artificial Intelligence · Digital transformation · Method · Sustainability · Sustainable development · Planet crisis · Climate change · Data obesity

1.1 Introduction

According to NASA [1] and some other sources, we entered into Anthropocene Epoch. Human activity has a significant impact on the planet's climate and ecosystems [2]. World Economic Forum addresses also this topic [3].

From the beginning of intensive industrialization in the early 1960s alerts have been raised by among others Antonina Lenkowa [4] and Erik Eckholm [5] trying to awake governments and industries about the various and heavy consequences of such intensive industrialization.

E. Mercier-Laurent (✉)
University of Reims Champagne Ardenne, Reims, France
e-mail: eunika.mercier-laurent@univ-reims.fr

G. Kayakutlu, M. Ö. Kayalica (eds.), *Decision Making Using AI in Energy and Sustainability*, Applied Innovation and Technology Management,
https://doi.org/10.1007/978-3-031-38387-8_1

However, we can observe the lack of birth control in developing countries. The "business first" attitude and related globalization amplified by technological progress lead to intensive production and transportation of goods around the globe just for sale. Investing in housing is very lucrative, but intensive destruction of nature just for high revenue produces disastrous consequences.

Despite introducing Corporate Social Responsibility, many companies still practice planned obsolescence [6] and shrinkflation to generate more revenues [7].

Is getting huge and quick revenue compatible with planet protection?

Technological progress facilitates communication, and contributes to growing sales online, but also generates a lot of visible and hidden waste [8]. Analytics and some AI techniques are used by marketing with the aim to sell more to more people. Collecting "big data" is fundamental for this activity. Ubiquitous and not targeted advertisements, SMS and emails pushing to buy generate unnecessarily carbon emissions [32]. Shopping online has become a world trend and influencer on social networks has been recognized as a new profession.

Many organizations such as NASA, European Space Agency (ESA), World Economic Forum (WEF), and some countries are concerned with climate change but consider only some factors. Numerous reports elaborated by the World Health Organization (WHO) [9], OECD [10], and GIEC [11] and conferences alert again about the state of our planet Earth. The United Nations defined the 17 goals of Sustainable Development but did not provided the recipe for how to achieve them. Intergovernmental Panel on Climate Change (IPCC), the United Nation Body has been created to provide policymakers with regular scientific assessments on climate change, its implications and potential future risks, as well as to put forward adaptation and mitigation options. It is composed of several working groups including one multidisciplinary. Through its assessments, the IPCC determines the state of knowledge on climate change. It identifies where there is an agreement in the scientific community on topics related to climate change, and where further research is needed [14]. The involved stakeholders focus mainly on carbon dioxide generation, but it is only one factor. Main conferences devoted to climate change blame coal and the automotive industry as the biggest emitters of carbon dioxide. They seem to ignore the other influencing factors such as globalization and miss to study the pollution systems as a whole.

UNESCO elaborated a Strategy for Climate Actions [12]. It is based on four pillars: education, scientific cooperation, culture, and social inclusion. Their video [13] mentions the need for a knowledge-based leader. How many such leaders do we have? As many, they produce reports that few read. Those of UNESCO contain 490 pages.

ADEME presents the main factors on only three pages [15].

The participants of the recent COP26 in Glasgow aim in accelerating action toward the goals of the Paris Agreement and the UN Framework Convention on Climate Change [16].

European initiatives EIT [17] include the Knowledge and Innovation Community (KIC), working to accelerate the transition to a zero-carbon economy. Supported by the European Institute of Innovation and Technology, EIT Climate-KIC identifies and supports innovation that helps society mitigate and adapt to climate change. EIT: "Climate-KIC believes that a decarbonized, sustainable economy is not only necessary to prevent catastrophic climate change but presents a wealth of opportunities for business and society." Other EIT communities work on Digital, Food, Health, Energy, Manufacturing, and Raw materials, but separately. A typical example of such separation is the program Horizon Europe addressing two interrelated topics Digital Transformation and Green Deal separately; submitted projects are evaluated by experts in related fields which limits the opportunity for connecting these two related fields [43]. However, the recent posts on Twitter show the emerging synergy between the two programs [18].

Despite all these initiatives, discussions, agreements, and recommendations, we still have to face global warming. The increasing frequency and strength of natural disasters is a complex consequence of decisions taken at governmental and other levels and of human behaviors as well.

In reality, the addressed problem of climate change is very complex and requires a deep understanding of all aspects before trying to improve the planet condition with adequate actions. The KPI (Key Progress Indicators) should be defined to measure the impact of such actions.

Paradoxically, the governments and worldwide organizations spent money on conferences and reports, but the investment in impactful actions is still very weak [19]. From the research point of view, there is a barrier to evaluating and publishing papers – multidisciplinary papers are weakly considered by reviewers specialized for the most in one area.

1.2 Complexity of Climate Change: Comprehension of the Influencing Factors

Improving the planet's condition requires bringing together specialists from various related fields, sharing knowledge and best practices at the global level, and the appropriate policies, funding, and incentives. Reducing the footprint cannot be achieved without awakening companies and people about sustainable business, "be happy with few" attitude, reusing instead of recycling and repairing (opposite of planned obsolescence), smart recycling, and applying environmental principles to design [20, 21].

Most human-made things and actions impact the environment because they are not conceived in harmony with nature. Such an approach requires understanding, using or inventing alternative solutions, and verifying the possible impacts before implementation [8].

1.2.1 Influencing Factors

Among the main influencing factors are greedy business, industry, technology, and human behaviors.

The intensive business produces waste with design just for sale, planned obsolescence, and shrinkflation.

The healthcare industry is probably among the most polluting [44].

The growing population raises the need for housing. This fact is explored also by investors, destroying natural spaces to build as many houses as possible.

Feeding the world population requires the right effort of both farmers and consumers without overproduction leading to food waste. Direct sales from producer to consumer avoid massive production just to pay intermediates. Agro-ecology is already practiced [22] and should be extended.

Human behaviors such as intensive shopping activity, choice of the mean of transportation, food waste, and excessive use of technology, poor thinking, lack of considering alternative solutions.

There are more, but we focus here on globalization effects and the impact of innovation and technology.

1.2.1.1 Globalization

Focus on profit the "World Companies" have changed the economic landscape. Opened to business and to quick development of China and other Asian and South American countries offering low labor cost increases the relocation-out from origin countries, mainly the USA and Europe, in search of quick and high revenues. Such business depends of transportation around the globe, which is mostly not optimized across the world by ships and airplanes, which increases pollution [23]. Besides, the end-buyers have to recycle products, often of poor quality, made just for sale. As product designers integrate embedded planned obsolescence, repairing such products costs often more than new products and requires the availability of spare pieces and related tools and expertise.

Huge global business amplified through communication technology, e-commerce, and transportation facilities. Each day of the recent Suez Canal blockage disrupted more than $9 billion worth of goods, according to *The Associated Press*, citing estimates from Lloyd's List. A large number of ships stuck in one place – at the northern end of the Suez Canal in the Mediterranean Sea – caused sulfur emissions to spike in the region [24]. The route through the North Pole will increase ice melting.

The knowledge of adequate techniques of AI and expertise in applying them can help find and optimize the local resources with the aim of avoiding or minimizing transportation of many goods and enhancing the local economy. AI is also useful for optimizing logistics. The Covid pandemic demonstrated clearly the need for relocation, especially for electronic components in Europe.

1.2.1.2 Innovation and Technology

Despite the provided benefits, the innovation contributes to planet disaster, because the inventors and designers think mostly about the functionalities, shape, look, attractiveness, and few about minimizing the use of raw materials, reducing the energy consumption, providing the remanufacturing and recycling facilities and other ways to reduce environmental impacts. "Innovation" in packaging by reducing the quantity of a given product (food, ink) in a more attractive box produces more waste.

The primary eco-innovation movement claims to generate new businesses and jobs, but the use of technology in related activities is moderate, except maybe for waste sorting [25]. Instead of sorting, it is smarter to produce less waste. The standards associated with product lifecycle management (PLM) software claim to manage the environmental impact of the design [20]. Autodesk is probably among the pioneers of industrial convergence including constraints in design [26]. However, they do not consider yet the environmental and recycling constraints.

Technological progress, considered a powerful engine of economy, produces a twofold effect. It provides benefits for humanity in many fields but contributes also to Planet Crisis with massive waste generation, energy consumption, the generation of heat by data centers situated mainly in the north, and depletion of scarce metals. The enthusiasm of the users for novelty and recently for AI lead to sometimes hidden waste.

Information technologies have provided a considerable help for humans in various contexts, such as administration, businesses, schools, research, health, cities, and many others. Everything is digital and connected is everywhere, even if it is not necessary. Thanks to IT, many activities went online during the Covid pandemic. This experience demonstrated that it is possible to reduce traveling and optimize between telework and work in an office. Although records from meeting and conferences are registered and stored contributing to data obesity.

Related to innovation, technology is able to enhance creativity (generator of ideas in [21]) and provide considerable help in eco-design, verifying constraints before implementation, optimizing design, reparability, and recyclability. However, the way of thinking compatible with such complex problem-solving should combine the global, holistic, and system approaches [8, 27]. For example, aerospace and automotive industries focus on lightening weight and reducing carbon footprint, while other related aspects such as route optimization for users are weakly considered. Current travel search engines work for their clients, not for travelers. The air transportation system "hub-and-spoke" in which local airports offer air transportation to a central airport where long-distance flights are available needs innovation to minimize the distance and related CO2 emission. The effort can be done through better collaboration between airlines. Three main alliances, Sky Team, Star Alliance and One World, do not accept combining the flight tickets between them for optimizing flight distance.

AI-based simulators can help check and optimize things. Such a system has been prototyped in the framework of the French national-funded project Convergence [20] that delivered a game-based guide for SMEs aiming in guiding them in applying ISO 26000 requirements related to eco-design. AI for optimizing the use of energy is described in [45].

Amazon claims to reduce carbon dioxide emissions by using technology in the whole cycle of sales [28]. Although, they also contribute to the C02 generation. However, total environmental balance is not published. Schneider elaborated the offer Smart city by optimizing the use of energy, water, smart transportation, and control of the whole [46].

Although, the majority of computers, smartphones, IoT, and other electronic devices are mostly not eco-designed, except Xiaomi Modular smartphone, and even integrate a principle of planned obsolescence. Consequently, they are thrown away and replaced by the latest models. Not conceived with a knowledge approach, the consecutive layers of software and frequent update for various reasons increase already existing "obesity" and generate a need for more powerful computers, smartphones, and other devices. The combination of various communicating software requires "up-to-date" hardware to run correctly. Software is so complex that it is impossible to remove non-used parts (e.g., Windows).

While web 2.0 brought useful services, the massive appropriation of social networks and various applications produce an exponential amount of data stored in data centers that need cooling. Fortunately, few apply circular energy to reduce environmental impact [29], but still, those in Scandinavian countries and North America clearly contribute to the rise of temperature and melting of ice [8, 45].

The third hype of AI has been mostly triggered by marketing needs to sale more and quicker to the largest public. The involved companies collect all available data with an aim to elaborate models and select algorithms to support the whole acceleration and amplification of business including "client experience." Covid-19 pandemic amplified also online shopping. . . .and the need for transportation.

The advertisement companies explore the navigation data and "client experience" with the aim of pushing not always targeted advertisements to the large population of Internet users. It generates not only a carbon footprint but also visual pollution and certainly influences concentration. According to [31, 32], simple email without attachments generates the equivalent of 4 g CO_2, an annual Google search of 10 kg by only one person.

Advertisement-based business model, invented by Google, influences and mass media promotes the "buy more" and "to have more and to show" mentality. "Buy more, throw away and buy new" are the engines of today's business. The demand for data scientists, especially in marketing, is increasing, which is good for employment. The general trend of "Big Data and machine learning" affected also the problem-solving capacity among the AI programmers to the point that many apply the Big Data + machine learning principle to solve all kinds of problems even if the alternative methods, greener and smarter, are available. Collection of data became a prerequisite. However, their quality should be adapted to the expected results without cheating aiming in influencing the results and producing the famous bias. Raw data need "cleaning" before processing.

In fact, the quality of data is not controlled – there is no "garbage collector" for outdated data, no verification of consistency, and no selection in function of the target. For example, many organizations collect our personal data that are in various databases instead of having just one (identity). By consequence the same data may be registered a dozen times, contributing again to data obesity.

Artificial Intelligence plays already a significant role in decision support systems, optimization, and simulation, but can do better at the condition to consider the mentioned elements as an ecosystem. It requires a capacity for different thinking without the barriers between fields [33], complex problem-solving, the art of choosing appropriate techniques, and managing available knowledge and experience that apply. AI techniques help understand the satellite images of burning forest, melting ice, urbanization, and others related to climate change.

Various models have been also elaborated for a better understanding of various components and their interrelations.

1.2.2 Models, Components, Relations

The climate portal of MIT [34] provides some explanations the state of current knowledge, limits, of course, authors' knowledge of the topic and explains what can be done. They elaborated a simulator exploring seven physical principles:

The dynamics of the climate system are governed by seven physical principles:

- Conservation of air mass
- Conservation of water mass
- Conservation of energy
- Conservation of momentum of air in three directions
- The ideal gas law applied to air

Climate models describe these principles as seven equations, which constrain seven variables:

- Air temperature
- Pressure
- Density
- Water vapor content
- Wind magnitude in three directions

By solving the equations, climate models can simulate all these variables in three dimensions and in time.

Other variables that affect the Earth's climate are hard to model directly. Clouds are a good example: a cloud is much smaller than the smallest unit of distance in a typical climate model, so the model cannot "see" individual clouds, but taken together they have big effects on the Earth's temperature. For these factors, climate

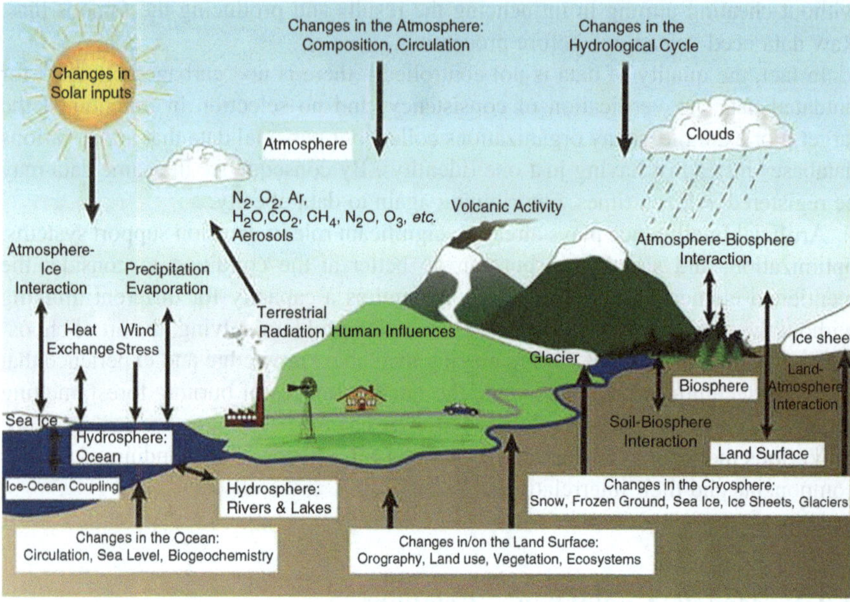

Fig. 1.1 A schematic view of many of the processes and interactions in the global climate system (From Le Treut et al., 2007). (Source NIVA [36])

models use "parameterizations," or simplified equations that behave roughly the same as the real thing. Rain, snow, and evaporation are other physical processes that have to be "parameterized" in climate models. These are important features of the Earth's climate, so getting the parameterizations right is a huge part of designing a good climate model. This model lacks in considering the impact of human activity in various contexts.

In the framework of project Clima, the researchers of Caltech are developing an "Earth system model that automatically learns from diverse data sources. Our model will exploit advances in machine learning and data assimilation to learn from observations and data generated on demand in targeted high-resolution simulations, for example, of clouds or ocean turbulence. This will allow us to reduce and quantify uncertainties in climate predictions" [35].

The National Science Center NIWA (New Zealand) focuses on Climate, Fresh Water, and Ocean Sciences [36]. Their model is presented in Fig.1.1.

Geophysical Fluid Dynamic Laboratory elaborated a global climate model (GCM) which is a complex mathematical representation of the major climate system components (atmosphere, land surface, ocean, and sea ice), and their interactions. Earth's energy balance between the four components is the key to long-term climate prediction. The main climate system components treated in a climate model are:

- The atmospheric component, which simulates clouds and aerosols, plays a large role in the transport of heat and water around the globe.

Fig. 1.2 Multiple interrelated eco-crises. (Source [37])

- The land surface component, which simulates surface characteristics such as vegetation, snow cover, soil water, rivers, and carbon storing.
- The ocean component, which simulates current movement and mixing, and biogeochemistry, since the ocean is the dominant reservoir of heat and carbon in the climate system.
- The sea ice component, which modulates solar radiation absorption and air-sea heat and water exchanges [48].

In 2010 Charles Savage [37] introduced multiple interrelated crises, as shown in Fig. 1.2.

He points out that all elements mentioned in the Fig. 1.3 are interrelated in Profit People Planet context.

Presented first to the students of German Business School, the question in the middle concerns all citizens in their respective contexts.

Are these models sufficient for a deep understanding of climate change at the point to help improve the Planet's conditions?

There is a need for the big picture, and a combination of system and holistic approaches for better understanding [27].

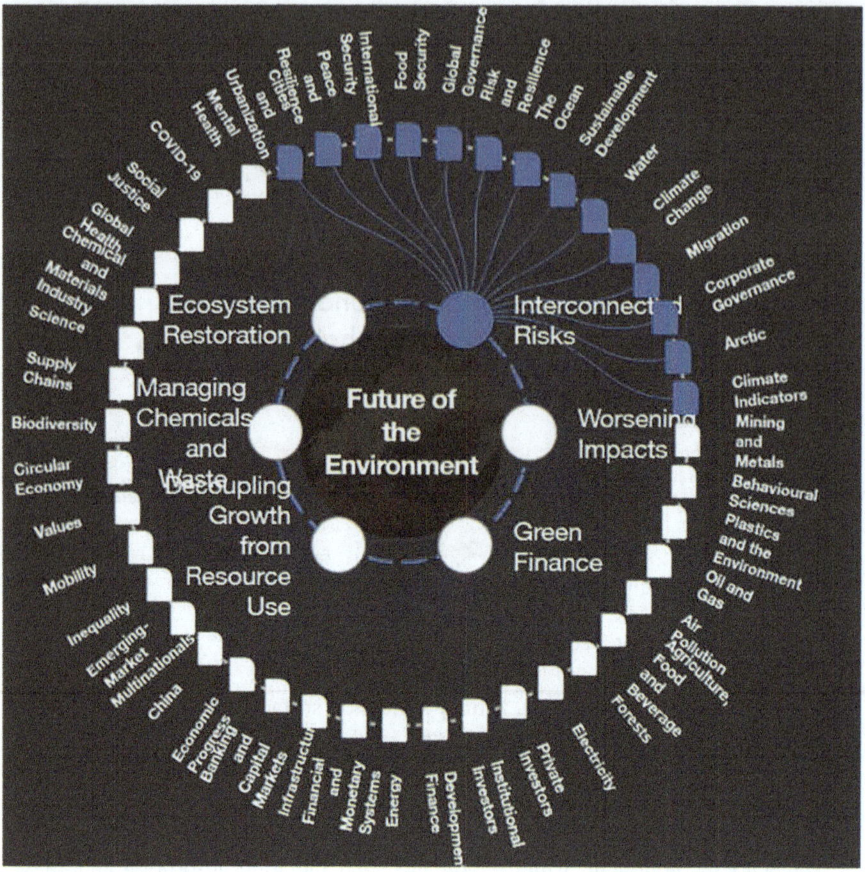

Fig. 1.3 Simulator of various impacts related to the Future of Environment. (Source [41])

1.3 Some Initiatives

The circular economy [38] suggests reusing and recycling; however, recycling uses energy and produces pollution.

Circular energy proposes to reuse, for example, the heat produced by data centers for heating schools, houses, and hotels in the north of Europe and Canada [45].

The initiatives as buy local or share knowledge on reparing via video are numerous. Repair cafés are present in a few cities for sharing expertise and tips [47]. We can already find some refurbished devices [49]. More incentives at the governmental level could improve this situation.

There are some efforts in eco-designing of products for easier repairing or smart recycling.

Do governments push for buying electric cars to reduce carbon emissions but is electricity less polluting?

Among the trends is smart buying and smart transportation, and smart and green programming as well. Technology and especially AI helps "greening and smarting" information technologies (IT) at the condition to apply the suited method (simple and effective).

Games and serious games (with AI) have also the potential to encourage people to think and act differently. For example, the game Nega-water developed by the association Du Flocon à la Vague calculates the use of water when doing daily activities [30].

Such initiatives deserve to be "copied."

However, there is still a need for evaluation and minimizing the impacts on Climate change.

1.4 Evaluation of Impacts

Since the official birth of AI in 1956 and before some approaches and techniques for problem-solving, such as General Problem solver (GPS), Knowledge Acquisition Design system (KADS) [39], Multi-agent Systems (MAS) [50], and Multi-strategy Machine Learning (MSML) [40] have been invented and applied in many fields.

Knowledge-based artificial intelligence focuses on problem-solving and provided some methods for deep problem understanding, such as FOREX KADS [39] aiming in first defining a goal, considering all related knowledge needed for given problem-solving and contextual knowledge.

In the case of climate change, the objective is to slow down or stop.

Context: growing population, distribution of population in cities, investors businesses, behaviors (shopping, ads, air conditioning...), technology.

Knowledge related to such problem-solving is ecosystems of climate in the solar system and the universe as well as interrelations between all elements.

After a deep understanding and elaboration of related knowledge models, ex ontologies, and reasoning models, adequate techniques have to be selected and combined for processing.

Simulation is also helpful to understand the effects of the condition to implement all related knowledge on a given topic.

1.4.1 Simulation

Well-conceived simulator including all possible knowledge is a powerful tool for understanding the various impacts. World Economic Forum proposes those shown in Fig. 1.3.

Clicking on one of five points shows the possible impacts. This simulator deserves to be largely used for better understanding of impacts, improving it by adding all additional knowledge and deciding about the actions.

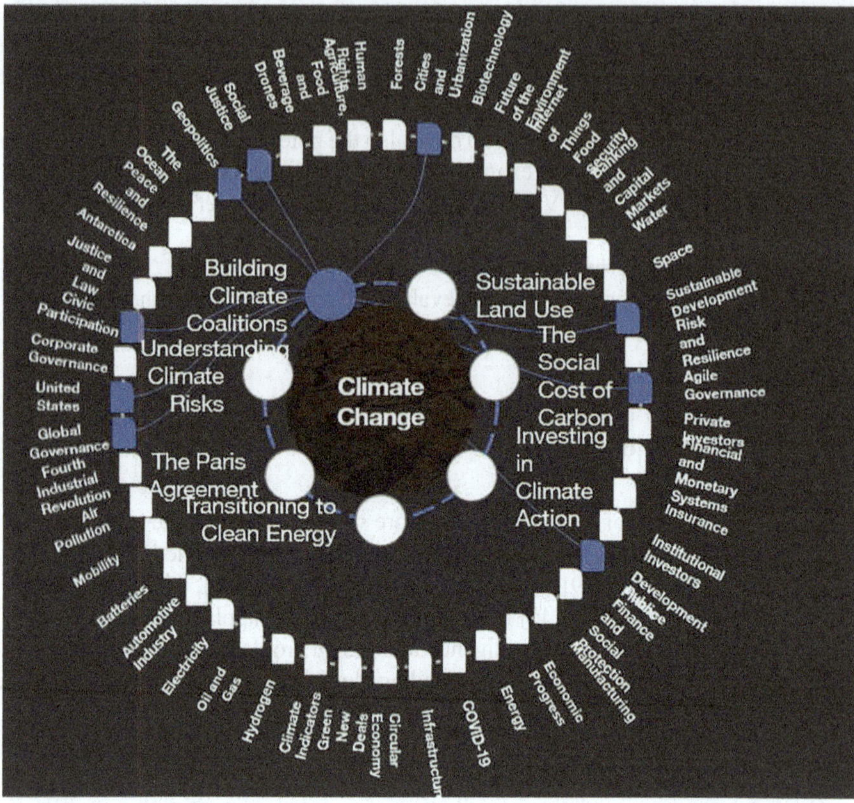

Fig. 1.4 Simulator of impacts Climate Change (impact of buildings). (Source [51])

WEF just published those shown in Fig. 1.4 on climate change [51], but we need to combine both for better comprehension.

Such a simulator can serve for defining the initial elements to assess and control the progress.

1.4.2 Assessment

In her book *The Innovation Strategy for the Knowledge Economy* [42], Debra Amidon proposed an assessment that can be applied for Strategic Knowledge Management and also for evaluation of the Innovative Capacity of a given organization (Fig. 1.5). This assessment practiced by the Entovation network of experts was enriched by adding among other environmental impacts [21].

Usually, the concerned people answer the questions related to each item and the radar chart expresses the current state and desired state. Each item represents a metric (KSI = Key Success Indicators), and all are interconnected. Analyzing the results

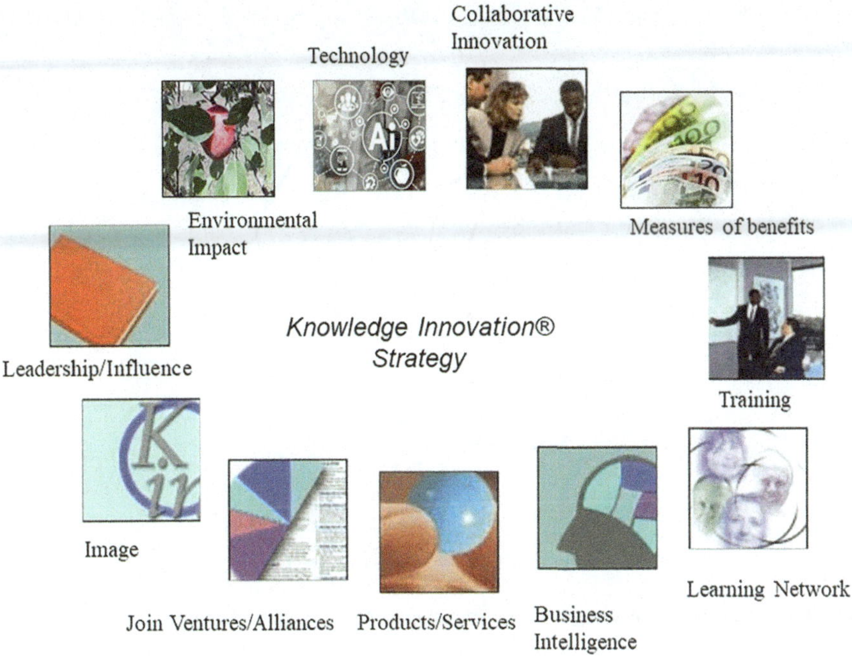

Fig. 1.5 Main elements for assessment of Strategic Knowledge Management or Innovative Capacity [21]

pointed out on the radar chart, the participants select the most important actions to achieve the desired state.

A similar but more complex assessment can be used for evaluating and managing the actions aimed at reducing our footprint. The axis should be carefully selected to reflect a given situation and objective. In our case, we want to minimize the impact on planet.

In a general manner, sustainability is about the integration of environmental health, social equity, and economic vitality in order to create thriving, healthy, diverse, and resilient communities for this generation and generations to come. The practice of sustainability recognizes that these complex issues are interconnected and require a systems approach. Indeed, sustainability depends on how human activities are in balance with natural ecosystems and how they affect biodiversity and our biosphere [8].

Figure 1.6 represents an example of a radar chart evaluating selected impacts.

All items are high (close to 10) and should be minimized

In the case of sustainability, the contextual knowledge of policies in a given country, the base of available actions, and practice/prototyping should be considered. Taking inspiration from games described in [52], we can imagine building a Sustainability Support System as a simulator containing various interrelated components to evaluate the impact of decisions for policymakers, planners,

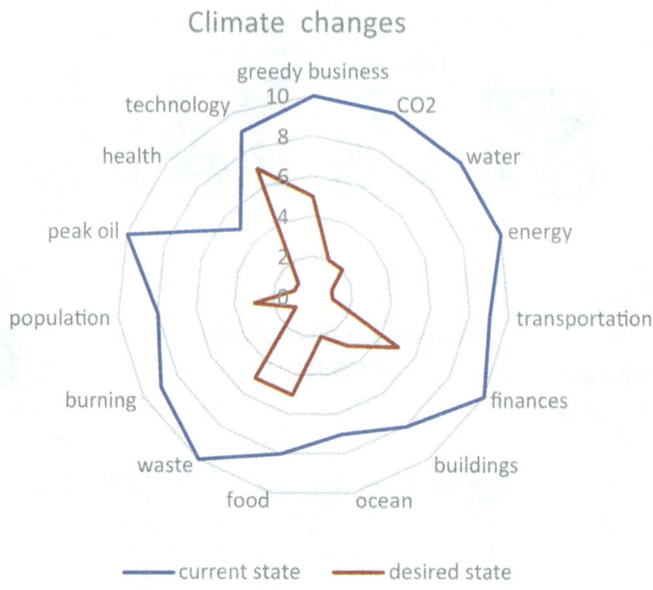

Fig. 1.6 Assessment of main system elements impacting the climate change

organizations, and companies. It can be conceived as a toolbox allowing easy configuration in function of the needs and allowing verifying various constraints before decision making.

1.5 Conclusion and Perspective

Planet Emergency, sustainability, and sustainable development are still big challenges for humanity, organizations, and businesses. This very complex problem requires a deep understanding of all system components and their interactions.

This chapter is probably the first to consider climate change with a knowledge-based AI approach. We tried to demonstrate that it is possible to gather main elements and understand the various impacts. To be honest, facing such a complex problem requires bringing together specialists from various fields to validate and enrich this work in a large project.

Amidst understanding and evaluating, the efforts have to be made on evolving human behaviors which is a tough task but worth trying to help preserve our Home and continue living in harmony with nature. The better use of our brain capacity is mandatory.

Instead of conferences and reports, we need to invest in various connected projects and measure the progress on regular bases.

To remain sustainable, IT development requires smart use of available AI techniques, not only trendy ones but also those effective for a given case. Combined with intensive use of available knowledge – individual and collective, from related domains, from the past, and currently gained from experience, such systems can really help minimize the impact of human activity.

AI way of thinking helps understand better such a complex problem, finding alternative solutions. AI is useful in many situations if we apply appropriate methods and techniques.

AI can help manage many aspects when used appropriately, but we still need to define the right metrics for the correct evaluation of the effects of our actions – individual and collective. Simulators can help evaluate the situation, testing the effect of influencing factors and consequences of decisions to take before doing. It applies in industry, city, and in our lives.

The balance between the use of technology and human capacity should be preserved. Technology producers have a tendency to produce software and devices that think instead of humans and take decisions for him/her. This kind of application may replace humans in the long run and reduce their cognitive capacity. That is why we have to design applications able to enhance human intelligence without switching it off.

Technology alone cannot save the planet but can help us to be smarter.

References

1. https://astrobiology.nasa.gov/news/the-anthropocene-humankind-as-a-turning-point-for-earth/
2. https://education.nationalgeographic.org/resource/anthropocene
3. https://www.weforum.org/agenda/2016/08/what-is-the-anthropocene-and-why-does-it-matter/
4. Lenkowa, A. (1961). Oskalpowana ziemia.
5. Eckholm, E. P. (1976). *Loosing losing ground: Environmental stress and world food prospect.*
6. Cueto, J. (2009, July). *Planned obsolescence – an economic and cultural anxiety*, source researchgate.net.
7. https://corporatefinanceinstitute.com/resources/knowledge/economics/shrinkflation
8. Mercier-Laurent, E. (2015). *The innovation biosphere –planet and brains in digital era*. Wiley.
9. WHO: Environment and Health https://www.euro.who.int/en/health-topics/environment-and-health/Climate-change/data-and-statistics. Last consulted 12/2021.
10. OECD: Work in support of Climate actions. (2019, Dec). https://www.oecd.org/environment/cc/OECD-work-in-support-of-climate-action.pdf
11. Changements climatiques: les éléments scientifiques du GIEC. (2021, Aug). https://www.ecologie.gouv.fr/sites/default/files/21144_GIEC-2.pdf
12. https://www.unesco.org/en/climate-change
13. https://www.youtube.com/watch?v=7ee3WSwPa_Y
14. IPCC 6th Assessment Report. (2021, March). https://www.ipcc.ch/report/sixth-assessment-report-working-group-3/.
15. https://librairie.ademe.fr/cadic/6748/exposition_changement_climatique_rac_ademe_web.pdf
16. https://unfccc.int/sites/default/files/resource/cp2021_12E.pdf
17. https://eit.europa.eu/our-communities/eit-climate-kic
18. Digital EU on Twitter. https://twitter.com/DigitalEU
19. https://www.youtube.com/watch?v=SA4_gGR3Vn4

20. Feng, Z., Rio, M., Allais, R., Zwolinski, P., Reyes, T., Roucoules, L., Mercier-Laurent, E., & Buclet, N. (2013). Toward a systemic navigation framework to integrate sustainable development into the company. *Journal of Cleaner Production, 54*, 199–214.
21. Mercier-Laurent, E. (2011). *Innovation ecosystems*. Wiley.
22. Côte, F. X., Bruno, R. B., Sourisseau, J. M., Francois, A. F., Andrieu, N., Bessou, C., Caron, P., Deguine, J. P., Faure, G., Hainzelin, E., Malezieux, E., Poirier-Magona, R. P., Scopel, E., Tixier, P., Toillier, A., & Perret, S. (2022). Levers for the agroecological transition of tropical agriculture. *Agronomy for Sustainable Development, 42*, 67. https://doi.org/10.1007/s13593-022-00799-z
23. Carbon footprint of global trade, OECD. (2015). https://www.itf-oecd.org/sites/default/files/docs/cop-pdf-06.pdf
24. https://www.dailymail.co.uk/sciencetech/article-9469541/Suez-canal-blockage-caused-spike-air-pollution.html. April 14, 2021.
25. Waste sorting. https://www.futura-sciences.com/tech/actualites/robotique-robot-dope-ia-ameliorer-tri-dechets-73184/
26. https://www.autodesk.com/industry/convergence. Last consulted 12/2021.
27. Mercier-Laurent, E. (2007). *Rôle de l'ordinateur dans le processus global de l'innovation à partir de connaissances, HDR*. Université Jean Moulin, Lyon.
28. https://sustainability.aboutamazon.com/environment/sustainable-operations/carbon-footprint. Last consulted 12/2021.
29. Capiluppi, A., & Baravalle, A. (2010, Oct). Matching demand and offer in on-line provision: A longitudinal study of monster.com, Web System Evolution.
30. Nega-water. https://waterfamily.org/water-academie
31. The carbon footprint of the Internet. https://www.climatecare.org/resources/news/infographic-carbon-footprint-internet/, April 2021.
32. https://communication-responsable.ademe.fr/sites/default/files/guide-ademe-tic-quels-impacts.pdf
33. Cavallucci, D., Cascini, G., Duflou, J., Livotov, P., & Vaneker, T. (2015). *TRIZ and knowledge-based innovation in science and industry*.
34. https://climate.mit.edu/explainers/climate-models
35. https://clima.caltech.edu/
36. https://niwa.co.nz/
37. Savage C. (2010). Grokking* the Multiple Interrelated Eco-Crises, Rev. 1.6 – 2010-01-11 – EuroNews: CommentVision - www.commentvisions.com/month/january/2010/visions_from#4
38. https://www.weforum.org/agenda/2022/06/what-is-the-circular-economy
39. Schreiber, G., Wielinga, B., & Breuker, J. (1993). *KADS – A principled approach to knowledge-based system development*. Academic Press.
40. Michalski, R. S., & Tecuci, G. (1994). *Machine learning: A multistrategy approach* (Vol. IV). Morgan and Kaufmann. ISBN-13: 978-1558602519.
41. https://intelligence.weforum.org/topics/a1Gb0000001hXikEAE?tab=publications
42. Amidon, D. (1997). *The innovation strategy for the knowledge economy*. Butterworth Heinemann.
43. https://ec.europa.eu/info/funding-tenders/find-funding/eu-funding-programmes/digital-europe-programme_en
44. https://news.yale.edu/2019/08/02/healthcare-industry-major-source-harmful-emissions
45. Kayakutlu, G., & Mercier-Laurent, E. (2017). *Intelligence in Energy*. Elsevier.
46. Schneider 2018–2019 Sustainability Report https://www.schneider-electric.com/en/about-us/sustainability/
47. Repair café https://www.repaircafe.org/fr/. Last consulted April 2021.
48. https://www.gfdl.noaa.gov/climate-modeling/
49. Back market https://www.backmarket.fr. Last consulted April 2021.

50. Julian, V., & Botti, V. (2019). Multi-agent systems. *Applied Science, 9*(7). https://www.mdpi.com/2076-3417/13/2/1021
51. https://intelligence.weforum.org/topics/a1Gb0000000LHVfEAO?emailType=Strategic+Intelligence+Weekly&ske=MDAxNjgwMDAwMENQQ1ZzQUFQ
52. Sustainability Simulation Game. https://www.gblsim.com/. Last accessed 27 Sep 2023.

Chapter 2
The European Green Deal and the 17 SDGs: Uncovering their Connection with a ML-based Approach

Phoebe Koundouri, Nicolaos Theodossiou, Charalampos Stavridis, Stathis Devves, and Angelos Plataniotis

Abstract The European Green Deal (EGD) was published in December 2019 with the ambition of being Europe's new growth strategy, making it climate neutral by 2050, and ensuring its citizens a sustainable, prosperous, and inclusive future. The energy sector is central to this ambition, as the European Commission's objectives are, among others, to increase the efficiency of energy production by establishing a fully integrated, interconnected, and digitalized EU energy market. The EGD was the starting point for the publication of a large number of Policy and Strategy documents for achieving Sustainability in Europe. One of the first attempts to

A Preprint working paper of this contribution exists in Athens University of Economics and Business' Working Papers, available here: https://ideas.repec.org/s/aue/wpaper.html

P. Koundouri (✉)
School of Economics and ReSEES Research Laboratory, Athens University of Economics and Business, Athens, Greece

Sustainable Development Unit, ATHENA RC, Athens, Greece

Department of Technology, Management and Economics, Technical University of Denmark, Denmark, Copenhagen

UN SDSN (Global Climate Hub, European Hub, Greek Hub), Athens, Greece
e-mail: pkoundouri@aueb.gr; http://www.phoebekoundouri.org
https://www.dept.aueb.gr/en/ReSEES; https://www.athenarc.gr/en/sdu
https://www.man.dtu.dk/english/; https://ae4ria.org

N. Theodossiou · C. Stavridis
Department of Civil Engineering, AUTh, UN SDSN Black Sea, Thessaloniki, Greece
e-mail: niktheod@civil.auth.gr; cstavrid@civil.auth.gr

S. Devves
School of Economics and ReSEES Research Laboratory, Athens University of Economics and Business, Athens, Greece

Sustainable Development Unit, ATHENA RC, Athens, Greece

UN SDSN (Global Climate Hub, European Hub, Greek Hub), Athens, Greece
e-mail: devves@aueb.gr; https://www.dept.aueb.gr/en/ReSEES; https://ae4ria.org

G. Kayakutlu, M. Ö. Kayalica (eds.), *Decision Making Using AI in Energy and Sustainability*, Applied Innovation and Technology Management,
https://doi.org/10.1007/978-3-031-38387-8_2

systematically correlate the policy areas of the European Green Deal with the 17 Sustainable Development Goals (SDGs) was made in the first report of the UN SDSN's Senior Working Group for the Joint Implementation of the SDGs and the EGD, which was published in February 2021, where the EGD framework was linked to each of the 17 SDGs using textual analysis. Building on this methodology, in this chapter we extend the manual linkage of policy texts to SDGs, by using Natural Language Processing and Machine Learning techniques to automate it, focusing on Energy-related documents derived by the EGD.

Keywords European Green Deal · Policies · Sustainable Development Goals · Deep Learning · Natural Language Processing · Semantics

2.1 Introduction

2.1.1 The European Green Deal (EGD)

Since the launching of the **European Green Deal (EGD)** [1] by the European Commission at the end of 2019, a significant number of Policy and Strategy documents have been introduced to support the effective implementation of it through the mobilization of funds and the engagement of the various actors from both public and private sector.

The EGD serves as Europe's long-term growth plan and as a roadmap for the establishment of new Policies and Strategies at both central European and national levels. It is a comprehensive strategy that aspires to make the European continent carbon-neutral by 2050, as well as resource-efficient, cutting-edge in terms of technology, and socially just.

The Energy sector is one of the fundamental Pillars to achieve the EGD objective of climate neutrality, considering that the production and consumption of energy are responsible for more than 75% of greenhouse emissions globally and buildings are responsible for about 40% of energy consumption [2].

Below we provide a list of the most important Policies and Strategies **related to Energy, that were** published in the past two years, and we accompany each one with a brief description, highlighting their main points. Then we provide a method to link this set of Policies to the 17 Sustainable Development Goals of the United

A. Plataniotis
Sustainable Development Unit, ATHENA RC, Athens, Greece

National and Kapodistrian University of Athens, Athens, Greece

UN SDSN Global Climate Hub, Athens, Greece
e-mail: aplataniotis@econ.uoa.gr; https://www.athenarc.gr/en/sdu; https://ae4ria.org

Nations Agenda 2030 [3]. Next, we provide a Machine Learning Method to make this linkage in an automated way. Finally, a discussion of the results is made and directions for future research are given.

2.1.2 Energy-Related Policies Derived from the EGD

2.1.2.1 A New Industrial Strategy for Europe

Europe's new industrial strategy was published in March 2020 [4]. Its key priorities are:

1. **Commonly created transition pathways across relevant industrial ecosystems**. The Annual Single Market Report 2021 highlights the challenges encountered by 14 of these ecosystems. Not only are these challenges identified, but the report also mentions transformative initiatives that have been started to achieve the Green and Digital (Twin) Transition. To implement this transition, targeted policies are essential. An example of such a policy is the promotion of clean and competitive European steel, which is supported by specific measures tailored to the industry.
2. **Supporting the business case for the green and digital transitions**. Real business cases, and a resilient regulatory framework, infrastructure, finance for innovation financing, raw materials innovation, clean energy, demand-side measures for climate neutral, production circularity, up-skilling in labor potential, synergies between the sustainable, and digital transitions. Digital solutions, such as digital twins in advanced manufacturing, can help optimization of processes in all ecosystems.
3. **Investment in skills** is a highly appreciated component of investing in labor. The European Skills Agenda supports the green and digital transitions. Mobilization of the private sector is a critical parameter. Large-scale skills partnerships per ecosystem, skilling commitments in the critical industry fields (automotive, microelectronics, aerospace, defense industry).

2.1.2.2 EU Hydrogen Strategy

Europe's Hydrogen Strategy was published in July 2020 [5]. The establishment of a resilient and Sustainable Eco-system for Hydrogen in Europe consists of:

- **Electricity-based hydrogen** through water electrolysis procedure. It is up to renewable electricity production to characterize as fully Green the Hydrogen Production.

- Renewable hydrogen may also be produced through the reforming of biogas (instead of natural gas) or biochemical conversion of biomass, if in compliance with sustainability requirements, therefore, "**Clean hydrogen**" refers to renewable hydrogen.
- "**Fossil-based hydrogen with carbon capture**" means that greenhouse gases emitted can be captured. The variable effectiveness of greenhouse gas capture (maximum 90%) needs to be taken into account.
- **"Hydrogen-derived synthetic fuels"** refer to a variety of gaseous and liquid fuels based on hydrogen and carbon. For synthetic fuels to be considered renewable, the hydrogen part of the syngas should be renewable. Synthetic fuels include for instance synthetic kerosene in aviation, synthetic diesel for cars, and various molecules used in the production of chemicals and fertilizers.

*The **main Key actions** are Strategic investments agenda; Boosting demand for and scaling up production (incentives and supporting regulations, such as a* **Carbon Contracts for Difference scheme;** Design enabling market rules to the deployment of hydrogen; *Promoting research and innovation in hydrogen technologies;* Strengthen EU leadership in international fora for technical standards, regulations and definitions on hydrogen.

2.1.2.3 The Annual Sustainable Growth Strategy of 2021 (7 Technology Flagship Areas)

The Annual Sustainable Growth Strategy of 2021 for Europe was published in September 2020 [6].

The green and digital transformations required unprecedented expenditures in re- and upskilling. Reforms to direct public and private investments toward climate and environmental initiatives are required to complete the green transition. The EU-ID and other important digital public services, including the legal and medical systems, need to be revitalized and be accessible by all and the public administration and services, in general, will be improved by Digitalization.

A Recovery and Resilience Facility will help rebuild and prepare the next generation, but the transition process must be equitable for all Europeans to avoid further inequalities, ensure societal support, and promote social, economic, and territorial cohesion.

Reskilling and upskilling are key to advancing the green and digital transitions, boosting innovation and growth potential, and assuring quality employment and social inclusion. The issues addressed by the Annual Sustainable Growth Strategy (ASGS) 2021 include employment, skills, health, and education. Building renovation projects will help the economy by creating jobs, lowering energy costs, improving living conditions, and reducing energy poverty. Also, by setting the European Green Deal as its cornerstone, aspires to maintain global health, prosperity, and happiness.

In the face of Energy, the EU should speed up the deployment of renewable energy and hydrogen, as well as intensify efforts to improve building energy efficiency. The promotion of future-proof clean technologies to accelerate the use of sustainable, accessible, and smart transportation will make European cities and regions cleaner. The EU's Recovery and Resilience Facility will help the Member States achieve climate neutrality and help to meet the Paris climate goals.

Cross-border and multi-country projects are necessary to promote specific types of investments, such as energy interconnectors, transportation networks, and forward-thinking digital and green projects.

2.1.2.4 Chemicals Strategy for Sustainability

Chemicals strategy for Sustainability was published in October 2020 [7].

Human Health and a **toxic-free environment** are the keywords for reflecting the Chemical Strategy. These two keywords should be protected within an ambitious approach to pollution tackling from all kinds of sources in our common life and activities.

During the full lifecycle of new chemicals and materials, their safety and sustainable utilization and disposal must be ensured. The complexity and global context of manufacturing and supply chains for critical chemicals (i.e., pharmaceuticals) should be treated with **resilient value chains** and diversified sourcing toward climate-neutral and circular economy pathways. Toxic-free material cycles and clean recycling from the EU should lead to a world-widely spread benchmark. Any **substances of concern in products and recycled materials** should be minimized.

Financial instruments with regard to Innovation Calls of EU Commission aspire to support the deployment of the necessary infrastructure to develop the **use, transport, and storage of electricity** from renewable/carbon-neutral energy sources for the production of chemicals. Another dimension is the priority for KPIs of measurement of the industrial transition. The reduction in waste exports, especially plastics and textiles, will be supported by cleaner production processes and technologies. Additionally, there will be a focus on promoting **innovative business models** and providing financial access for **startups**.

2.1.2.5 EU Strategy to Reduce Methane Emissions

EU Strategy to reduce methane emissions was published in October 2020 [8] to address the challenge of Agricultural Transition. Dietary changes will be promoted through the actions relevant to Farm to Fork (F2F) Strategy. Biogas production will be supported for agricultural waste via National Strategic Plans under the Common Agricultural Policy. The gas market regulatory framework will be under review, to promote the facilitation of distributed and locally connected production of biogas. Best practices and technologies, feed and breeding changes, will be promoted for the reduction of agricultural emissions. Improvement of leak detection and repair (LDAR), regarding leaks on fossil gas infrastructure, production, transport, and use, will be an obligation.

Future legislation will include venting, flaring and standards for the overall supply chain, and support to the World Bank "Zero Flaring' initiative. An expert group will develop life-cycle methane emissions methods, especially for the agricultural sector. Partner countries shall be engaged in the global coordination of energy-sector methane emissions, with regard to diplomatic action of the EU. Methane emissions from international partners will be reduced, under standards, targets (of emission reduction), and incentives for any kind of (fossil) energy carrier imported to the EU.

2.1.2.6 A Renovation Wave for Europe

Europe's Renovation Plan was published in October 2020 [9]. The main Priorities in Renovation Wave and initiations on Energy Poverty are: Waste and renewable heat and cool into energy systems; Extension of the use of emission trading to emissions from buildings; Minimum proportions of renewable energy in buildings; Green public procurement criteria related to life cycle and climate resilience; Eco-design and energy labelling; Creating green jobs, upskilling workers and attracting new talent within the Frame of European Skills Agenda; Deep renovation standard; Horizon Europe and the R&I co-creation space; Building Information Modelling; 2050 whole life-cycle performance roadmap; Public buildings and social infrastructure showing the way; Tackling energy poverty and worst-performing buildings; Launching the Affordable Housing Initiative piloting 100 renovation districts; Reviewing the General Block Exemption Regulation and Energy and Environmental Aid Guidelines; European Bauhaus platform (combination of sustainability with art and design); Digitalization in the construction sector.

2.1.2.7 EU Commission Recommendation on Energy Poverty

Europe's Recommendation on Energy Poverty was published in October 2020 [10]. Energy Poverty has become a real challenge for the European Union, since almost 34 million citizens cannot withstand their lack of access to affordable energy for warming their households, feeding their appliances, isolating their buildings from extreme weather/temperature conditions, and finally living in a comfortable home environment.

European Commission Recommendation on Energy Poverty outlines specific indicators for the proper assessment of energy-poor households, such as (a) indicators for the comparison of energy expenditures versus income, (b) a self-assessment on which is the current access of each household in energy services (ability on heating, cooling, warming), (c) explicit monitoring by real measurements of physical parameters/figures, such as room temperatures, as a feedback of the current situation and potential/or implemented interventions, (d) monitoring of the ability for households to pay for their energy bills, the quality of their home, the number of energy disconnection, as a measurement of people weakness to finance their energy consumption/ needs, (e) last but not least, the electricity/fuels prices charging the end-users.

Long-Term Renovation Strategies in every Member State should take into account to erase barriers to investments relevant to energy efficiency, according to the residential profiles that are seeking an energy upgrade.

EU funding programs that are enhancing cohesion policies should be included to develop their distributional effects with regard to energy transition models and the support of vulnerable society profiles.

Finally, Energy Saving Companies (ESCO) and the relevant contracts should be mentioned as key elements to anticipate the upfront costs that are raising barriers to the vulnerable groups and to provide reliable financing solutions for the necessary energy performance interventions.

2.1.2.8 EU Strategy to Harness the Potential of Offshore Renewable Energy for a Climate-Neutral Future

The offshore renewable energy Strategy of Europe was published in November 2020 [11]. The EU has the great advantage of having the biggest variety in sea basins, so the EU is in a position to develop offshore renewable energy. Spatial planning in maritime space is necessary, and there are successful pilot projects specialized in the identification of environmental benefits of offshore wind and aquaculture. Mollusks, algae, and multi-performed offshore platforms are these examples running in Belgium, Germany, France, the Netherlands, and Portugal (MERMAID, TROPOS, Edulis, Wier en Wind).

On the other hand, coordination between Transmission System Operators and Regulatory Authorities is necessary to support offshore renewables interconnection between local and central grids. Integration of large-scale or distributed offshore energy systems into the grids should follow the regulatory rules. EU Sustainable Taxonomy and Grids development, innovation applications, and technical maturity can support private capital investment.

Research and innovation can also support the choice of the proper materials (i.e., corrosion resistance, etc.) and ensure high capacity rates and long-term efficiency and "sufficiency." Re-skill and Up-skill is an important challenge with the potential of high-quality employment opportunities to skilled workers affected by the transition. The Circular Economy approach is important to focus on recycling and re-use of renewable machinery equipment components (i.e., wind turbine blades), affecting the modern design and provoking niche and non-generic applications.

New technologies and methods should be developed and deployed for this, increasing the value retention of products and services within the manufacturing sector. Last but not least, through Green Deal diplomacy, the EU is engaged with its international partners to develop an ecosystem of offshore renewables in cooperation with low-income countries and emerging markets.

2.1.2.9 Smart Mobility Strategy

The Strategy of Europe for Smart Mobility was published in December 2020 [12]. Mass transportation and mobility is a crucial parameter in the social life of citizens. But it has to cut off the costs derived by air-water, greenhouse emission pollution. Environmental impacts affect human health, well-being, and biodiversity. To protect these public goods, especially the non-market goods, the mobility sector has to become sustainable, green, and digitalized and serve the energy transition toward clean, low carbon, green fuels, or other energy carriers that are facilitating modern transportation solutions.

Charging and fuel infrastructure for zero-emission vehicles, the supply chain for renewable and low-carbon fuels, can support developed green mobility in cities with more healthy and environmentally friendly conditions.

On the other hand, digitalization and Artificial Intelligence services can enhance circularity and efficiency in mass production and travel for vehicles, aircraft, and shipping, considering mobility as a service (MAaS). Therefore, safe, comfortable, reliable, interconnected (for both urban and rural areas), and affordable for all mobility, especially for the vulnerable groups of people, should be maximized via digitally powered competitive advantages and logistic chains with increased resilience. All these provide good chances for all in social life, the development of new skills, and, as a result, new and better opportunities for attractive jobs.

Commission has settled down certain milestones for European Smart and Sustainable Mobility:

(1) By 2030, at least 30 million zero-emission cars and 80,000 zero-emission heavy-duty vehicles should be in operation. By 2050, only these categories should be in traffic. (2) Zero-emission ocean-going vessels and large zero-emission aircraft will be introduced in the market by 2030 and 2035, respectively. (3) Traffic on High-speed rail utilization should be twice by 2030 and three times by 2050. (4) Transport by inland waterways and short-sea transport systems should be increased to 25% by 2030 and to 50% by 2050.

2.1.2.10 Updating the 2020 New Industrial Strategy: Building a Stronger Single Market for Europe's Recovery

The European Commission published an update for the Industrial Strategy in May 2021 [13]. The Upgrade of the New Industrial Strategy relies on (a) the monitoring of industry trends and the **Key Performance Indicators (KPIs)** of competitiveness (b) making the Single Market more resilient to anticipate crises and support the recovery. (c) The toolbox development for the release of strategic dependencies. (d) The boost of Digital and Energy Transition.

Commission in combination with the current tools (i.e., Single Market Scoreboard) will monitor main indicators of the competitiveness of the European economy such as:

1. Single Market integration, based on indicators on intra-EU trade across the Member States, to support the innovation ecosystems.

2. Productivity growth.
3. International competitiveness, for open and fair access to export markets.
4. Public and private investment, as a percentage of GDP, performs with regard to the capabilities of both digital and energy potential transformation of the economy.
5. R&D expenditure as a percentage of GDP, to indicate the innovations support.

For the Single Market Resilience, the Key Actions are (1) The Single Market Emergency Instrument, (2) The Annual Single Market Report, (3) A potential consolidated template for the declaration of workers posting, (4) The Surveillance of European and imported products, (5) The Capital or Equity financing of the SME's (under InvestEU).

The Commission is preparing the Alliance on processors and semiconductor technologies and the Alliance for Industrial Data, Edge, and Cloud.

The Commission is also considering the preparation of an Alliance on Space Launchers for the autonomous European access to space, as well as an Alliance on Zero Emission Aviation to provide the future fuels and energy carriers for aircraft (e.g., hydrogen, electric), with regard to the European 2050 climate neutrality objective, as a complementary Group to the Renewable and Low-Carbon Fuels Alliance.

The Commission will continue to support Important Projects of Common European Interest (IPCEIs), especially when the market on its own cannot achieve groundbreaking innovations. This is particularly evident in sectors like batteries and microelectronics.

The Commission will strengthen the synergies between the Green and Digital transitions. Digital twins for the development of the digital copies, historic data, predictive analytics, and all Asset Management Performance of Critical Industrial Equipment will reinforce the reliability of all involved ecosystems.

European Green Digital Coalition will implement the monitoring of green/digital combination solutions and the derived impact. Furthermore, the carbon neutrality goal for ICT technologies will be pursued in alignment with the commitments set out by the Digital Compass, aiming to reduce the environmental footprint of digital technologies.

The twin transition will be supported by the Investment in skills. On the other hand, the twin transition will enhance the Just recovery. The Pact for Skills is a tool of the European Skills Agenda. The European Labor force will be improved in upskilling and reskilling throughout the mobilization of Private Enterprises, Institutes, Academia, and all the relevant Stakeholders.

2.1.3 The European Green Deal and the 17 SDGs

The **Agenda 2030** for Sustainable Development, agreed by all UN Member States in 2015, contains **17 Sustainable Development Goals (SDGs)** and 169 objectives

(Table 2.1). The Agenda is a pledge to eradicate poverty and achieve sustainable development on a global scale by 2030, considering three pillars of sustainable development – *economic, social,* and *environmental.* The SDGs are global in scope and universal in application, considering the different national specificities, capacities, and stages of development, as well as specific difficulties. Thus, all countries share responsibility for achieving the SDGs and each has a critical role to play locally, nationally, and globally, under the United Nations dogma of "leaving no one behind" [3].

The EGD and the 17-SDGs context share common objectives, meaning that the implementation of EGD policies would at the same time support actions that will contribute to achieving various SDGs. In [14], a methodology is presented to link the objectives of the EGD Policy Areas with those of the 17 SDGs, which is based on a text-mining exercise to match specific parts of the EGD document to all relevant SDGs. Their results are very vibrant on the relationship between the two frameworks. Figure 2.1 demonstrates the relation between SDGs and EGD policy areas. Dark green represents a straight connection between EGD and SDG, according to the number of the EGD text extracts that are conceptually similar to the SDG ambition. Light green-colored cells illustrate indirectly associations between EGD and SDGs, and white cells show a weak or no obvious linkage.

The Sustainable Development Goals (SDGs), like the Paris Climate Agreement, call for profound changes in every country, requiring coordinated efforts by governments, civil society, research, and business. However, stakeholders lack a broad consensus on how to operationalize the 17 SDGs, but [15], suggest six major categories of transformations that integrate SDGs in public policy interventions, namely: (1) education, gender, and inequality; (2) health, well-being, and demography; (3) energy decarbonization and sustainable industry; (4) sustainable food, land, water, and oceans; (5) sustainable cities and communities; and (6) digital revolution for sustainable development. Each transformation is accompanied by recommended key investments and regulatory concerns, and as a result, government structures can be used to operationalize transformations while still respecting the 17 SDGs' interdependencies (Fig. 2.2).

2.2 Alignment Between Energy-Related Policies and the 17 SDGs

One of the political guidelines of Ursula von der Leyen [16], the President of the European Commission, is that the process of **European Semester**[1] must be reoriented and become an instrument that integrates the **United Nations Sustainable Development Goals (SDGs) derived from Agenda 2030** [3], which constitute the most widely accepted framework for sustainable development, globally.

[1] The European Semester serves as a framework for the integrated monitoring and coordination of economic and employment policies across the European Union. Since its inception in 2011, it has become a well-established forum for discussing the fiscal, economic, and employment policy challenges confronting EU countries on a yearly basis

Table 2.1 The 17 Sustainable Development Goals of UN Agenda 2030

Goal	The ambition of the Goals
Goal 1 - No Poverty	End poverty in all its forms everywhere
Goal 2 - Zero Hunger	End hunger, achieve food security and improve nutrition and promote sustainable agriculture
Goal 3 - Good Health & Well Being	Ensure healthy lives and promote well-being for all at all ages
Goal 4 - Quality Education	Ensure inclusive and equitable quality education and promote lifelong learning opportunities for all
Goal 5 - Gender Equality	Achieve gender equality and empower all women and girls
Goal 6 - Clean Water & Sanitation	Ensure availability and sustainable management of water and sanitation for all
Goal 7 - Affordable & Clean Energy	Ensure access to affordable, reliable, sustainable, and modern energy for all
Goal 8 - Decent Work & Economic Growth	Promote sustained, inclusive, and sustainable economic growth, full and productive employment, and decent work for all
Goal 9 - Industry, Innovation & Infrastructure	Build resilient infrastructure, promote inclusive and sustainable industrialization, and foster innovation
Goal 10 - Reduced Inequalities	Reduce inequality within and among countries

(continued)

Table 2.1 (continued)

Goal 11 - Sustainable Cities & Communities	Make cities and human settlements inclusive, safe, resilient, and sustainable
Goal 12 - Response Consumption & Production	Ensure sustainable consumption and production patterns
Goal 13 - Climate Action	Take urgent action to combat climate change and its impacts
Goal 14 - Life Below Water	Conserve and sustainably use the oceans, seas, and marine resources for sustainable development
Goal 15 - Life On Land	Protect, restore and promote sustainable use of terrestrial ecosystems, sustainably manage forests, combat desertification, halt and reverse land degradation and halt biodiversity loss
Goal 16 - Peace Justice & Strong Institutions	Promote peaceful and inclusive societies for sustainable development, provide access to justice for all, and build effective, accountable, and inclusive institutions at all levels
Goal 17 - Partnerships for the Goals	Strengthen the means of implementation and revitalize the Global Partnership for Sustainable Development

It is believed that the integration of SDGs into the European Policy Framework will help ensure that Europe's pathway in achieving climate neutrality will happen within a comprehensive economic framework that gives equal opportunities to everyone, and properly.

Below, we present a method used to identify how each Policy mentioned above is related to the 17 Sustainable Development Goals and, thus, contributes to their achievement in various levels.

The linkage is made by identifying phrases or sentences in each Policy text that are conceptually related to specific SDGs. Then, assuming that the greater the number of relevant phrases or sentences, the greater the influence of the Policy on the SDGs, we assign a score to show the level of impact, using a 4-point scale, as follows:

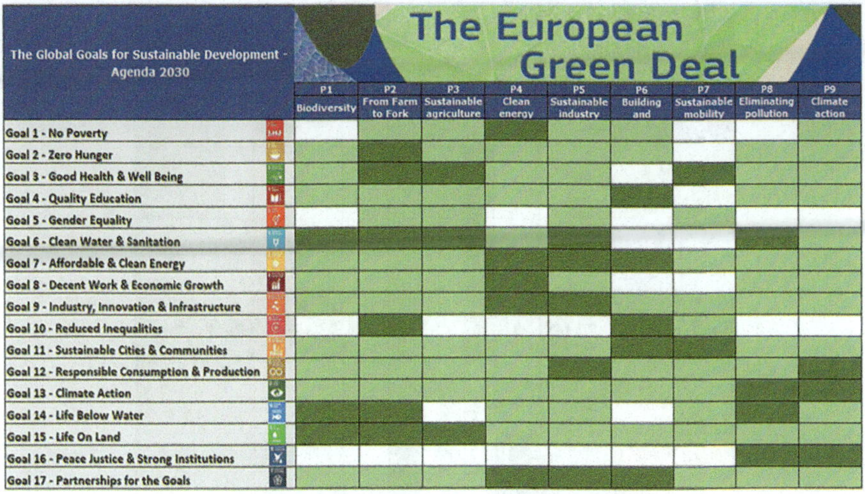

Fig. 2.1 The correlation between the 17 SDGs and the European Green Deal Policy areas. (Source: [14])

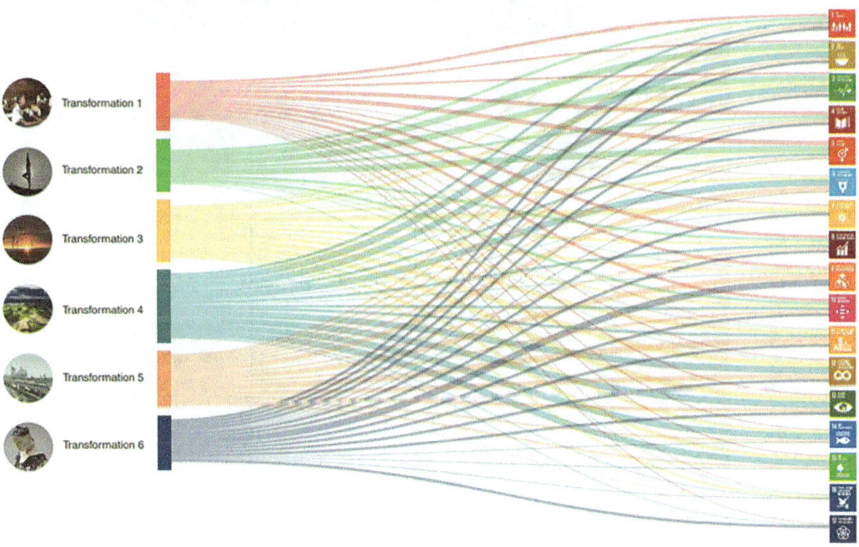

Fig. 2.2 The translation of the 17 SDGs into six transformations. (Source: [15])

- 3, the Policy document directly affects the SDG outcomes
- 2, the Policy document reinforces the SDG outcomes
- 1, the Policy document enable the SDG outcomes
- 0, the Policy document do not interact with the SDG (Table 2.2)

Table 2.2 Connection between Policies and the 17 SDGs

	SDG 1-No poverty	SDG 2-Zero hunger	SDG 3-Good health and well-being	SDG 4-Quality education	SDG 5-Gender equality	SDG 6-Clean water and sanitation	SDG 7-Affordable and clean energy	SDG 8-Decent work and economic growth	SDG 9-Industry, innovation and infrastructure	SDG 10-Reduced inequalities	SDG 11-Sustainable cities and communities	SDG 12-Responsible consumption and production	SDG 13-Climate action	SDG 14-Life below water	SDG 15-Life on land	SDG 16-Peace, justice and strong institutions	SDG 17-Partnerships for the goals
A New Industrial Strategy for Europe (COM/2020/102)	1	2	1	2	0	0	3	2	3	0	1	2	2	1	2	2	2
EU Hydrogen Strategy	1	0	0	2	0	0	3	2	3	1	2	2	3	0	0	2	1
7 technology flagship Areas, ASGS for 2021	0	0	2	1	1	0	2	3	3	3	3	2	2	0	1	2	1
Chemicals strategy for Sustainability	0	1	3	0	0	0	1	0	3	0	1	2	3	3	3	1	0
EU Strategy to reduce methane emissions	1	3	1	1	0	0	2	1	2	0	1	2	1	1	1	1	1
A Renovation Wave for Europe	1	0	0	1	0	0	3	1	2	0	3	2	3	1	1	1	1
EU Commission Recommendation on Energy Poverty	3	0	0	0	0	0	2	2	0	3	1	1	2	0	0	0	0
EU Strategy to harness the potential of offshore renewable energy for a climate neutral future	0	0	0	1	0	0	3	2	3	0	2	1	3	2	0	2	2
Smart Mobility Strategy	0	1	2	0	0	0	3	0	3	2	2	2	3	2	0	0	1
Updating the 2020 New Industrial Strategy: Building a stronger Single Market for Europe's recovery	1	2	1	2	0	0	3	2	3	0	1	2	2	1	2	2	2

2.3 A Machine Learning Method to Evaluate the Connection Between Policy Documents and the 17 SDGs

Within the scope of work undertaken by the authors was the utilization of Machine Learning for three main reasons: (a) to validate the linkages found by the classical approach, (b) to create a tool that will serve as a fast classifier for present and future work, and (c) to discover any new possible connections between the SDGs and the policy documents scanned that were not identified during the classical approach in the first place.

As [17] suggested, Machine Learning is the "fundamental way to make the computer intelligent" and it offers a fast accumulation of knowledge. Thus, it is widely used in Natural Language Processing problems to aid in text analysis.

Natural Language Processing (or NLP for short) is a set of methods for making the human language accessible to computers [18] or else a set of methods that allow the analysis, the understanding and translating of the meaning of human language from computers. The core idea is that for a machine to "understand" what is written or spoken, the message is necessary to break down into a numerical format.

While NLP is widely used in our everyday lives, such as spam e-mail detection, machine translation [19], chatbots and virtual agents development, text classification and summarization, there are major challenges in its implementation. The Natural Language is quite ambiguous. For example, a given text may consist of contextual words and phrases, homonyms and synonyms, errors and slang, and may express irony or sarcasm. The above cannot be easily understood and deciphered by a machine.

For our study, we have developed and deployed two different models. The first one is a simple Information Retrieval Model using Bag-of-Words, while the second one is a more developed and complicated one using Deep Learning Techniques (BERT).

2.3.1 *Information Retrieval*

Information Retrieval (IR), as the term reveals, is the field involved with the search and retrieval of information. It is concerned with all the activities related to the organization, processing, and access of information of all forms and formats [20]. The purpose of IR is the rapid retrieval of documents, texts, and information in general, based on a user's query. An example of IR, and a rather advanced one, is the Google Search Engine.

Bag-of-Words (BoW), a term first used linguistically by [21], is among the prevalent techniques used for textual Information Retrieval [22]. It is a simplifying representation of a document, in which words of a text are presented as a multiset (Bag) disregarding grammar and syntax while keeping simplicity.

The BoW technique was used for the study, to quickly find similarities between energy policy documents and the 17 Sustainable Development Goals (SDGs). Specifically, 17 different vocabularies were built, each consisting of words and phrases describing the essence of each SDG. For instance, SDG 4("Quality Education") was described with keywords such as "primary education", "secondary education", "reading", "mathematics", "organized learning", "information technology", "equal access", "training", "literacy", "numeracy", "fixed level of proficiency", "learner", "curricula", "scientific programs", "teachers", "Learning opportunities", "School enrolment", "Secondary education", "Teacher training", "Universal education"," Vocational training", while SDG 6 ("Clean Water and Sanitation") was described with keywords such as "drinking water", "safe water", "water service", "water sanitation", "hygiene", "defecation", "hand-washing facilities", "shower", "wastewater treatment", "water-use", "freshwater", "water management", etc.

Next, policy documents were read in a .html form and were preprocessed by lowercasing the documents, removing numbers, punctuation, stop words, contractions, and whitespaces. Both the vocabularies and the pre-processed policy documents were transformed into vectors by using CountVectorizer provided by the sci-kit-learn.

To determine the similarity between each policy token and each vocabulary token, we used cosine similarity, which disregards the magnitude of the vectors being compared.

2.4 Results 1

In Table 2.3, the correlation between each Energy Policy with the SDGs is presented. Higher scores present higher similarities.

Table 2.3 Correlation between each Energy Policy with the SDGs

Short Name	SDG1	SDG2	SDG3	SDG4	SDG5	SDG6	SDG7	SDG8	SDG9	SDG10	SDG11	SDG12	SDG13	SDG14	SDG15	SDG16	SDG17
A New Industrial Strategy	0,451	0,333	0,361	0,417	0,278	0,352	0,552	0,512	0,533	0,471	0,423	0,520	0,398	0,427	0,307	0,422	0,556
Hydrogen Strategy	0,468	0,410	0,375	0,385	0,259	0,410	0,659	0,485	0,556	0,494	0,473	0,512	0,545	0,418	0,346	0,394	0,541
7 technology flagship Areas, ASGS for 2021	0,538	0,367	0,402	0,468	0,344	0,401	0,544	0,525	0,590	0,485	0,522	0,525	0,497	0,388	0,346	0,422	0,504
Chemicals strategy for Sustainability	0,500	0,422	0,489	0,447	0,359	0,491	0,528	0,499	0,596	0,480	0,494	0,568	0,484	0,436	0,453	0,441	0,551
Strategy to reduce methane emissions	0,468	0,455	0,402	0,385	0,287	0,468	0,582	0,431	0,508	0,456	0,502	0,545	0,585	0,388	0,453	0,408	0,470
Renovation Wave	0,523	0,410	0,415	0,454	0,337	0,452	0,666	0,534	0,596	0,508	0,605	0,584	0,568	0,436	0,403	0,422	0,556
Recommendation on Energy Poverty	0,459	0,294	0,280	0,278	0,238	0,296	0,475	0,392	0,368	0,375	0,393	0,343	0,390	0,309	0,226	0,316	0,379
Strategy for offshore renewable energy	0,508	0,404	0,368	0,454	0,313	0,401	0,639	0,508	0,579	0,485	0,486	0,557	0,497	0,527	0,392	0,387	0,546
Smart Mobility	0,552	0,427	0,451	0,425	0,352	0,444	0,604	0,555	0,606	0,542	0,577	0,565	0,528	0,480	0,392	0,477	0,566
Updating the 2020 Industrial Strategy	0,451	0,360	0,368	0,393	0,313	0,392	0,544	0,530	0,527	0,485	0,469	0,525	0,398	0,418	0,307	0,435	0,561

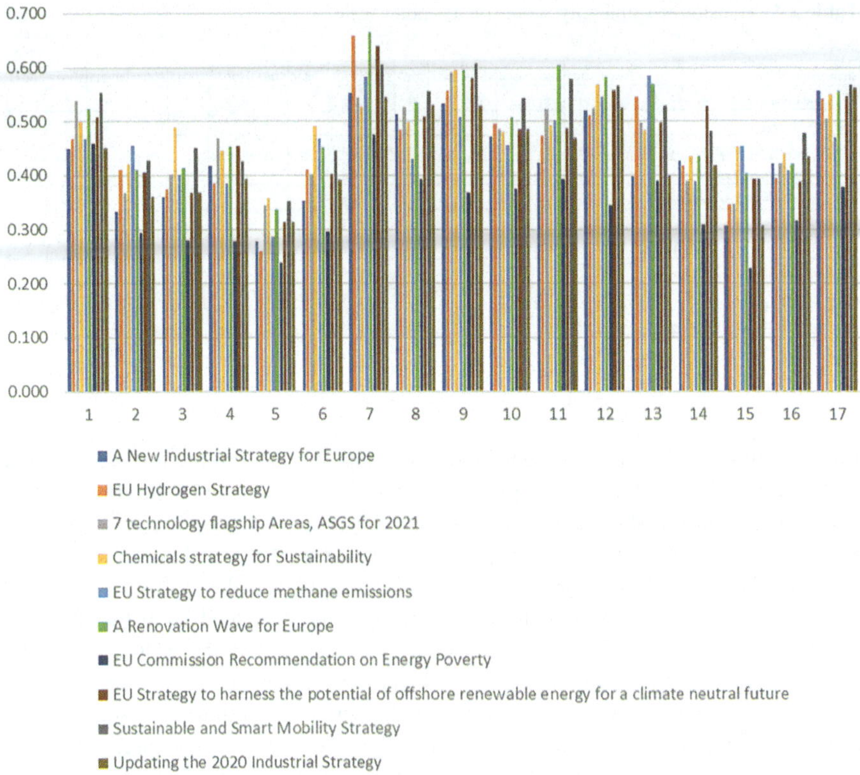

Fig. 2.3 Correlation between each Energy Policy with the SDGs in Bar Chart

The highest similarity score of each policy does not exceed 0.70, meaning that the algorithm did not find a high correlation with a certain SDG. This can be due to the preprocessing of the documents and the vocabularies' construction. Although this might seem disappointing at a first sight, one can draw several conclusions after carefully examining the results (Fig. 2.3).

To further understand the correlation and to easily compare the results with the "human" approach described in the previous chapter, we have assigned the values of 0, 1, 2, and 3 according to a range of similarity scores, as follows:

- Similarity scores below 0.3 are transformed to 0
- Similarity scores between 0.3 and 0.4 are assigned to 1
- Similarity scores between 0.4 and 0.5 are assigned to 2
- Similarity scores greater than 0.5 are assigned to 3

By implementing the aforementioned allocation, Table 2.3 is transformed as Table 2.4.

As one can easily notice, all energy strategies seem to be highly connected to SDG 7 ("Affordable and Clean Energy"), following SDG 9 ("Industry and

Table 2.4 Transformed similarity scores on a 4-point scale

Short Name	SDG1	SDG2	SDG3	SDG4	SDG5	SDG6	SDG7	SDG8	SDG9	SDG10	SDG11	SDG12	SDG13	SDG14	SDG15	SDG16	SDG17
A New Industrial Strategy	2	1	1	2	0	1	3	3	3	2	2	3	1	2	1	2	3
Hydrogen Strategy	2	2	1	1	0	2	3	2	3	2	2	3	3	2	1	1	3
7 technology flagship Areas, ASGS for 2021	3	1	2	2	1	2	3	3	3	2	3	3	2	1	1	2	3
Chemicals strategy for Sustainability	2	2	2	2	1	2	3	2	3	2	2	3	2	2	2	2	3
Strategy to reduce methane emissions	2	2	2	1	0	2	3	2	3	2	3	3	3	1	2	2	2
Renovation Wave	3	2	2	2	1	2	3	3	3	3	3	3	3	2	2	2	3
Recommendation on Energy Poverty	2	0	0	0	0	0	2	1	1	1	1	1	1	1	0	1	1
Strategy for offshore renewable energy	3	2	1	2	1	2	3	3	3	2	2	3	2	3	1	1	3
Smart Mobility	3	2	2	2	1	2	3	3	3	3	3	3	3	2	1	2	3
Updating the 2020 Industrial Strategy	2	1	1	1	1	1	3	3	3	2	2	3	1	2	1	2	3

Infrastructure"), SDG 12 ("Responsible Consumption and Production"), and SDG 8 ("Decent Work and Economic Growth"), goals that are expected to be connected to energy production and management. On the other hand, energy policies seem to be less related to SDG 5 ("Gender Equality"), SDG 15 ("Life on Land"), and SDG 3 ("Good Health and Well-Being"), which seems again sensible.

Comparing and contrasting the table with the one produced from the classical approach of reading and rating each policy's connection to the SDGs, it can be observed that the algorithm, in general, did well. Of course, due to the use of similar keywords in some vocabularies and due to the semantic similarity of some of the SDGs, the similarity scores produced by the algorithm are, in most cases, higher.

A great difference is observed in SDG 6 ("Clean Water and Sanitation"), in which there was no correlation found following the classical approach, but some connection with most policies was produced by the algorithm. This can mean either that different and maybe more targeted keywords should be used as a vocabulary for describing SDG 6, or that there might be a connection, found by the algorithm, that was not detected by the classical approach of reading through the policies and scoring their similarity.

Another odd and unexpected result is the high similarity scores produced in SDG 1 ("No poverty"), as the Classical approach produced none or little correlation between SDG 1 and the scanned energy policies. Again, this might have happened due to the words and expressions chosen as the Goal's describing vocabulary.

Of course, one must not forget that a machine is not human as it lacks semantic capabilities, which are crucial for understanding what a text communicates. Considering that the SDGs are closely related and interconnected, a machine cannot easily capture the essence of each SDG. For instance, "No poverty" and "Zero Hunger" are two very interrelated Goals that, in some cases, relevant texts or phrases cannot be easily labelled or even classified based on their correspondence with the first or the second one.

After we compared and contrasted the results with the classical approach, we decided to take a step further and try to build an even better and more developed algorithm that would perform better and have semantic capabilities.

2.4.1 Deep Learning

As said in the previous chapter, the similarity tool developed showed possible correlations between the SDGs and the scanned policy documents. To make a step further, we decided to deploy a model that has the semantic capability, since our goal was to find the semantic similarity[2] of the EU energy policies and the SDGs. To do so, we used deep learning.

Deep learning refers to very large neural networks with many layers (deep), that "allow computational models that are composed of multiple processing layers to learn representations of data with multiple levels of abstraction" [23]. Simply put, machines learn from experience by representations that are expressed in terms of other, simpler representations [24]. In 2017, Google Research introduced The Transformer, a deep learning model, which is based on attention mechanisms, dispensing with recurrence and convolutions entirely [25]. This innovation led to the development of a wide range of models based on transformers, allowing the processing of entire sequences without the need for labeled data in pre-training.

As a result, and taking particularly into account the ambiguity of the Natural Language (Lexically, Syntactically, Semantically, Anaphorically, Pragmatically), we have fine-tuned a pre-trained transformer-based model to find the similarity score of each policy document with each SDG.

BERT, which stands for "Bidirectional Encoder Representations from Transformers," was firstly introduced by Google Research in 2018 [26] and revolutionized the approach of modeling. Standard Language Models are unidirectional, limiting the architectures that can be used for pre-training, while BERT is a bidirectional transformer pre-trained by using masked language modelling objective and next sentence prediction. Data used for its pretraining are the Toronto Book Corpus (approx. 800 million words) and Wikipedia (approx. 2500 million words). For the study, the "bert-base-uncased" (12-layer, 768-hidden, 12-heads, 110M parameters) model was used.

To fine-tune the pre-trained BERT model, we have used the latest available OSDG Community Dataset, along with the description of the targets and indicators of each Goal. As the OSDG Community Dataset does not include texts for SDG16 and SDG17 yet, we have used expressions and sentences linked to those specific goals from the humanely scanning related policies.

The OSDG Community Dataset is the result of the work of more than 1000 volunteers from all over the world using the OSDG Community Platform, who label sentences and paragraphs according to their relationship with each SDG. Each labeling exercise is a binary decision problem, in which the volunteer has to answer positively or negatively if the presented excerpt is connected to the presented SDG or not.

---

[2] Semantic Similarity is the task of determining how similar two sentences are, in terms of what they mean

The dataset is formed after text excerpts of paragraphs deriving from public documents, such as reports, policies, and publication abstracts. Furthermore, some documents originate from UN-related sources (e.g., SDG-Pathfinder and SDG Library). The released dataset [27] constitutes 32,115 labeled document excerpts, and it contains the referred SDG, the number of volunteers that classified the connection to the SDG as negative, the number of volunteers that classified the connection to the SDG as positive, and the agreement score based on the formula:

$$\text{Agreement} = \frac{|\,\text{labels}_{\text{positive}} - \text{labels}_{\text{negative}}\,|}{|\,\text{labels}_{\text{positive}} + \text{labels}_{\text{negative}}\,|}$$

For the presented research, data used were pre-selected using the following criteria:

1. $\text{labels}_{\text{positive}} > \text{labels}_{\text{negative}}$, as we needed to only use data related to an SDG
2. Agreement >0.6, as we needed to be sure that the volunteers agreed to the labeling

This pre-selection process produced 14,280 excerpts, in which minor corrections occurred, such as joined words separation and wrong letter replacing (e.g., in some words the letter "r" was "i"), etc. In the final set of excerpts, we have added the indicators and the descriptions for each of the 169 targets of the SDGs, as retrieved from the website sdg-tracker.org, leading to a total of 15,083 text excerpts for fine-tuning the model.

The number of text excerpts used for each SDG is presented in Table 2.5.

The model was developed in Python, using PyTorch and Scikit-Learn. Eighty percent of the text excerpts were used as training data and 20% as testing data. Adam Optimizer was used as an optimizer, while the Cross-Entropy Loss was chosen as the optimization criterion. Next, the model was trained for 10 Epochs with a learning rate of 10-5, resulting in an accuracy score of 0.889.

Considering that the training data used for each SDG are not of the same size and the fact that a text excerpt is most probably linked to more than one SDG, the accuracy score is acceptable. In other words, the algorithm simply answers the question "how probable is the X energy policy document to be related to the Y SDG."

By deploying the trained model, the same ten energy policy documents were classified, in the form of .html files as published by the European Commission, giving the results presented below.

Table 2.5 Number of text excerpts used for each SDG

SDG	Number of excerpts used for fine-tuning
1	970
2	726
3	1639
4	1993
5	1948
6	1132
7	1391
8	759
9	605
10	385
11	1081
12	232
13	991
14	613
15	480
16	66
17	72

Table 2.6 Total Similarity Scores

POLICY DOCUMENT	SDG Correlation																
Column1	SDG1	SDG2	SDG3	SDG4	SDG5	SDG6	SDG7	SDG8	SDG9	SDG10	SDG11	SDG12	SDG13	SDG14	SDG15	SDG16	SDG17
A New Industrial Strategy	0,00%	0,04%	0,02%	0,04%	0,03%	0,02%	0,11%	0,21%	99,11%	0,02%	0,03%	0,19%	0,04%	0,03%	0,03%	0,02%	0,05%
Hydrogen Strategy	0,05%	0,09%	0,09%	0,03%	0,02%	0,11%	92,81%	0,07%	0,14%	0,05%	0,12%	1,19%	2,36%	0,11%	0,04%	0,62%	2,08%
7 technology flagship Areas, ASGS for 2021	0,05%	0,47%	1,58%	0,14%	0,08%	0,34%	1,64%	0,19%	92,25%	0,17%	0,21%	0,16%	0,94%	0,23%	0,10%	0,26%	1,17%
Chemicals strategy for Sustainability	0,04%	0,15%	1,81%	0,08%	0,12%	1,60%	0,97%	0,06%	0,21%	0,08%	0,27%	92,41%	0,25%	0,81%	0,31%	0,43%	0,39%
Strategy to reduce methane emissions	0,13%	0,07%	0,16%	0,03%	0,05%	0,41%	84,73%	0,13%	0,08%	0,13%	0,38%	4,44%	4,25%	0,12%	0,09%	1,39%	3,39%
Renovation Wave	0,01%	0,01%	0,02%	0,03%	0,02%	0,04%	0,08%	0,03%	0,07%	0,03%	98,96%	0,26%	0,24%	0,03%	0,03%	0,09%	0,03%
Recommendation on Energy Poverty	0,03%	0,03%	0,03%	0,02%	0,01%	0,15%	98,30%	0,05%	0,03%	0,02%	0,07%	0,41%	0,22%	0,06%	0,01%	0,20%	0,35%
Strategy for offshore renewable energy	0,01%	0,01%	0,01%	0,01%	0,01%	0,02%	99,20%	0,05%	0,03%	0,01%	0,04%	0,26%	0,13%	0,03%	0,01%	0,06%	0,11%
Smart Mobility	0,01%	0,01%	0,03%	0,02%	0,01%	0,02%	0,04%	0,01%	0,07%	0,01%	99,64%	0,04%	0,02%	0,02%	0,01%	0,03%	0,01%
Updating the 2020 Industrial Strategy	0,03%	0,10%	0,09%	0,13%	0,12%	0,11%	1,16%	3,69%	87,95%	0,11%	0,12%	5,61%	0,16%	0,12%	0,12%	0,10%	0,28%

2.5 Results 2

Similarity scores calculated by the model are presented in Table 2.6. The higher the score, the bigger the probability for a scanned policy to be linked to a certain SDG. It is noted that to get better insight, one should view the presented results with a qualitative and not with a quantitative look (Fig. 2.4).

Surprisingly, not all Energy Policies scanned are closely related to SDG 7 ("Affordable and Clean Energy"), as was expected. Only "Hydrogen Strategy," "Strategy to reduce methane emissions," "Recommendation on Energy Poverty," and "Strategy for offshore renewable energy" found to be closely related to SDG 7. "A New Industrial Strategy," "7 Technology flagship Areas," and "Updating the 2020 Industrial Strategy" were found to be probably more related to SDG 9 "Industry, Innovation and Infrastructure." On the other hand, "Renovation Wave" and "Smart Mobility" seem to be more related to SDG 11 "Sustainable Cities and Communities," while "Chemicals Strategy for Sustainability" seems to be closer to SDG 12 "Responsible Consumption and Production."

To understand the correlation between the policies and the SDGs, and to draw meaningful conclusions, the rest of the percentages should be examined as well. To do so, we excluded the highest score of each row (policy document) and by re-adjusting the percentages, we get Table 2.7:

The differences between each policy can be now clearly understood. For instance, energy policies that are related more to industry, such as "A New Industrial Strategy"

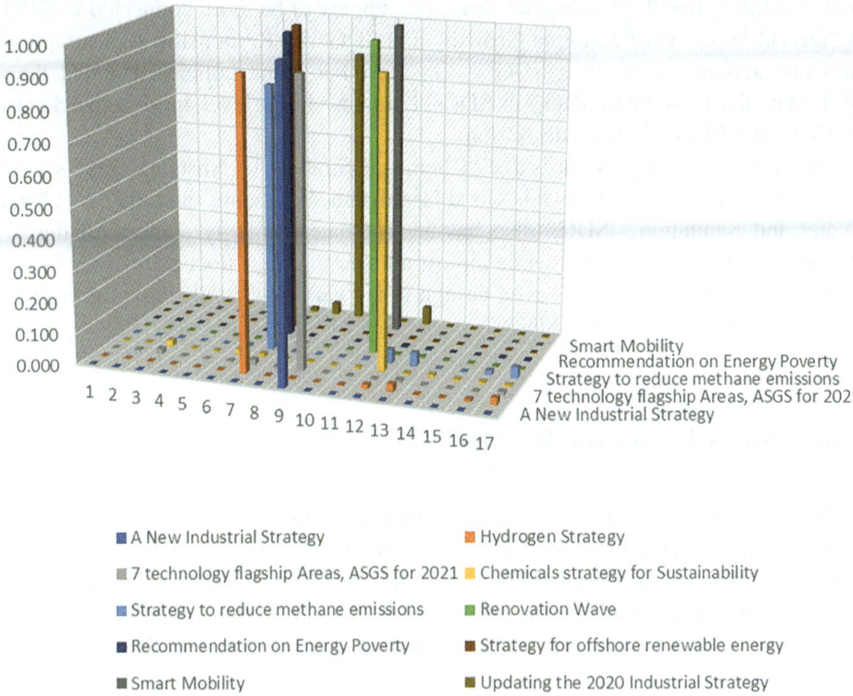

Fig. 2.4 Total Similarity Scores in Bar Chart

Table 2.7 Correction table

Short Name	SDG1	SDG2	SDG3	SDG4	SDG5	SDG6	SDG7	SDG8	SDG9	SDG10	SDG11	SDG12	SDG13	SDG14	SDG15	SDG16	SDG17
A New Industrial Strategy	0,54%	4,04%	2,31%	4,86%	3,61%	2,50%	12,25%	23,41%		2,13%	3,89%	21,38%	4,21%	3,15%	3,83%	2,08%	5,80%
Hydrogen Strategy	0,76%	1,27%	1,19%	0,39%	0,34%	1,58%		0,97%	2,02%	0,66%	1,71%	16,52%	32,81%	1,58%	0,59%	8,65%	28,97%
7 technology flagship Areas, ASGS for 2021	0,70%	6,04%	20,37%	1,85%	0,97%	4,44%	21,13%	2,50%		2,19%	2,74%	2,09%	12,15%	3,02%	1,29%	3,38%	15,15%
Chemicals strategy for Sustainability	0,58%	2,04%	23,87%	1,08%	1,60%	21,09%	12,72%	0,79%	2,75%	1,08%	3,60%		3,24%	10,65%	4,13%	5,65%	5,13%
Strategy to reduce methane emissions	0,87%	0,48%	1,07%	0,19%	0,34%	2,72%		0,85%	0,53%	0,86%	2,52%	29,09%	27,81%	0,81%	0,58%	9,08%	22,20%
Renovation Wave	1,11%	1,04%	2,24%	2,95%	2,03%	4,31%	7,97%	3,24%	6,55%	2,52%		25,18%	22,84%	2,95%	3,32%	8,46%	3,29%
Recommendation on Energy Poverty	1,55%	2,01%	1,66%	0,90%	0,75%	8,99%		3,00%	2,05%	1,02%	4,15%	24,03%	13,17%	3,38%	0,81%	11,79%	20,74%
Strategy for offshore renewable energy	1,33%	0,92%	1,64%	1,16%	1,06%	2,95%		6,15%	3,98%	1,63%	4,39%	32,83%	16,39%	4,17%	0,91%	7,12%	13,35%
Smart Mobility	1,56%	1,67%	8,49%	5,26%	3,96%	6,58%	10,11%	2,96%	20,42%	3,22%		10,09%	6,09%	5,42%	3,52%	8,37%	2,28%
Updating the 2020 Industrial Strategy	0,25%	0,82%	0,74%	1,10%	0,97%	0,87%	9,62%	30,67%		0,95%	1,03%	46,54%	1,30%	0,99%	0,99%	0,86%	2,30%

and "Updating the 2020 Industrial Strategy," appear to be more connected to SDG 8 "Decent Work and Economic Growth" and SDG 12 "Responsible Consumption and Production," rather than SDG 7. On the other hand, policies that had their highest similarity score to SDG 7 or SDG 11 are, on a lower level, more connected to SDG 12 and SDG 13 "Climate Action."

Another interesting insight is that "Chemicals Strategy for Sustainability" seems to be closely connected to SDG 3 "Good Health and Well-being" and SDG 6 "Clean Water and Sanitation." SDG3 also has an adequate similarity score with the "7 Technology Flagship Areas," while almost all energy policy documents seem to be closely connected to SDG 17, even though it had very few labelled text excerpts in comparison to the rest of the SDGs.

2.6 Discussion on the Results

In this study, three different approaches were used to find a correlation between EU energy policy documents and the SDGs. The first one referred to a "classical approach," which consists of reading the policy documents, derived from the EGD and relate to energy, and finding phrases and extracts related to each SDGs. The second one uses a simple Information Retrieval Bag-of-Words Technique, which seeks correlation between the energy policy documents and the SDGs using cosine similarity of vectors, formed by the aforementioned documents and sets of keywords describing each SDGs, referred to as vocabularies. The third one involves the fine-tuning of a pre-trained BERT model (deep learning), using over 15,000 excerpts from documents, which were labeled by the OSDG Volunteer Community.

Assuming that the "classical approach" is a standard and concrete way of finding connections between two documents or a document and a concept, although time-consuming, its results were used as a benchmark for the effectiveness of the two other methods.

Overall, both machine approaches worked, providing a quick way of classifying a policy document regarding the 17 SDGs. The first algorithmic approach performs well as it captures the correlation of energy-related Policies with SDG 7, SDG 9, and SDG 12 as expected. However, some deviations were observed, namely unexpected connections with SDG 1, SDG 5, and SDG 6. Most probably this is due to the vocabularies used, or the fact that the algorithm does not have semantic capabilities to decipher the true meaning and the use of words and phrases. Nevertheless, by focusing on each policy document separately, it is obvious that although there are a few cases where the algorithm produced connections to the related SDGs, none was found through the classical method of scanning the policy text and scoring its relativity to the SDGs. But we must have in mind that, since the algorithm is incapable of semantic capabilities, the results are based purely on the frequency of terms appearing in each policy document. This explains why no policy was classified with a 0,00-similarity score in Table 2.6. Results that may seem quite peculiar are the strong connection between all policies with SDG 1 and SDG 6. These discrepancies may occur again due to the words used to form the vocabularies describing SDG 1 and SDG 6 and their frequency in the examined policy documents. In Conclusion, the 1st algorithmic approach performs well in capturing the overall connection between policies and the SDGs, but not so well in identifying which SDGs are mostly related to the documents under examination. As a result, the 1st algorithm could be used for a quick first glimpse of the overall relativeness between SDGs and the policy documents.

On the other hand, the second algorithmic approach, which utilizes deep learning techniques to calculate the similarity between SDGs and energy policy documents, performs much better. Overall, its results seem to agree with the "classical approach," but, unlike both the latter and the 1st algorithmic approach, they provide a better insight on scoring the identified relationship. This scoring can be of great use, for a quick but adequate evaluation of policy documents, thus providing a useful tool for highlighting the interlinkages between Policy and Strategy documents, like the ones examined here, and the SDGs. Moreover, what seems quite interesting is the fact that some energy-related policy documents are somewhat relevant to SDG 6. This correlation can be also found in the results of the first algorithmic approach but not with the "classical approach." This means that there might be a connection between some policies and "Clean Water and Sanitation" that one must look more closely to find out. After all, one of the advantages of utilizing Machine Learning is the fact that they quickly identify insights and correlations, not easily observed by the human eye. For instance, SDG 6 (Clean Water and Sanitation) targets safe drinking for all; therefore, it is necessary to review the progress in Water-Use efficiency and the reduction of water-stress levels. These two important indicators are dependent on the maintenance of the water distribution systems and the investments in innovative technologies within the industrial, manufacturing, and agricultural systems. Moreover, water-use efficiency and savings are highly associated with energy savings as well, due to the water-use intensity in the energy systems. Similarly, relief of water stress is essential for the activities associated with Energy-Agriculture-Food Industry Nexus. To sum up, there are real technical initiatives that are proving hidden relationships and implicit correlations between Policies and SDGs, and ML utilizing can support this in such accuracy, as the correspondent technical libraries can be embedded within the ML approach and process.

On the other hand, similarity scores with SDG 1 were low, confirming the results of the classical method. Finally, one must highlight the fact that policies are somewhat connected to SDGs 16 and 17, which were the two goals that had the fewest labeled text excerpts used for training, showing that without "Peace, Justice and Strong Institutions" and without building "Partnerships for the Goals" little can be achieved.

2.7 Conclusions—Ideas for Further Research

The presented work gives oversight on how the main Energy Policies of the European Union connect to the 17 SDGs and roughly answers, in short, the question "Where is the EU heading, considering the announcements and the promises made for a sustainable future." The work included three different methodologies targeting grading the relativity of each policy with each SDG.

The first methodology, referred to as the "classical approach," included the reading of each document and grading its SDG-relativity using a scale of 0-3, where 0 stands for "no relativity" and 3 for "great relativity." The second methodology included using BoW Techniques (Information Retrieval) to compare the similarity of any given document to a vocabulary of keywords, formatted by the research group. Each document was preprocessed and transformed into a vector. The document-vector and each vocabulary-vector were compared using the Cosine Similarity and their similarity scores were produced. The third methodology included the fine-tuning of a deep learning BERT model, using more than 15,000 labeled texts in total.

By utilizing Natural Language Processing techniques, one can find a quick and adequate correlation between the examined policy documents and the 17 SDGs, but further examination is needed to validate the results produced. A clearer view with better accuracy is given after utilizing Deep Learning Techniques with semantic capabilities, which produced insights that need further investigation to be confirmed.

Overall, this study utilized Natural Language Processing Techniques and Machine Learning to calculate the interconnectivity of policy documents and SDGs, setting a concrete work that can be used in future research.

For instance, one could further improve the results by including more labelled texts in the training data set, especially for Goals SDGs 12, 16, and 17. Furthermore, considering the interconnectivity and relativity of the 17 SDGs with each other, one could use multi-labelling and voting ensembles, to get better results for each scanned document, or deploy a Neural Information Retrieval Model for better queries. This work could help in utilizing Machine Learning in the Six Transformational Pathways toward Sustainability that base upon the interconnectivity of the SDGs.

Additionally, to improve the public dissemination and understanding of the connection between policies and the SDGs, a web platform could be set up, open to anyone interested in finding more information regarding a policy document and its connection to the SDGs. The platform could host the aforementioned developed tools, that would be regularly maintained and upgraded to provide more accurate results from time to time.

The classical approach will be still more popular since human working groups will prefer with no doubt to discuss and exchange ideas or even have arguments about the relevance between policy documents and ideas or commonly agreed targets (such as SDGs). But this concrete and the fair process can be supported in the future by ML or similar tools since these tools can be upgraded, to mention and remind explicit and implicit relationships between policies and targets. For instance, an upgraded ML tool, as it presents the result of relevance, can include also a short description, proving and giving feedback (by a short example) why they give a "point 3 or point 2" relevance between the x SDG and the A Policy, presenting a short text when computer censor is over this specific score assessment. This means that the algorithm should exploit all possible correlations between crucial document keywords and synonymous ideas that are leading to specific targets and their semantic relations.

References

1. European Commission. (2019, 11 Dec). *The European Green Deal* (COM/2019/640 final), Brussels, https://eur-lex.europa.eu/legal-content/EN/TXT/?qid=1576150542719&uri=COM%3A2019%3A640%3AFIN
2. World Resources Institute (By Mengpin Ge, Johannes Friedrich and Leandro Vigna), (2020, Feb 6). *4 charts explain greenhouse gas emissions by countries and sectors*, https://www.wri.org/insights/4-charts-explain-greenhouse-gas-emissions-countries-and-sectors
3. United Nations. (2015, Oct). *Transforming our world: The 2030 Agenda for Sustainable Development (A/RES/70/1)*, https://sdgs.un.org/2030agenda
4. European Commission. (2020, 10 March). *A new industrial strategy for Europe (COM(2020) 102 final)*, Brussels https://eur-lex.europa.eu/legal-content/EN/TXT/HTML/?uri=CELEX:52020DC0102&from=EN
5. European Commission. (2020, 8 July). *A hydrogen strategy for a climate-neutral Europe (COM (2020) 301 final)*, Brussels, https://eur-lex.europa.eu/legal-content/EN/TXT/HTML/?uri=CELEX:52020DC0301&qid=1641991695109&from=EN
6. European Commission. (2020, 17 Sept). *Annual sustainable growth strategy 2021 (COM(2020) 575 final)*, Brussels, https://eur-lex.europa.eu/legal-content/EN/TXT/HTML/?uri=CELEX:52020DC0575&from=en
7. European Commission. (2020, 14 Oct). *Chemicals strategy for sustainability (COM(2020) 667 final)*, Brussels, https://eur-lex.europa.eu/legal-content/EN/TXT/HTML/?uri=CELEX:52020DC0667
8. European Commission. (2020, 14 Oct). *On an EU strategy to reduce methane emissions (COM (2020) 663 final)*, Brussels, https://eur-lex.europa.eu/legal-content/EN/TXT/HTML/?uri=CELEX:52020DC0663
9. European Commission. (2020, 14 Oct). *A Renovation Wave for Europe - greening our buildings, creating jobs, improving lives (COM(2020) 662 final)*, Brussels, https://eur-lex.europa.eu/legal-content/EN/TXT/HTML/?uri=CELEX:52020DC0662&from=EN
10. European Commission. (2020, Oct). *Official Journal of the European Union (L 357/35)*, COMMISSION RECOMMENDATION (EU) 2020/1563 on energy poverty, https://eur-lex.europa.eu/legal-content/EN/TXT/HTML/?uri=CELEX:32020H1563&from=EN
11. European Commission. (2020, 19 Nov). *An EU Strategy to harness the potential of offshore renewable energy for a climate neutral future (COM(2020) 741 final)*, Brussels, https://eur-lex.europa.eu/legal-content/EN/TXT/HTML/?uri=CELEX:52020DC0741&from=EN
12. European Commission. (2020, 9 Dec). *Sustainable and Smart Mobility Strategy – putting European transport on track for the future (COM(2020) 789 final)*, Brussels, https://eur-lex.europa.eu/legal-content/EN/TXT/HTML/?uri=CELEX:52020DC0789&from=EN
13. European Commission. (2021, 5 May). *Updating the 2020 New Industrial Strategy: Building a stronger Single Market for Europe's recovery (COM(2021) 350 final)*, Brussels.
14. Sachs, J., Koundouri, P., et al. (2021). *Transformations for the joint implementation of Agenda 2030 for Sustainable Development and the European Green Deal - A Green and Digital, job-based and inclusive recovery from the COVID-19 pandemic*. Report of the UN Sustainable

Development Solutions Network, https://resources.unsdsn.org/transformations-for-the-joint-implementation-of-agenda-2030-the-sustainable-development-goals-and-the-european-green-deal-a-green-and-digital-job-based-and-inclusive-recovery-from-covid-19-pandemic
15. Sachs, J. D., Schmidt-Traub, G., Mazzucato, M., et al. (2019). Six transformations to achieve the sustainable development goals. *Nature Sustainability, 2*, 805–814. https://doi.org/10.1038/s41893-019-0352-9
16. von der Leyen, U. (2019). *Political guidelines for the next Commission (2019-2024) - "A Union that strives for more: My agenda for Europe"*, https://ec.europa.eu/info/files/political-guidelines-new-commission_en
17. Wang, H., Ma, C., & Zhou, L. (2009). *A brief review of machine learning and its application*. 2009 international conference on information engineering and computer science, pp. 1–4. https://doi.org/10.1109/ICIECS.2009.5362936
18. Eisenstein, J. (2019). *Introduction to natural language processing*. MIT Press.
19. Wu, Y., Schuster, M., Chen, Z., Le, Q. V., Norouzi, M., Macherey, W., Krikun, M., Cao, Y., Gao, Q., Macherey, K., et al. (2016). Google's neural machine translation system: Bridging the gap between human and machine translation. arXiv preprint arXiv:1609.08144, .
20. Chowdhury, G. G. (2010). *Introduction to modern information retrieval*. Facet Publishing.
21. Zellig, S. H. (1954). Distributional structure. *WORD, 10*(2-3), 146–162. https://doi.org/10.1080/00437956.1954.11659520
22. Passalis, N., & Tefas, A. (2018). Learning bag-of-embedded-words representations for textual information retrieval. *Pattern Recognition, 81*, 254–267.
23. LeCun, Y., Bengio, Y., & Hinton, G. (2015). Deep learning. *Nature, 521*, 436–444. https://doi.org/10.1038/nature14539
24. Goodfellow, I., Bengio, Y., & Courville, A. (2016). *Deep learning*. MIT Press. Available at: www.deeplearningbook.org
25. Vaswani, A., Shazeer, N. M., Parmar, N., Uszkoreit, J., Jones, L., Gomez, A. N., Kaiser, L., & Polosukhin, I. (2017). Attention is all you need. ArXiv, abs/1706.03762.
26. Devlin, J., Chang, M. W., Lee, K., & Toutanova, K. (2018). Bert: Pre-training of deep bidirectional transformers for language understanding. arXiv preprint arXiv:1810.04805.
27. OSDG, UNDP IICPSD SDG AI Lab, & PPMI. (2021). *OSDG Community Dataset (OSDG-CD) (2021.09)* [Data set]. Zenodo. https://doi.org/10.5281/zenodo.5550238

Chapter 3
Single-Valued Neutrosophic CRITIC-Based ARAS Method for the Assessment of Sustainable Circular Supplier Selection

Arunodaya Raj Mishra, Raghunathan Krishankumar, Fausto Cavallaro, Ram Kishun Lodhi, and K. S. Ravichandran

Abstract A *"Circular Supplier Selection (CSS)"* is a procedure of choosing suppliers in a *"Closed-Loop Supply Chain (CLSC)"*. A *"Sustainable Circular Supplier Selection (SCSS)"* discusses the environmental and social issues in the region of *"Circular Economy (CE)"* and *"Sustainable Supplier Selection (SSS)"*. Due to the subjectivity of human judgment, the best SCSS often comprises ambiguous and indeterminate data, and the *"Single-Valued Neutrosophic Sets (SVNSs)"* have a vast proficiency to treat uncertain, indeterminate, and inaccurate information in the *"Multi-Criteria Decision-Making (MCDM)"* process. This chapter develops an extended methodology with the *"Additive Ratio Assessment (ARAS)"* and *"Criteria*

The original version of the chapter has been revised. A correction to this chapter can be found at https://doi.org/10.1007/978-3-031-38387-8_20

A. R. Mishra
Department of Mathematics, Government College Raigaon, Satna, MP, India

R. Krishankumar
Information Technology Systems and Analytics Area, Indian Institute of Management Bodh Gaya, Bodh Gaya, Bihar, India
e-mail: raghunathan.k@iimbg.ac.in

F. Cavallaro (✉)
Department of Economics, University of Molise, Campobasso, Italy
e-mail: cavallaro@unimol.it

R. K. Lodhi
Department of Applied Science, Symbiosis Institute of Technology, Symbiosis International University, Pune, India
e-mail: ramkishun.lodhi@sitpune.edu.in

K. S. Ravichandran
Department of Mathematics, Amrita School of Physical Sciences, Amrita Vishwa Vidyapeetham, Coimbatore, Tamil Nadu, India
e-mail: ks_ravichandran@cb.amrita.edu

G. Kayakutlu, M. Ö. Kayalica (eds.), *Decision Making Using AI in Energy and Sustainability*, Applied Innovation and Technology Management,
https://doi.org/10.1007/978-3-031-38387-8_3

Importance Through Intercriteria Correlation (CRITIC)" under SVNSs called SVN-CRITIC-ARAS method and utilizes it to treat the SCSS in the manufacturing companies. The comparison with extant models shows that the SVN-CRITIC-ARAS tool is capable and proficient to select the best sustainable circular supplier. Hence, the developed framework can be implemented by companies to evaluate and find the appropriate supplier in the SCSS procedure in the CE.

Keywords Circular economy · Single-valued neutrosophic sets · Sustainable circular supplier · MCDM · CRITIC · ARAS

3.1 Introduction

The selection of supplier in *"Circular Supply Chain (CSC)"* system is a concern of such huge significance that it can have a massive influence on falling environmental degradation and the chain costs [18]. Alternatively, the earning of ecological and recyclable raw products indicates to the reduction of waste and the use of raw materials dependent on appropriate *"Circular Supplier Selection (CSS)"*. Though the *"Green Supplier Selection (GSS)"* or CSS issue incorporates both economical and environmental features, social accountability is additional significant feature that deliberates worker rights and encourages safe and secure operational surroundings for all employees, but it obtains little responsiveness [17, 55]. This deficiency of academic consideration gives a substantial task to reach *"Sustainable Development Goal (SDG)"*. Consequently, SCSS that considers economic, environmental, social, and circular features leads to a theoretical explanation of a *"Sustainable Circular Supplier Selection (SCSS)"* concern that is more inclusive than the CSS/GSS concern.

In this setting, numerous works have discussed the environmental issued *"Supplier Selection (SS)"* concern in *"Forward Chain (FC)"* whereas there is always an investigation gap in CSS (SS in *"Forward and Reverse Supply Chains (FRSCs)"*) [12]. The SS issue indicates the development of *"Multi-Criteria Decision-Making (MCDM)"* models [48]. There are various MCDM models and their generalizations in the associated studies reveal both pros and cons [3]. Thus, a model proposed by Kusi-Sarpong et al. [24] is using fuzzy *"Best-Worst Method (BWM)"*, and VIKOR models for *"Sustainable Supplier Selection (SSS)"* in CSC problem are discussed. To extract attributes and sub-attribute's weights in MCDM procedure, Tavana et al. [48] used the *"Analytic Hierarchy Process (AHP)"* and *"Analytic Network Process (ANP)"*.

Though, the idea of *"Fuzzy sets (FSs)"* and *"Intuitionistic Fuzzy Sets (IFSs)"* can only tackle uncertain and ambiguous data but not indeterminate and inconsistent data describes in realistic settings. To treat the data more precisely, the *"Neutrosophic sets (NSs)"* was initiated by Smarandache [45]. NS is described by the independent *"Membership Functions (MFs)"* which designate the *"Truth-membership Function (TF)"*, *"Indeterminacy-membership Function (IF)"*, and *"Falsity-membership Function (FF)"* [45]. It is complicated to utilize the *NSs* in realistic circumstances from a scientific viewpoint. In 2010, the doctrine of *"Single-Valued Neutrosophic Sets (SVNSs)"* was discussed by Wang et al. [53] with several elegant and inventive postulates for better implementations in realistic science and engineering disciplines. An SVNS involves of a TF $t_S(z_i)$, an IF $i_S(z_i)$, and an FF $f_S(z_i)$ in the interval [0,1] and holds the condition $0 \leq t_S(z_i) + i_S(z_i) + f_S(z_i) \leq 3$. SVNSs show huge prospective to treat ambiguous, indeterminate, and inconsistent data which arises in realistic concerns [40, 41].

Owing to the idea of SVNSs, the paper develops an integrated MCDM method for treating SCSS problem in the manufacturing companies on SVNSs. In handling with decision-making problems, the computation of criteria weights and priority order of options are the vital and precarious stages for *"Decision-Experts (DEs)"*. Here, the *"Criteria Importance Through Intercriteria Correlation (CRITIC)"* method is an objective weighting assessment procedure, which evaluates the contrast intensity of attribute with the help of *"Standard Deviation (SD)"* and assesses the conflicts among the attributes using the *"Correlation Coefficient (CRC)"*. The *"Additive Ratio Assessment (ARAS)"* model makes use of the idea of an optimality degree in order to achieve the prioritization. It includes direct and proportional relationship with attribute weights, having the ability of solving complicated problems and involving some simple and direct steps for the assessment of a number of options or choices based on their performance in comparison with the chosen evaluation criteria that obtained suitable, sensible, and comparatively accurate results.

Inspired by the aforementioned studies, the SVN-CRITIC-ARAS methodology is utilized in this chapter to do SCSs' assessment. To the best of authors' knowledge, this is the first work to assess the SCSS and sustainable parameters to achieve the goal. Alternatively, the assessment criteria are distributed into four types: economic, environmental, social and circular perspectives. This chapter combines the CRITIC and ARAS approaches under SVNSs setting. Though, few scholars have integrated both models on diverse settings, but no study has been discussed on SVNS setting. The main contributions of the chapter are given by

- A decision-making model is introduced with the use of CRITIC and ARAS methods on SVNSs setting.
- To obtain the attribute weights of SCSS, the CRITIC tool is applied.
- To prioritize the different SCSs in the manufacturing companies, the ARAS method is used.
- A case study of SCSS with SVNSs is discussed to illustrate the practicality of the discussed integrated tool.
- Comparative study is made to show the consistency and strength of the decision outcomes.

The remaining work of the study is discussed as Sect. 3.2 presents the literature work. Section 3.3 discusses the elementary ideas on SVNSs. Section 3.4 shows the SVN-CRITIC-ARAS model for treating decision-making problems on SVNSs setting. Section 3.5 presents a case study of SCSS in manufacturing companies and further makes comparative analysis. Finally, Section 3.6 concludes the work and shows the future direction of further study.

3.2 Literature Review

As FSs doctrine [58] is more accomplished in replicating human rational related to the crisp values, it has been found huge interest in MCDM analysis. Next, the idea of IFSs [4] is an extension of FSs, which provides each member to the *"Membership Degree (MD)"* and *"Non-membership Degree (ND)"*. Thus, it shows more efficient procedure to treat the ambiguous and uncertain data in realistic situations. For example, if an expert provides their judgment about an expression, then he/she may express that 0.5 being the "possibility that the statement is true", 0.6 being the "possibility that the statement is false", and 0.2 being the "possibility that he or she is on the fence". This concern is omitted from the range of FSs and IFSs, and hence, the concept of SVNSs is essential.

To treat the issues, Smarandache [45] announced the NSs, which models with the *"Truth-membership Function (TF)"*, *"Indeterminacy-membership Function (IF)"*, and *"Falsity-membership function (FF)"*. It can treat ambiguous, imprecise, incomplete, and inconsistent information, which the FSs, IFSs, *"Pythagorean Fuzzy Sets (PFSs)"*, and *"Interval-Valued Intuitionistic Fuzzy Sets (IVIFSs)"* cannot precisely treat. Afterward, NSs have broadly been applied in various realistic applications to tackle the uncertain data [49, 61]. In 2010, Wang et al. [53] revealed the doctrine of SVNSs as a particular case of NS, SVNSs, which models by the *TF*, *IF*, and *FF* to treat with indeterminate and inconsistent data. For example, Luo et al. [26] gave an MCDM tool with the use of SVN-similarity measure for handling the appointment registering structures problems. Garg and Nancy [15] discussed a decision-making model for treating the MCDM problems. In recent times, Mishra et al. [32] gave an *"Additive Ratio Assessment (ARAS)"* approach to assess and rank the sites for *"Electric Vehicle Charging Station (EVCS)"*. Jana and Pal [20] presented the Dombi operators on SVNSs to treat the MCDM problems. Mishra and Rani [30] initiated an integrated tool with the CRITIC and CoCoSo methods for assessing sustainable 3PRLP selection problem under SVNS settings.

In the MCDM system, the evaluation of attribute weights is a momentous task for DEs. Various academicians have presented numerous frameworks for computing the attributes weights [29]. The CRITIC approach was initiated by Diakoulaki et al. [13] to estimate the objective weight of attributes. To use CRITIC tool, the attribute's weight is estimated by the contrast intensity of attribute applying the *"Standard Deviation (SD)"*, whereas conflicts among the attributes are estimated with the *"Correlation Coefficient (CRC)"*. It has effectively been utilized to obtain the weights of attributes in several realistic MCDM issues [1, 38]. For example, Ghorabaee et al. [16] gave an integration of CRITIC-WASPAS tool to ruminate the *"Third Party Reverse Logistic Provider (3PRLP)"* with IT2FSs. Wei et al. [54] discussed CRITIC-GRA model to treat the EVCS assessment problem. Mishra and Rani [30] gave an integrated CRITIC-CoCoSo approach for handling the *"sustainable 3PRLPs (S3PRLPs)"* assessment on SVNSs. Mishra et al. [34] presented the CRITIC-EDAS tool to treat the S3PRLPs on *"Fermatean Fuzzy Sets (FFSs)"*. Baidya et al. [7] developed BCF-Archimedean power-weighted operator-based

CRITIC-MULTIMOORA method to solve 3PRLP selection problem. Mishra and Rani [30] initiated the CRITIC-CoCoSo tool for handling logistic service providers on SVNSs. Rani et al. [43] offered a combined framework using the CRITIC-MULTIMOORA method on SVNSs to handle the food waste treatment technique assessment problem. Mishra et al. [35] applied the CRITIC-VIKOR tool and applied to treat the Kharif crop evaluation of Satna in India.

During the past few decades, MCDM was taken into consideration as a key process of people's daily lives. In realistic circumstances, it is not easy to treat the MCDM problems [10]. Owing to the increasing complexity and extensive alterations to today's environments, the conventional MCDM methods are generally inapplicable to the MCDM problems. Turskis and Zavadskas [51] introduced the ARAS method indicating that the actions of this intricate domain may be implied using simple relative assessments. ARAS makes use of the concept of an optimality degree in order to achieve the prioritization. The most important benefits of ARAS include: (1) direct and proportional association with attribute weights [19], (2) having the ability of solving complex problems [9], (3) involving some simple and direct steps for the assessment of a number of options or choices based on their performance in comparison with the chosen evaluation criteria that obtained suitable, sensible, and comparatively accurate results. Most situations where the conventional ARAS has been recently utilized have been aimed at the ranking of firms on the basis of indicators of corporate social accountability [21], and the evaluation of drug selections [33]. In recent years, this tool has been elaborated in diverse uncertain disciplines. A popular instance is the ARAS Grey model [52], which was extended on interval-valued triangular fuzzy numbers [47]. ARAS was used by Mishra et al. [31] for the assessment and selection of desired IT personnel for a firm on IFSs. Rani et al. [42] used the SWARA-ARAS tool to obtain the best *"healthcare waste treatment (HCWT)"* tool.

3.3 Preliminaries

Consider a discourse set U and $u \in U$. $\widetilde{N}$ as a *"neutrosophic sets (NS)"* is given as a *TF* $T_{\widetilde{N}}(u)$, an *IF* $I_{\widetilde{N}}(u)$ and an *FF* $F_{\widetilde{N}}(u)$ and is shown like $\widetilde{N} = \left\{ \left(T_{\widetilde{N}}(u), I_{\widetilde{N}}(u), F_{\widetilde{N}}(u) \right) : u \in U \right\}$. Also the *MFs* of $T_{\widetilde{N}}(u)$, $I_{\widetilde{N}}(u)$ and $F_{\widetilde{N}}(u)$ are real standard or nonstandard subsets of $]0^{-}, 1^{+}[$ such that $T, I, F : U \rightarrow]0^{-}, 1^{+}[$ with the constraint as $0^{-} \leq \sup T_{\widetilde{N}}(u) + \sup I_{\widetilde{N}}(u) + \sup F_{\widetilde{N}}(u) \leq 3^{+}$.

As aforesaid, it is tough to employ the NSs to tackle various concerns. Henceforth, Wang et al. [53] gave an SVNS as follows:

Definition 1 [53] A *"Single Valued Neutrosophic Set (SVNS)"* is originated by Wang et al. [53] for handling uncertain MCDM problems. An SVNS N on U is defined as a TF $t_N(u)$, an IF $i_N(u)$, and an FF $f_N(u)$ such that $t_N : U \rightarrow [0, 1]$, $i_N : U \rightarrow [0, 1]$, and $f_N : U \rightarrow [0, 1]$ with the constraint as $0 \leq t_N(u) + i_N(u) + f_N(u) \leq 3$.

For an SVNS N in U, the triple (t_N, i_N, f_N) is termed as *"Single Valued Neutrosophic Number (SVNN)"*.

Definition 2 [45] Let $\alpha = (t_\alpha, i_\alpha, f_\alpha)$, $\beta = (t_\beta, i_\beta, f_\beta) \in SVNNs(U)$ and $\gamma > 0$, then the same operations are described as

- $\alpha^c = (f_\alpha, 1 - i_\alpha, t_\alpha)$
- $\alpha \subseteq \beta$ if $t_\alpha \leq t_\beta$, $i_\alpha \geq i_\beta$ and $f_\alpha \geq f_\beta$,
- $\alpha = \beta$ if and only if $\alpha \subseteq \beta$ and $\beta \subseteq \alpha$
- $\alpha \cup \beta = (\max\{t_\alpha, t_\beta\}, \min\{i_\alpha, i_\beta\}, \min\{f_\alpha, f_\beta\})$
- $\alpha \cup \beta = (\min\{t_\alpha, t_\beta\}, \max\{i_\alpha, i_\beta\}, \max\{f_\alpha, f_\beta\})$
- $\alpha + \beta = (t_\alpha + t_\beta - t_\alpha t_\beta, i_\alpha i_\beta, f_\alpha f_\beta)$
- $\alpha \cdot \beta = (t_\alpha t_\beta, i_\alpha + i_\beta - i_\alpha i_\beta, f_\alpha + f_\beta - f_\alpha f_\beta)$
- $\gamma\alpha = \left(1 - (1 - t_\alpha)^\gamma, i_\alpha^\gamma, f_\alpha^\gamma\right)$
- $\alpha^\gamma = \left(t_\alpha^\gamma, 1 - (1 - i_\alpha)^\gamma, 1 - (1 - f_\alpha)^\gamma\right)$

Definition 3 [46] Let $\alpha = (t_\alpha, i_\alpha, f_\alpha) \in SVNNs(U)$, then the score function of α is given by

$$\mathbb{S}(\upsilon) = \frac{2 + t_\upsilon - i_\upsilon - f_\upsilon}{3}; \mathbb{S}(\upsilon) \in [0, 1] \tag{3.1}$$

On the similar line, accuracy and certainty values are given by

$$a(\upsilon) = t_\upsilon - f_\upsilon \text{ and } c(\upsilon) = t_\upsilon \tag{3.2}$$

Definition 4 [56] Let $\alpha_j = (t_j, i_j, f_j) \in SVNNs(U)$, $j = 1, 2, \ldots, n$ and $w = (w_1, w_2, \ldots, w_n)^T$ with $w_j \in [0, 1]$ and $\sum_{j=1}^{n} w_j = 1$ be the weights of α_j. Then the *"Single-Valued Neutrosophic Weighted Average (SVNWA)"* and *"Single-Valued Neutrosophic Weighted Geometric (SVNWG)"* operators of α_j are given as

$$\begin{aligned} \text{SVNWA}(\alpha_1, \alpha_2, \ldots, \alpha_n) &= \bigoplus_{j=1}^{n} (w_j \alpha_j) \\ &= \left(1 - \prod_{j=1}^{n} (1 - t_j)^{w_j}, \prod_{j=1}^{n} (i_j)^{w_j}, \prod_{j=1}^{n} (f_j)^{w_j}\right), \end{aligned} \tag{3.3}$$

$$\begin{aligned} \text{SVNWG}(\alpha_1, \alpha_2, \ldots, \alpha_n) &= \bigotimes_{j=1}^{n} (w_j \alpha_j) \\ &= \left(\prod_{j=1}^{n} (t_j)^{w_j}, 1 - \prod_{j=1}^{n} (1 - i_j)^{w_j}, 1 - \prod_{j=1}^{n} (1 - f_j)^{w_j}\right). \end{aligned} \tag{3.4}$$

3.4 SVN-CRITIC-ARAS Method

The main purpose of the work is to use SVN-CRITIC-ARAS framework to select the best SCSS. Here, the weights of SCSS criteria are computed with the SVN-CRITIC tool. After the evaluation of the criteria and expert weights, SVN-CRITIC-ARAS framework is implemented first time on SVNSs to select the SCSs. The brief descriptions of these methods are given below:

Step I: *Originate the MCDM procedure*

The aim of MCDM structure is to obtain an appropriate choice from a set of m options $S = \{s_1, s_2, \ldots, s_m\}$ over the criteria set $F = \{f_1, f_2, \ldots, f_n\}$. Consider a committee of ℓ experts $D = \{d_1, d_2, \ldots, d_\ell\}$ has been created to get the best option (s).

Step II: *Estimate the DEs' weights.*

Significance degree of DEs is defined by in terms of *"Linguistic Variables (LVs)"* and is articulated in SVNNs. For computation of kth DE, let $D_k = (t_k, i_k, f_k)$ be the SVNN, then the weight value is obtained as

$$\lambda_k = \left(\frac{\left(t_k + i_k \left(\frac{t_k}{t_k + f_k} \right) \right)}{\sum_{k=1}^{\ell} \left(t_k + i_k \left(\frac{t_k}{t_k + f_k} \right) \right)} \right), k = 1(1)\ell. \tag{3.5}$$

Here, $\sum_{k=1}^{\ell} \lambda_k = 1$ and $\lambda_k \geq 0$.

Step III: *Obtain the "Aggregated SVN-decision-matrix (A-SVN-DM)".*

To create the A-SVN-DM, we apply the SVNWAO and then $Z = \left(\widehat{\xi}_{ij} \right)_{m \times n}$, where

$$\begin{aligned} \widehat{\xi}_{ij} &= \text{SVNWA}\left(h_{ij}^{(1)}, h_{ij}^{(2)}, \ldots, h_{ij}^{(\ell)} \right) \\ &= \left(1 - \prod_{k=1}^{\ell} (1 - t_k)^{\lambda_k}, \prod_{k=1}^{\ell} (i_k)^{\lambda_k}, \prod_{k=1}^{\ell} (f_k)^{\lambda_k} \right). \end{aligned} \tag{3.6}$$

Step IV: *Estimate the attribute weights by CRITIC method*

Let $w = (w_1, w_2, \ldots, w_n)^T$ with $\sum_{j=1}^{n} w_j = 1$ and $w_j \in [0, 1]$ be criteria weight. Here, the CRITIC tool is used to obtain the attribute weights as

Step IV-A: Applying (3.7) to obtain the *"SVN-score matrix (SVN-SM)"* $Z = \left(\widehat{\xi}_{ij} \right)_{m \times n}, \forall i, j$ of each SVNN $\widehat{\xi}_{ij}$

$$\widehat{\xi}_{ij} = \frac{2 + t_{ij} - i_{ij} - f_{ij}}{3} \tag{3.7}$$

Step IV-B: Utilizing Eq. (3.8), convert the SVN-DM $Z = \left(\widehat{\xi}_{ij}\right)_{m \times n}$ into standard SVN-DM $\widetilde{S} = \left(\widetilde{\varsigma}_{ij}\right)_{m \times n}$

$$\widetilde{\varsigma}_{ij} = \begin{cases} \dfrac{\widehat{\xi}_{ij} - \varsigma_j^-}{\varsigma_j^+ - \varsigma_j^-}, & j \in F_b \\ \dfrac{\varsigma_j^+ - \widehat{\xi}_{ij}}{\varsigma_j^+ - \varsigma_j^-}, & j \in F_n \end{cases} \tag{3.8}$$

where $\varsigma_j^+ = \max_i \widehat{\xi}_{ij}$ and $\varsigma_j^- = \min_i \widehat{\xi}_{ij}$.

Step IV-C: Obtain the SDs of each attribute form Eq. (3.9)

$$\sigma_j = \sqrt{\frac{\sum_{i=1}^{m} \left(\widetilde{\varsigma}_{ij} - \overline{\varsigma}_j\right)^2}{m}}, \text{ where } \overline{\varsigma}_j = \sum\nolimits_{i=1}^{m} \widetilde{\varsigma}_{ij}/m. \tag{3.9}$$

Step IV-D: Obtain the CRC between attributes by Eq. (3.10)

$$r_{jt} = \frac{\sum_{i=1}^{m} \left(\widetilde{\varsigma}_{ij} - \overline{\varsigma}_j\right)\left(\widetilde{\varsigma}_{it} - \overline{\varsigma}_t\right)}{\sqrt{\sum_{i=1}^{m} \left(\widetilde{\varsigma}_{ij} - \overline{\varsigma}_j\right)^2 \left(\widetilde{\varsigma}_{it} - \overline{\varsigma}_t\right)^2}}, \tag{3.10}$$

Step IV-E: Examine the quantity of information as

$$c_j = \sigma_j \sum_{t=1}^{n} \left(1 - r_{jt}\right). \tag{3.11}$$

Step IV-F: Obtain the attribute weight by Eq. (3.13)

$$w_j = \frac{c_j}{\sum\nolimits_{j=1}^{n} c_j}. \tag{3.12}$$

Step V: *Evaluate "Normalized A-SVN-DM (NA-SVN-DM)"*

In MCDM system, the NA-SVN-DM $\overline{N} = \left(\varsigma_{ij}\right)_{m \times n}$ is obtained, where

$$\varsigma_{ij} = \begin{cases} \widehat{\xi}_{ij} = \left(\widehat{t}_{ij}, \widehat{i}_{ij}, \widehat{f}_{ij}\right), & j \in f_b \\ \left(\widehat{\xi}_{ij}\right)^c = \left(\widehat{f}_{ij}, 1 - \widehat{i}_{ij}, \widehat{t}_{ij}\right), & j \in f_n \end{cases} \tag{3.13}$$

such that f_b and f_n are benefit and cost-types attributes, respectively.
Step VI: *Assessment the "Optimal Significance Rating (OSR)"*

$$s_0 = \begin{cases} \max \widehat{\xi}_{ij}, \ j \in f_b \\ \min \widehat{\xi}_{ij}, \ j \in f_n \end{cases} \tag{3.14}$$

Step VII: *Create "Weighted Normalized A-SVN-DM (WNA-SVN-DM)"*
Let $w = (w_1, w_2, \ldots, w_n)^T$ be the attribute weight of $f_j : j = 1(1)n$, then the WNA-SVN-DM $\overline{N}_w = \left(\widetilde{\varsigma}_{ij}\right)_{m \times n}$ is given by

$$\widetilde{\varsigma}_{ij} = \langle \widetilde{\mu}_{ij}, \widetilde{\nu}_{ij} \rangle = \bigoplus_{j=1}^{n} w_j \varsigma_{ij} = \left(1 - \prod_{j=1}^{n} (1 - t_{ij})^{w_j}, \prod_{j=1}^{n} (i_{ij})^{w_j}, \prod_{j=1}^{n} (f_{ij})^{w_j}\right). \tag{3.15}$$

Step VIII: *Calculation of score values*
Using Eq. (3.10), the score values of WNA-SVN-DM $\overline{N}_w = \left(\widetilde{\varsigma}_{ij}\right)_{m \times n}$ are computed as

$$\mathbb{S}\left(\widetilde{\varsigma}_{ij}\right) = \left[\frac{2 + \widetilde{t}_{ij} - \widetilde{i}_{ij} - \widetilde{f}_{ij}}{3}\right]; i = 1(1)m, j = 1(1)n. \tag{3.16}$$

Step IX: *Calculate the "Overall Performance Score (OPS)" and "Utility Degree (UD)"*
The OPSs can be obtained by using the expression:

$$M_i = \sum_{j=1}^{n} \mathbb{S}\left(\widetilde{\varsigma}_{ij}\right), i = 1(1)m. \tag{3.17}$$

The best choice has the utmost *UD*, while the worst option has the lowermost degree of M_i. In order to the different *UD* is evaluated using the observed variant with OSR s_0. The *UD* (u_i) of option s_i is given as

$$u_i = \frac{M_i}{S_0}; i = 1(1)m. \tag{3.18}$$

Hence, $u_i \in [0, 1]$and UD u_i prioritizes in an ascending order, which offers the essential preference order.

Step X: *Select the best option*

The determined options are prioritized by ascending u_i, i.e., the maximum value u_i of option is the optimal one. Therefore, the most appropriate option can be obtained as

$$M^* = \left\{ M_i | \ \max_i u_i; i = 1(1)m \right\}, \tag{3.19}$$

where M^* is most desirable option.

Step XI: *End.*

3.5 Case Study: Evaluation of "Sustainable Circular Supplier Selection (SCSS)"

Here, the SVN-CRITIC-ARAS methodology is utilized to treat the SCSS assessment in the manufacturing companies. To assess, prioritize, and examine the SCS, a committee of three DEs is created in the manufacturing companies, comprising one supply chain executives, one *CE* manager, and one manufacturing supervisor. All nominated DEs have had experience to SSC and *CE* processes for over ten years. Based on the previous reviews, the four key SCS (s_1, s_2, s_3, and s_4) options are chosen over various sustainable parameters after a preliminary assessment.

Next, certain DEs aided researchers in preparing the questionnaires to rank the SCSs in the manufacturing companies in the *CE*. Initially, a survey method with a conference with DEs and existing studies has been made to select the sustainable parameters to evaluate SCSs. Then, we have provided a questionnaire involving 20 parameters to DEs to choose and evaluate these selected parameters. Lastly, the DEs considered 13 sustainable parameters to select the SCSs in the *CE*. The DEs offer the evaluation ratings of SCSs in terms of SVNNs over 13 different sustainable parameters (Table 3.1). The selected sustainable parameters are characterized into the four dimensions of sustainable viewpoints. The details of the different sustainable parameters and dimensions are given in Table 3.1 and Fig. 3.1.

Steps I and II Tables 3.2 and 3.3 depict the significance ratings of the DEs and parameters in *"linguistic values (LVs)"* and then changed into SVNNs. Table 3.4 presents the DEs weight based on Table 3.2 and Eq. (3.5). Table 3.5 shows the importance ratings of SCSs over considered sustainable parameters by DEs.

Step III The ratings given by three DEs are combined using Eq. (3.6) into an A-SVN-DM $Z = \left(\widehat{\xi}_{ij}\right)_{m \times n}$, with the consideration the individual DEs' importance; then, they are presented in Table 3.6.

Table 3.1 Detailed explanation of criteria for SCSS

Dimensions	Criteria	Type	Sources
Economic	Quality (f_1)	Max	[6, 8, 14, 27, 28]
	Costs (f_2)	Min	[6, 8, 27, 28, 36]
	Flexibility (f_3)	Max	[6, 8, 14]
	Technology capability (TC) (f_4)	Max	[8, 36]
Environmental	Green technology innovation (f_5)	Max	[5, 23, 32, 59, 60]
	Environmental management system (EMS) (f_6)	Max	[2, 11, 34, 50, 57]
	Eco-design (f_7)	Max	[5, 23, 32, 59, 60]
Social	Occupational health & safety (f_8)	Max	[12, 14, 27, 36]
	Job creation (f_9)	Max	[12, 27, 44]
	Reputation (f_{10})	Max	[12, 22, 25, 28]
Circular	Air pollution caused from production and recycling products (f_{11})	Min	[14, 18, 27]
	Developing eco-friendly and recyclable raw materials (f_{12})	Max	[18, 27, 28]
	Using systematic and clean technologies in production and recycling products (f_{13})	Max	[12, 14, 18, 39]

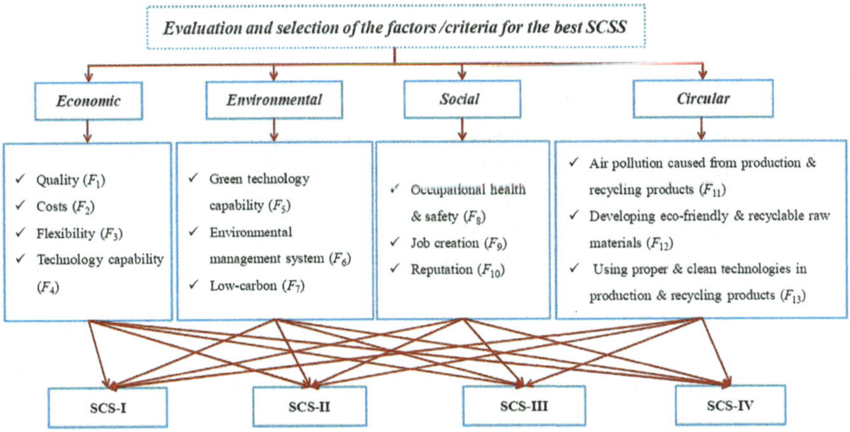

Fig. 3.1 Hierarchical procedure for the assessment of sustainable parameters of SCSS

Table 3.2 LVs for estimating the DEs weights

LVs	SVNNs
Extremely eligible (EE)	(0.90,0.10,0.10)
Very very eligible (VVE)	(0.80,0.25,0.20)
Very eligible (VE)	(0.70,0.35,0.30)
Eligible (E)	(0.60,0.45,0.40)
Less eligible (LE)	(0.50,0.65,0.60)
Very less eligible (VLQ)	(0.40,0.75,0.70)

Table 3.3 LVs for the evaluation of SCSS

LVs	SVNNs
Absolutely good (AG)	(1.00, 0.00, 0.00)
Very very good (VVG)	(0.90, 0.10, 0.10)
Very high (VG)	(0.80, 0.15, 0.20)
Good (G)	(0.70, 0.25, 0.30)
Moderately good (MG)	(0.60, 0.35, 0.40)
Good (G)	(0.50, 0.50, 0.50)
Moderately low (ML)	(0.40, 0.65, 0.60)
Low (L)	(0.30, 0.75, 0.70)
Very low (VL)	(0.20, 0.85, 0.80)
Very very low (VVL)	(0.10, 0.90, 0.90)
Absolutely low (AL)	(0.00, 1.00, 1.00)

Table 3.4 DEs' weights for assessing of SCSS

Des	LVs	SVNNs	Weights
d_1	EE	(0.90, 0.10,0.10)	0.3548
d_2	VVE	(0.75, 0.25, 0.20)	0.3396
d_3	VE	(0.60, 0.40, 0.35)	0.3056

Table 3.5 LVs with diverse DEs for SCSS

	s_1	s_2	s_3	s_4
F1	(G,MG,G)	(F,G,VG)	(G,F,G)	(F,MG,G)
F2	(L,VL,F)	(F,VL,MG)	(F,L,F)	(F,ML,F)
F3	(G,MG,F)	(G,G,VG)	(F,MG,L)	(G,F,F)
F4	(MG,G,F)	(F,G,ML)	(F,F,L)	(F,G,MG)
F5	(F,MG,MG)	(VG,G,MG)	(F,F,MG)	(G,VG,MG)
F6	(G,F,MG)	(F,MG,VG)	(G,F,F)	(F,MG,F)
F7	(G,MG,VG)	(G,G,F)	(F,L,MG)	(F,VG,F)
F8	(VG,G,MG)	(MG,G,F)	(F,F,ML)	(ML,F,MG)
F9	(VG,G,VG)	(ML,MG,VG)	(F,ML,MG)	(ML,MG,G)
F10	(F,G,ML)	(MG,F,VG)	(ML,MG,VG)	(F,ML,VG)
F11	(ML,L,F)	(L,L,F)	(F,ML,ML)	(F,ML,L)
F12	(VG,ML,MG)	(MG,G,F)	(VG,F,ML)	(F,G,VG)
F13	(F,MG,G)	(VG,MG,F)	(G,MG,F)	(G,MG,G)

Table 3.6 The A-SVN-DM for SCSS

	s_1	s_2	s_3	s_4
f_1	(0.674, 0.280, 0.331)	(0.688, 0.273, 0.318)	(0.647, 0.316, 0.357)	(0.609, 0.358, 0.397)
f_2	(0.341, 0.691, 0.661)	(0.454, 0.537, 0.548)	(0.443, 0.574, 0.561)	(0.472, 0.547, 0.532)
f_3	(0.619, 0.346, 0.387)	(0.740, 0.214, 0.265)	(0.493, 0.501, 0.514)	(0.587, 0.391, 0.417)
f_4	(0.619, 0.348, 0.388)	(0.564, 0.428, 0.444)	(0.452, 0.566, 0.554)	(0.614, 0.354, 0.393)
f_5	(0.573, 0.397, 0.433)	(0.721, 0.231, 0.284)	(0.538, 0.448, 0.467)	(0.721, 0.233, 0.285)
f_6	(0.614, 0.351, 0.390)	(0.655, 0.307, 0.350)	(0.587, 0.391, 0.417)	(0.543, 0.443, 0.464)
f_7	(0.712, 0.240, 0.292)	(0.656, 0.309, 0.351)	(0.479, 0.515, 0.524)	(0.643, 0.332, 0.366)
f_8	(0.721, 0.251, 0.284)	(0.619, 0.348, 0.388)	(0.477, 0.542, 0.529)	(0.507, 0.492, 0.498)
f_9	(0.775, 0.178, 0.230)	(0.631, 0.337, 0.374)	(0.507, 0.490, 0.497)	(0.583, 0.393, 0.423)
f_{10}	(0.564, 0.428, 0.444)	(0.655, 0.305, 0.349)	(0.631, 0.337, 0.374)	(0.279, 0.762, 0.726)
f_{11}	(0.405, 0.630, 0.598)	(0.372, 0.663, 0.632)	(0.442, 0.592, 0.562)	(0.415, 0.619, 0.590)
f_{12}	(0.644, 0.320, 0.359)	(0.619, 0.348, 0.388)	(0.622, 0.353, 0.382)	(0.688, 0.273, 0.318)
f_{13}	(0.609, 0.358, 0.397)	(0.670, 0.289, 0.335)	(0.619, 0.346, 0.387)	(0.674, 0.280, 0.331)

Step IV The CRITIC tool is extended under SVNSs setting with the aim of determining the criteria weights. We, first, used Table 3.6 and Eq. (3.7) to calculate the score matrix $Z = \left(\widehat{\xi}_{ij}\right)_{m \times n}$. After that, the standard q-ROF-matrix $\widetilde{S} = \left(\widetilde{\varsigma}_{ij}\right)_{m \times n}$ needs to be computed by means of Eq. (3.8). In addition, SD, CRC, and the quantity of information of each parameter of SCSS are assessed with the use of Eqs. (3.9), (3.10), (3.11). Finally, Eq. (3.12) is used to obtain the weights of parameters of SCSS in the manufacturing companies and the results are provided in Table 3.7.

$$w_j = \begin{array}{l}(0.0720, 0.0919, 0.0489, 0.0667, 0.0944, 0.0667, 0.0499, 0.0563, 0.0547,\\ 0.1069, 0.0496, 0.1227, 0.1193).\end{array}$$

Step V Since parameter f_2 and f_{11} are cost-type and rest of the benefit-type, the NA-q-ROF-DM is obtained by Eq. (3.13) and shown in Table 3.8.

Step VI Afterward, the OSR of SCSs is determined using Eq. (3.14). The obtained OSR of the companies is given as

Table 3.7 The weight value for each SCS factor by CRITIC tool

	s_1	s_2	s_3	s_4	σ_j	c_j	w_j
*f*1	0.859	1.000	0.493	0.000	0.387	3.256	0.0720
*f*2	1.000	0.060	0.211	0.000	0.401	4.158	0.0919
*f*3	0.520	1.000	0.000	0.384	0.358	2.211	0.0489
*f*4	1.000	0.653	0.000	0.973	0.403	3.019	0.0667
*f*5	0.206	1.000	0.000	0.994	0.453	4.269	0.0944
*f*6	0.658	1.000	0.395	0.000	0.366	3.017	0.0667
*f*7	1.000	0.751	0.000	0.681	0.371	2.258	0.0499
*f*8	1.000	0.610	0.000	0.141	0.395	2.546	0.0563
*f*9	1.000	0.474	0.000	0.291	0.364	2.472	0.0547
*f*10	0.744	1.000	0.934	0.000	0.398	4.834	0.1069
*f*11	0.524	1.000	0.000	0.500	0.358	2.243	0.0496
*f*12	0.386	0.000	0.021	1.000	0.404	5.550	0.1227
*f*13	0.000	0.919	0.151	1.000	0.446	5.395	0.1193

Table 3.8 The NA-SVN-DM for SCSS

	*s*0	s_1	s_2	s_3	s_4
*f*1	(0.688, 0.273, 0.318)	(0.674, 0.280, 0.331)	(0.688, 0.273, 0.318)	(0.647, 0.316, 0.357)	(0.609, 0.358, 0.397)
*f*2	(0.661, 0.309, 0.341)	(0.341, 0.691, 0.661)	(0.454, 0.537, 0.548)	(0.443, 0.574, 0.561)	(0.472, 0.547, 0.532)
*f*3	(0.740, 0.214, 0.265)	(0.619, 0.346, 0.387)	(0.740, 0.214, 0.265)	(0.493, 0.501, 0.514)	(0.587, 0.391, 0.417)
*f*4	(0.619, 0.348, 0.388)	(0.619, 0.348, 0.388)	(0.564, 0.428, 0.444)	(0.452, 0.566, 0.554)	(0.614, 0.354, 0.393)
*f*5	(0.721, 0.231, 0.284)	(0.573, 0.397, 0.433)	(0.721, 0.231, 0.284)	(0.538, 0.448, 0.467)	(0.721, 0.233, 0.285)
*f*6	(0.655, 0.307, 0.350)	(0.614, 0.351, 0.390)	(0.655, 0.307, 0.350)	(0.587, 0.391, 0.417)	(0.543, 0.443, 0.464)
*f*7	(0.712, 0.240, 0.292)	(0.712, 0.240, 0.292)	(0.656, 0.309, 0.351)	(0.479, 0.515, 0.524)	(0.643, 0.332, 0.366)
*f*8	(0.721, 0.251, 0.284)	(0.721, 0.251, 0.284)	(0.619, 0.348, 0.388)	(0.477, 0.542, 0.529)	(0.507, 0.492, 0.498)
*f*9	(0.775, 0.178, 0.230)	(0.775, 0.178, 0.230)	(0.631, 0.337, 0.374)	(0.507, 0.490, 0.497)	(0.583, 0.393, 0.423)
*f*10	(0.655, 0.305, 0.349)	(0.564, 0.428, 0.444)	(0.655, 0.305, 0.349)	(0.631, 0.337, 0.374)	(0.279, 0.762, 0.726)
*f*11	(0.632, 0.337, 0.372)	(0.405, 0.630, 0.598)	(0.372, 0.663, 0.632)	(0.442, 0.592, 0.562)	(0.415, 0.619, 0.590)
*f*12	(0.688, 0.273, 0.318)	(0.644, 0.320, 0.359)	(0.619, 0.348, 0.388)	(0.622, 0.353, 0.382)	(0.688, 0.273, 0.318)
*f*13	(0.674, 0.280, 0.331)	(0.609, 0.358, 0.397)	(0.670, 0.289, 0.335)	(0.619, 0.346, 0.387)	(0.674, 0.280, 0.331)

Table 3.9 The WNA-SVN-DM for SCSS

	$s0$	s_1	s_2	s_3	s_4
$f1$	(0.080, 0.911, 0.921)	(0.077, 0.912, 0.923)	(0.080, 0.911, 0.921)	(0.072, 0.920, 0.928)	(0.065, 0.929, 0.936)
$f2$	(0.095, 0.898, 0.906)	(0.095, 0.898, 0.906)	(0.070, 0.932, 0.930)	(0.073, 0.925, 0.928)	(0.067, 0.930, 0.933)
$f3$	(0.064, 0.927, 0.937)	(0.046, 0.949, 0.955)	(0.064, 0.927, 0.937)	(0.033, 0.967, 0.968)	(0.042, 0.955, 0.958)
$f4$	(0.062, 0.932, 0.939)	(0.062, 0.932, 0.939)	(0.054, 0.945, 0.947)	(0.039, 0.963, 0.961)	(0.062, 0.933, 0.940)
$f5$	(0.114, 0.871, 0.888)	(0.077, 0.917, 0.924)	(0.114, 0.871, 0.888)	(0.070, 0.927, 0.931)	(0.114, 0.872, 0.888)
$f6$	(0.069, 0.924, 0.932)	(0.062, 0.932, 0.939)	(0.068, 0.924, 0.932)	(0.057, 0.939, 0.943)	(0.051, 0.947, 0.950)
$f7$	(0.060, 0.931, 0.940)	(0.060, 0.931, 0.940)	(0.052, 0.943, 0.949)	(0.032, 0.967, 0.968)	(0.050, 0.946, 0.951)
$f8$	(0.069, 0.925, 0.932)	(0.069, 0.925, 0.932)	(0.053, 0.942, 0.948)	(0.036, 0.966, 0.965)	(0.039, 0.961, 0.962)
$f9$	(0.078, 0.910, 0.923)	(0.078, 0.910, 0.923)	(0.053, 0.942, 0.948)	(0.038, 0.962, 0.962)	(0.047, 0.950, 0.954)
$f10$	(0.108, 0.881, 0.894)	(0.085, 0.913, 0.917)	(0.107, 0.881, 0.894)	(0.101, 0.890, 0.900)	(0.034, 0.971, 0.966)
$f11$	(0.048, 0.947, 0.952)	(0.044, 0.952, 0.956)	(0.048, 0.948, 0.952)	(0.040, 0.956, 0.960)	(0.043, 0.953, 0.957)
$f12$	(0.133, 0.853, 0.869)	(0.119, 0.869, 0.882)	(0.112, 0.879, 0.890)	(0.113, 0.880, 0.889)	(0.133, 0.853, 0.869)
$f13$	(0.125, 0.859, 0.876)	(0.106, 0.885, 0.896)	(0.124, 0.862, 0.878)	(0.109, 0.881, 0.893)	(0.125, 0.859, 0.876)

s_0 = {(0.688, 0.273, 0.318), (0.341, 0.691, 0.661), (0.740, 0.214, 0.265), (0.619, 0.348, 0.388), (0.721, 0.231, 0.284), (0.655, 0.307, 0.350), (0.712, 0.240, 0.292), (0.721, 0.251, 0.284), (0.775, 0.178, 0.230), (0.655, 0.305, 0.349), (0.372, 0.663, 0.632), (0.688, 0.273, 0.318), (0.674, 0.280, 0.331)}.

Step VII Applying Eq. (3.15) and Table 3.8, the WNA-q-ROF-DM is obtained and presented in Table 3.9.

Steps VIII–X Applying Table 3.9 and Eq. (3.1), the score degree $\mathbb{S}(\widetilde{\varsigma}_{ij})$ of SVNNs are calculated. Using Eq. (3.16), the OPS (M_i) of each SCS alternative is obtained and given in Table 3.10. From Eq. (3.17), the *UD* (u_i) is computed as $u_1 = 0.8822$, $u_2 = 0.8982$, $u_3 = 0.7212$ and $u_4 = 0.7798$, then, the ranking order for the SCS alternative is determined as $s_2 \succ s_1 \succ s_4 \succ s_3$. Hence, the desirable SCS alternative is SCS-*II* (s_2) is the best SCS alternative. Results are presented in Table 3.10.

Table 3.10 The outcomes of SVN-CRITIC-ARAS framework for SCSS

	f_1	f_2	f_3	f_4	f_5	f_6	f_7	f_8	f_9	f_{10}	f_{11}	f_{12}	f_{13}	M_i
s_0	0.083	0.097	0.066	0.064	0.118	0.071	0.063	0.071	0.082	0.111	0.050	0.137	0.130	1.1424
s_1	0.081	0.097	0.047	0.064	0.079	0.063	0.063	0.071	0.082	0.085	0.045	0.123	0.109	1.0078
s_2	0.083	0.070	0.066	0.054	0.118	0.071	0.053	0.054	0.054	0.111	0.050	0.114	0.128	1.0261
s_3	0.074	0.073	0.033	0.038	0.071	0.058	0.032	0.035	0.038	0.104	0.041	0.115	0.112	0.8239
s_4	0.067	0.068	0.043	0.063	0.118	0.051	0.051	0.039	0.047	0.032	0.044	0.137	0.130	0.8908

3.5.1 *Comparison and Discussion*

Here, a comparison is made with the results obtained from the SVN-CRITIC-ARAS and those extant approaches. To illustrate the efficacy and the irreplaceable virtues of the SVN-CRITIC-ARAS methodology, the SVN-TOPSIS method [37] and SVN-VIKOR [40, 41] model are used to assess the abovementioned MCDM problem. The structure SVN-TOPSIS tool is described in Sect. 3.5.1.1 as follows:

3.5.1.1 SVN-TOPSIS Model

Steps I–IV: *Similar to previous approach.*

Step V: *Compute the "SVN-Ideal Solution (SVN-IS)" and "SVN- Anti-Ideal Solution (SVN-AIS)".*

Let ξ_j^+ and ξ_j^- signify the "SVN-IS" and "SVN-AIS" and are obtained as

$$\xi_j^+ = \left(t_j^+, i_j^+, f_j^+\right) = \begin{cases} \left(\max_i \widehat{t}_{ij}, \min_i \widehat{i}_{ij}, \min_i \widehat{f}_{ij}\right), & j \in f_b \\ \left(\min_i \widehat{t}_{ij}, \max_i \widehat{i}_{ij}, \max_i \widehat{f}_{ij}\right), & j \in f_n \end{cases} \tag{3.19}$$

$$\xi_j^- = \left(t_j^-, i_j^-, f_j^-\right) = \begin{cases} \left(\min_i \widehat{t}_{ij}, \max_i \widehat{i}_{ij}, \max_i \widehat{f}_{ij}\right), & j \in f_b \\ \left(\max_i \widehat{t}_{ij}, \min_i \widehat{i}_{ij}, \min_i \widehat{f}_{ij}\right), & j \in f_n \end{cases} \tag{3.20}$$

Step VI: *Estimate the distances of option from "SVN-IS" and "SVN-AIS".*

Next, we obtain the weighted distances of each option are defined by

$$J_w\left(\widehat{\xi}_{ij}, \xi_j^+\right) = \frac{1}{(\sqrt{2}-1)} \sum_{i=1}^{n} w_j \left[\begin{array}{l} \left(\frac{\left(\widehat{t}_{ij}\right)^2 + \left(t_j^+\right)^2}{2}\right)^{1/2} - \frac{\widehat{t}_{ij} + \left(t_j^+\right)}{2} \\ + \left(\frac{\left(\widehat{i}_{ij}\right)^2 + \left(i_j^+\right)^2}{2}\right)^{1/2} - \frac{\widehat{i}_{ij} + \left(i_j^+\right)}{2} \\ + \left(\frac{\left(\widehat{f}_{ij}\right)^2 + \left(f_j^+\right)^2}{2}\right)^{1/2} - \frac{\widehat{f}_{ij} + \left(f_j^+\right)}{2} \end{array} \right], \tag{3.21}$$

Table 3.11 Results achieved by SVN-TOPSIS model for SCSS

SCS option	$J_w\left(\widehat{\xi}_{ij}, \xi_j^+\right)$	$J_w\left(\widehat{\xi}_{ij}, \xi_j^-\right)$	C_i	Ranking
s_1	0.051	0.125	0.711	2
s_2	0.051	0.134	0.726	1
s_3	0.125	0.051	0.289	4
s_4	0.107	0.068	0.390	3

$$J_w\left(\widehat{\xi}_{ij}, \xi_j^-\right) = \frac{1}{(\sqrt{2}-1)} \sum_{i=1}^{n} w_j \left[\begin{array}{l} \left(\frac{\left(\widehat{t}_{ij}\right)^2+\left(t_j^-\right)^2}{2}\right)^{1/2} - \frac{\widehat{t}_{ij} + \left(t_j^-\right)}{2} \\ +\left(\frac{\left(\widehat{i}_{ij}\right)^2+\left(i_j^-\right)^2}{2}\right)^{1/2} - \frac{\widehat{i}_{ij} + \left(i_j^-\right)}{2} \\ +\left(\frac{\left(\widehat{f}_{ij}\right)^2+\left(f_j^-\right)^2}{2}\right)^{1/2} - \frac{\widehat{f}_{ij} + \left(f_j^-\right)}{2} \end{array} \right]. \quad (3.22)$$

***Step VII:** Evaluate the "Closeness Index (CI)".*

Now, we obtain the CI of each option as follows:

$$C_i = \frac{J_w\left(\widehat{\xi}_{ij}, \xi_j^-\right)}{J_w\left(\widehat{\xi}_{ij}, \xi_j^-\right) + J_w\left(\widehat{\xi}_{ij}, \xi_j^+\right)}, C_i \in [0, 1]. \quad (3.23)$$

A set of options that can be prioritized using the decreasing order of $C_i \in [0, 1]$; and maximum rating shows the better choice.

***Step VIII:** End.*

From Table 3.6 and Eqs. (3.19) and (3.20), "SVN-IS" and "SVN-AIS" are evaluated.

ξ_j^+ = {(0.688, 0.273, 0.318), (0.341, 0.691, 0.661), (0.740, 0.214, 0.265), (0.619, 0.348, 0.388), (0.721, 0.231, 0.284), (0.655, 0.307, 0.350), (0.712, 0.240, 0.292), (0.721, 0.251, 0.284), (0.775, 0.178, 0.230), (0.655, 0.305, 0.349), (0.372, 0.663, 0.632), (0.688, 0.273, 0.318), (0.674, 0.280, 0.331)}

ξ_j^- = {(0.609, 0.358, 0.397), (0.472, 0.547, 0.532), (0.493, 0.501, 0.514), (0.452, 0.566, 0.554), (0.538, 0.448, 0.467), (0.543, 0.443, 0.464), (0.479, 0.515, 0.524), (0.477, 0.542, 0.529), (0.507, 0.490, 0.497), (0.279, 0.762, 0.726), (0.442, 0.592, 0.562), (0.619, 0.348, 0.388), (0.609, 0.358, 0.397)}.

The whole computational results using Eqs. (3.21) (3.22) and (3.23) of SVN-TOPSIS [37] method are depicted in Table 3.11. Thus, SCS-II (s_2) is the best SCS option.

3.5.1.2 SVN-VIKOR Method

Step I–V: *Same as the previous model.*

Step VI: The *"Group Utility Measure (GUM)"*, *"Individual Regret Measure (IRM)"* and *"Compromise Measure (CM)"* of each SCSS option are obtained in Table 3.12.

Step VII: Obtain the prioritization of the alternatives. The lowest degree of *CM* provides the best option.

The complete results of the SVN-VIKOR tool are obtained in Table 3.12. Thus, SCS-IV (s_4) is the best choice. The priority orders of the four SCS alternative obtained by the SVN-CRITIC-ARAS method is similar to the SVN-TOPSIS model, while is slightly differ with those obtained by the SVN-VIKOR model. Here, we discuss the outcomes of the developed SVN-CRITIC-ARAS approach as follows:

(i) In proposed method, the weights of DEs are computed with the developed procedure providing in more accurate results unlike randomly elected weights in SVN-TOPSIS model.
(ii) The criteria weights in developed approach are found using the CRITIC tool, which can tackle the inconsistency arise in weighting procedure; hence, the obtained outcomes are more precise and optimum, whereas in SVN-TOPSIS, the criteria weights are computed by discrimination measure on SVNNs, while in SVN-VIKOR, the subjective weights of attributes are estimated by SWARA tool.
(iii) The proposed SVN-CRITIC-ARAS approach only obtains the SVN-IS, while the SVN-TOPSIS, SVN-VIKOR models need to define the SVN-IS and SVN-AIS. This shows that the SVN-CRITIC-ARAS method can improve the operative usefulness to certain extent.
(iv) For SVN-TOPSIS and SVN-VIKOR models, it is essential to obtain the discriminations between each option and the SVN-IS, which is an extensive process and weakens the exactness of results. Alternatively, the computation formula of the SVN-CRITIC-ARAS approach is easier, and thus the correctness and consistency of the outcomes is higher (Figs. 3.2 and 3.3).

Table 3.12 The degrees of *GUM*, *IRM*, and *CM* for SCSS

SCS option	*GUM*	Priority order	*IRM*	Priority order	*CM*	Priority order
s_1	0.451	2	0.114	3	0.379	2
s_2	0.307	1	0.123	4	0.500	3
s_3	0.729	4	0.108	2	0.538	4
s_4	0.505	3	0.107	1	0.234	1

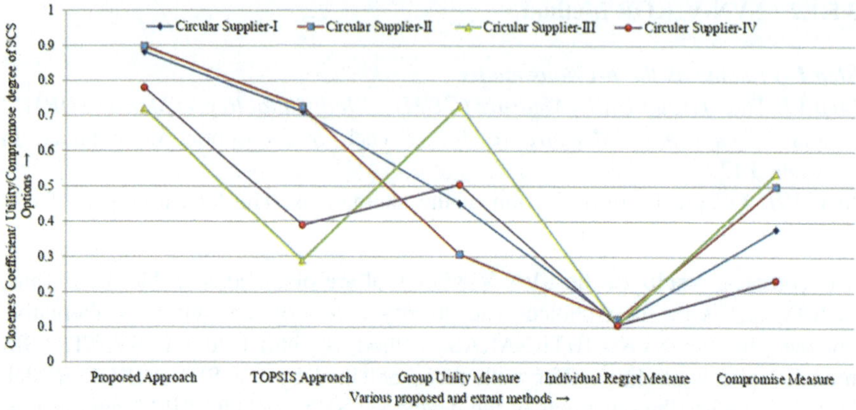

Fig. 3.2 Comparison of UDs of developed with extant methods for SCSS

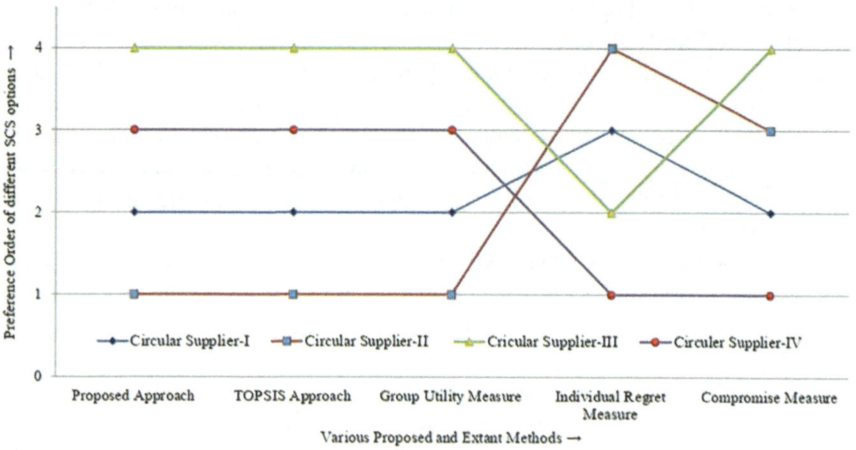

Fig. 3.3 Variation of preference order of developed with extant models for SCSS

3.5.2 *Managerial Implication*

This work describes an appropriate and effective MCDM system for helping managers in selecting SCSs using the SVN-CRITIC-ARAS method. One of the benefits of the developed method is that it is likely to add more attributes or subtract extant ones based on DEs and framework being discussed. Also, the SVN-CRITIC-ARAS method utilizes not only to SCSS but also to diverse regions namely preferences of supply chain risks, healthcare business, and among others. For instance, "Airbnb firm" is one of the major firms in the arena of peer-to-peer accommodation hires. This firm and other same companies are constantly observing for systems to increase consumer facility and fulfillment. These firms can, by utilizing the developed

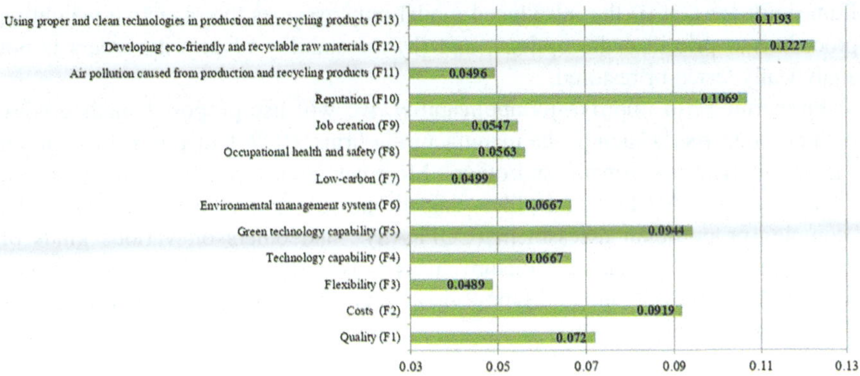

Fig. 3.4 Illustration of attribute weights with SVN-CRITIC-ARAS method

method, deliver them with appropriate accommodation options in relation to the prioritization so as to customers can find the appropriate options in the least feasible duration. Hence, the introduced model is a practical, efficient, and broad method that can be effortlessly employed in diverse application settings.

In this study, a survey approach with the use of the discussions with DEs and analysis has been discussed to classify and elect the key criteria to assess and examine SCSs in the *CE*. Here, after arguing with DEs in manufacturing companies and academician, we consider 13 parameters for SCSS assessment. We have obtained all the sustainable parameters form the previous literature review and mentioned in Table 3.1. Based on the outcomes of the developed methodology and from Fig. 3.4, it is found that sustainable parameter f_{12} takes the rank one (0.1227) followed by sustainable parameter f_{13} (0.1193), Reputation (0.1069), Green technology innovation (0.0944), costs (0.0919), Quality (0.072), EMS (f_6) (0.0667), TC (f_4) (0.0667), and other parameters which are obtained to utilize the SCS towards the accomplishments of sustainability perspectives.

3.6 Conclusions

The goal of the study is to develop a methodology to choose and evaluate the SCS in the *CE*. To do so, an extended methodology with the use of the ARAS and CRITIC methods for SVNSs has been proposed. A case study of SCSS has been discussed to illustrate the efficacy and practicality of the SVN-CRITIC-ARAS method. Lastly, an evaluation degree for SCS options in the *CE* is made, which involves four dimensions namely economic, social, environmental, and circular. The subsequent dimensions comprise four, three, three, and three parameters, respectively, and the parameters are widely considered from the literature. To certify the outcomes, a comparative discussion is made. The results certify that the introduced methodology has good efficacy and applicability with extant tools. Some flaws of the developed framework are that (i) the introduced model only assumes the objective weighting procedure to prioritize the options, and (ii) the stability of priority orders is not analytically tested or restored.

Furthermore, to improve its applicability, we will use proposed methodology with different issues such as the manufacturing firms, HCWT methods, low-carbon strategy tourism assessment, and others. Moreover, future studies can recommend various types of MCDM tools like "gained and lost dominance score (GDLS)", "complex proportional assessment (COPRAS)", and others on various kinds of "fuzzy sets (FSs)", namely "Picture fuzzy sets", "interval-valued Pythagorean fuzzy sets (IVPFSs)", "interval type-2 fuzzy sets (IT2FSs)", "Fermatean fuzzy sets (FFSs)", and others.

References

1. Adali, E. A., & Tus, A. (2019). Hospital site selection with distance-based multi-criteria decision-making methods. *International Journal of Healthcare Management*. https://doi.org/10.1080/20479700.2019.1674005
2. Alrasheedi, M., Mardani, A., Mishra, A. R., Rani, P., & Loganathan, N. (2021). An extended framework to evaluate sustainable suppliers in manufacturing companies using a new Pythagorean fuzzy entropy- SWARA-WASPAS decision-making approach. *Journal of Enterprise Information Management*. https://doi.org/10.1108/JEIM-07-2020-0263
3. Amini, A., Alinezhad, A., & Yazdipoor, F. (2019). A TOPSIS, VIKOR and DEA integrated evaluation method with belief structure under uncertainty to rank alternatives. *International Journal of Advanced Operations Management, 11*(3), 171–188.
4. Atanassov, K. (1986). Intuitionistic fuzzy sets. *Fuzzy Sets and Systems, 20*, 87–96.
5. Awasthi, A., Chauhan, S. S., & Goyal, S. K. (2010). A fuzzy multicriteria approach for evaluating environmental performance of suppliers. *International Journal of Production Economics, 126*, 370–378.
6. Awasthi, A., Govindan, K., & Gold, S. (2018). Multi-tier sustainable global supplier selection using a fuzzy AHP-VIKOR based approach. *International Journal of Production Economics, 195*, 106–117.
7. Baidya, J., Garg, H., Saha, A., Mishra, A. R., Rani, P., & Dutta, D. (2021). Selection of third party reverses logistic providers: An approach of BCF-CRITIC-MULTIMOORA using Archimedean power aggregation operators. *Complex & Intelligent Systems*. https://doi.org/10.1007/s40747-021-00413-x
8. Büyüközkan, G., & Çifçi, G. (2011). A novel fuzzy multi-criteria decision framework for sustainable supplier selection with incomplete information. *Computers in Industry, 62*(2), 164–174.
9. Büyüközkan, G., & Güler, M. (2020). Smart watch evaluation with integrated hesitant fuzzy linguistic SAW-ARAS technique. *Measurement, 153*, 107353.
10. Cavallaro, F. (2010). Fuzzy TOPSIS approach for assessing thermal-energy storage in concentrated solar power (CSP) systems. *Applied Energy, 87*(2), 496–503.
11. Chen, L., Duan, D., Mishra, A. R., & Alrasheedi, M. (2021). Sustainable third-party reverse logistics provider selection to promote circular economy using new uncertain interval-valued intuitionistic fuzzy-projection model. *Journal of Enterprise Information Management*. https://doi.org/10.1108/JEIM-02-2021-0066

12. Devika, K., Mina, H., Nosrati-Abarghooee, S., & Khosrojerdi, G. (2020). Sustainable circular supplier selection: A novel hybrid approach using streaming data. *Science of the Total Environment, 137936*. https://doi.org/10.1016/j.scitotenv.2020.137936
13. Diakoulaki, D., Mavrotas, G., & Papayannakis, L. (1995). Determining objective weights in multiple criteria problems: The CRITIC method. *Computers and Operations Research, 22*, 763–770.
14. Fallahpour, A., Olugu, E. U., Musa, S. N., Wong, K. Y., & Noori, S. (2017). A decision support model for sustainable supplier selection in sustainable supply chain management. *Computers and Industrial Engineering, 105*, 391–410.
15. Garg, H., & Nancy. (2020). Algorithms for single-valued neutrosophic decision making based on TOPSIS and clustering methods with new distance measure. *AIMS Mathematics, 5*(3), 2671–2693.
16. Ghorabaee, M. K., Amiri, M., Zavadskas, E. K., & Antuchevicience, J. (2017). Assessment of third-party logistics providers using a CRITIC–WASPAS approach with interval type-2 fuzzy sets. *Transport, 32*(1), 66–78.
17. Govindan, K., Shankar, M., & Kannan, D. (2018). Supplier selection based on corporate social responsibility practices. *International Journal of Production Economics, 200*, 353–379.
18. Govindan, K., Mina, H., Esmaeili, A., & Gholami-Zanjani, S. M. (2020). An integrated hybrid approach for circular supplier selection and closed-loop supply chain network design under uncertainty. *Journal of Cleaner Production, 242*, 118317. https://doi.org/10.1016/j.jclepro.2019.118317
19. Iordache, M., Schitea, D., Deveci, M., Akyurt, İ. Z., & Iordache, I. (2019). An integrated ARAS and interval type-2 hesitant fuzzy sets method for underground site selection: Seasonal hydrogen storage in salt caverns. *Journal of Petroleum Science and Engineering, 175*, 1088–1098.
20. Jana, C., & Pal, M. (2021). Multi-criteria decision making process based on some single-valued neutrosophic Dombi power aggregation operators. *Soft Computing, 25*, 5055–5072.
21. Karabasevic, D., Zavadskas, E. K., Turskis, Z., & Stanujkic, D. (2016). The framework for the selection of personnel based on the SWARA and ARAS methods under uncertainties. *Informatica, 27*(1), 49–65.
22. Kaya, R., & Yet, B. (2019). Building Bayesian networks based on DEMATEL formultiple criteria decision problems: A supplier selection case study. *Expert Systems with Applications, 134*, 234–248.
23. Kumari, R., & Mishra, A. R. (2020). Multi-criteria COPRAS method based on parametric measures for intuitionistic fuzzy sets: Application of green supplier selection. *Iranian Journal of Science and Technology, Transactions of Electrical Engineering, 44*, 1645–1662. https://doi.org/10.1007/s40998-020-00312-w
24. Kusi-Sarpong, S., Gupta, H., Khan, S. A., Jabbour, C. J. C., Rehman, S. T., & Kusi-Sarpong, H. (2019). Sustainable supplier selection based on industry 4.0 initiatives within the context of circular economy implementation in supply chain operations. *Production Planning and Control*. (in press).
25. Lo, H. W., Liou, J. J., Wang, H. S., & Tsai, Y. S. (2018). An integratedmodel for solving problems in green supplier selection and order allocation. *Journal of Cleaner Production, 190*, 339–352.
26. Luo, M., Wu, L., Zhou, K., & Zhang, H. (2019). Multi-criteria decision making method based on the single valued neutrosophic sets. *Journal of Intelligent & Fuzzy Systems, 37*(2), 2403–2417.
27. Memari, A., Dargi, A., Jokar, M. R. A., Ahmad, R., & Rahim, A. R. A. (2019). Sustainable supplier selection: A multi-criteria intuitionistic fuzzy TOPSISmethod. *Journal of Manufacturing Systems, 50*, 9–24.
28. Mina, H., Mirabedin, S. N., & Pakzad-Moghadam, S. H. (2014). An integrated fuzzy analytic network process approach for green supplier selection: A case study of petrochemical industry. *Management Science and Practice, 2*(2), 31–47.

29. Mishra, A. R., & Rani, P. (2019). Shapley divergence measures with VIKOR method for multi-attribute decision-making problems. *Neural Computing and Applications, 31*, 1299–1316.
30. Mishra, A. R., & Rani, P. (2021). Assessment of sustainable third party reverse logistic provider using the single-valued neutrosophic combined compromise solution framework. *Cleaner and Responsible Consumption*. https://doi.org/10.1016/j.clrc.2021.100011
31. Mishra, A., Sisodia, G., Raj Pardasani, K., & Sharma, K. (2020). Multi-criteria IT personnel selection on intuitionistic fuzzy information measures and ARAS methodology. *J Iranian Journal of Fuzzy Systems., 17*(4), 55–68.
32. Mishra, A. R., Rani, P., & Saha, A. (2021). Single-valued neutrosophic similarity measure-based additive ratio assessment framework for optimal site selection of electric vehicle charging station. *International Journal of Intelligent Systems, 36*, 5573–5604. https://doi.org/10.1002/int.22523
33. Mishra, A. R., Rani, P., Krishankumar, R., Ravichandran, K. S., & Kar, S. (2021). An extended fuzzy decision-making framework using hesitant fuzzy sets for the drug selection to treat the mild symptoms of Coronavirus Disease 2019 (COVID-19). *Applied Soft Computing, 103*, 107155. https://doi.org/10.1016/j.asoc.2021.107155
34. Mishra, A. R., Rani, P., & Pandey, K. (2021). Fermatean fuzzy CRITIC-EDAS approach for the selection of sustainable third-party reverse logistics providers using improved generalized score function. *Journal of Ambient Intelligence and Humanized Computing*. https://doi.org/10.1007/s12652-021-02902-w
35. Mishra, A. R., Rani, P., & Bharti, S. (2021). Assessment of agriculture crop selection using Pythagorean Fuzzy CRITIC–VIKOR decision-making framework. In H. Garg (Ed.), *Pythagorean fuzzy sets*. Springer. https://doi.org/10.1007/978-981-16-1989-2_7
36. Mohammed, A., Harris, I., & Kannan, G. (2019). A hybridMCDM-FMOO approach for sustainable supplier selection and order allocation. *International Journal of Production Economics, 217*, 171–184.
37. Nancy, & Garg, H. (2019). A novel divergence measure and its based TOPSIS method for multi criteria decision-making under single-valued neutrosophic environment. *Journal of Intelligent & Fuzzy Systems, 36*, 101–115.
38. Peng, X. (2019). New multiparametric similarity measure and distance measure for interval neutrosophic set with IoT industry evaluation. *IEEE Access, 7*, 28258–28280.
39. Qazvini, Z. E., Haji, A., & Mina, H. (2019). A fuzzy solution approach for supplier selection and order allocation in green supply chain considering location-routing problem. Scientia Iranica. *Transaction E Industrial Engineering*. https://doi.org/10.24200/sci.2019.50829.1885. (in press).
40. Rani, P., & Mishra, A. R. (2020). Single-valued Neutrosophic SWARA-VIKOR framework for performance assessment of eco-industrial thermal power plants. *ICSES Transactions on Neural and Fuzzy Computing, 3*, 1–9.
41. Rani, P., & Mishra, A. R. (2020). Novel single-valued neutrosophic combined compromise solution approach for sustainable waste electrical and electronics equipment recycling partner selection. *IEEE Transactions on Engineering Management*. https://doi.org/10.1109/TEM.2020.3033121
42. Rani, P., Mishra, A. R., Krishankumar, R., Ravichandran, K. S., & Gandomi, A. H. (2020). A new Pythagorean fuzzy based decision framework for assessing healthcare waste treatment. *IEEE Transactions on Engineering Management, 13*, 549–556. https://doi.org/10.1109/TEM.2020.3023707
43. Rani, P., Mishra, A. R., Krishankumar, R., Mardani, A., Ravichandran, K. S., & Kar, S. (2021). Multi-criteria food waste treatment method selection using single-valued neutrosophic-CRITIC-MULTIMOORA framework. *Applied Soft Computing, 111*, 107657. https://doi.org/10.1016/j.asoc.2021.107657
44. Rashidi, K., & Cullinane, K. (2019). A comparison of fuzzy DEA and fuzzy TOPSIS in sustainable supplier selection: Implications for sourcing strategy. *Expert Systems with Applications, 121*, 266–281.

45. Smarandache, F. (1999). *A unifying field in logics. Neutrosophy: Neutrosophic probability, set and logic*. American Research Press.
46. Smarandache, F. (2020). The score, accuracy, and certainty functions determine a total order on the set of neutrosophic triplets (T, I, F). *Neutrosophic Sets and Systems, 38*, 1–14. https://doi.org/10.5281/zenodo.4300354
47. Stanujkic, D. (2015). Extension of the ARAS method for decision-making problems with interval-valued triangular fuzzy numbers. *Informatica, 26*(2), 335–355.
48. Tavana, M., Khosrojerdi, G., Mina, H., & Rahman, A. (2019). A hybrid mathematical programming model for optimal project portfolio selection using fuzzy inference system and analytic hierarchy process. *Evaluation and Program Planning, 77*, 101703.
49. Topal, S., Çevik, A., & Smarandache, F. (2020). A new group decision making method with distributed indeterminacy form under Neutrosophic environment: An introduction to Neutrosophic Social Choice theory. *IEEE Access, 8*, 42000–42009.
50. Tseng, M. L., & Chiu, A. S. F. (2013). Evaluating firm's green supply chain management in linguistic preferences. *Journal of Cleaner Production, 40*, 22–31.
51. Turskis, Z., & Zavadskas, E. K. (2010). A new additive ratio assessment (ARAS) method in multicriteria decision-making. *Technological and Economic Development of Economy, 16*(2), 159–172.
52. Turskis, Z., & Zavadskas, E. K. (2010). A novel method for multiple criteria analysis: Grey additive ratio assessment (ARAS-G) method. *Informatica, 21*(4), 597–610.
53. Wang, H., Smarandache, F., Zhang, Y. Q., & Sunderraman, R. (2010). Single valued neutrosophic sets. *Multispace and Multistructure, 4*, 410–413.
54. Wei, G., Lei, F., Lin, R., Wang, R., Wei, Y., Wu, J., & Wei, C. (2020). Algorithms for probabilistic uncertain linguistic multiple attribute group decision making based on the GRA and CRITIC method: Application to location planning of electric vehicle charging stations. *Economic Research-Ekonomska Istraživanja, 33*(1), 828–846.
55. Xu, Z., Qin, J., Liu, J., & Martínez, L. (2019). Sustainable supplier selection based on AHPSort II in interval type-2 fuzzy environment. *Information Sciences, 483*, 273–293.
56. Ye, J. (2014). A multicriteria decision-making method using aggregation operators for simplified neutrosophic sets. *Journal of Intelligent & Fuzzy Systems, 26*, 2459–2466.
57. Yu, Q., & Hou, F. (2016). An approach for green supplier selection in the automobile manufacturing industry. *Kybernetes, 45*, 571–588.
58. Zadeh, L. A. (1965). Fuzzy sets. *Information and Control, 8*, 338–353.
59. Zarbakhshnia, N., Wu, Y., Govindan, K., & Soleimani, H. (2020). A novel hybrid multiple attribute decision making approach for outsourcing sustainable reverse logistics. *Journal of Cleaner Production, 242*, 118461.
60. Zhang, X., & Su, T. (2020). The dominance degree-based heterogeneous linguistic decision-making technique for sustainable 3PRLP selection. *Complexity*. https://doi.org/10.1155/2020/6102036. Article ID 6102036, 18 pages.
61. Zhang, H. Y., Ji, P., Wang, J.-Q., & Chen, X.-H. (2016). A neutrosophic normal cloud and its application in decision-making. *Cognitive Computation, 8*, 649–669.

Part II
Climate Change

Chapter 4
Linguistic-Based MCDM Approach for Climate Change Risk Evaluation Methodology

Gülçin Büyüközkan, Deniz Uztürk, and Yağmur Karabulut

Abstract Conventional production networks that focus on optimizing budgets and increasing productivity often give way to vulnerabilities, which are under threat by external shocks that can find and exploit these weaknesses. Climate change (CC) is increasingly becoming a main stressor for such shocks in supply chain management (SCM). In this chapter, significant risks and their interrelations in supply chains are explored. The risk evaluation is carried out with a Multi-Criteria Decision-Making (MCDM) approach, integrated with the Group Decision-Making (GDM) technique. This hybrid model is selected to better understand and analyze SCM risks that arise from CC. Multi-granular linguistic values are recommended for this assessment and prioritization. Multi-granular information helps gather evaluations from different experts. Additionally, linguistic assessments provide a more straightforward analysis closer to the human cognitive processes. For the risk assessment process, the most affected socio-economic systems are chosen as the first level, then the risks related to SCM are assigned as the second level. Risks are evaluated with the 2-Tuple-GDM framework. Multi-granular data is handled with the 2-Tuple Linguistic model, where its aggregation tools are applied to prioritize the risks stemming from the CC. A case study is provided to test the plausibility of the suggested risk assessment model.

Keywords Multi-granular linguistic values · MCDM · Climate change · Risk evaluation · Supply chain management

---

G. Büyüközkan (✉)
Galatasaray University, Industrial Engineering Department, Ortaköy-Istanbul, Türkiye
e-mail: gbuyukozkan@gsu.edu.tr

D. Uztürk
Galatasaray University, Department of Management, Business Research Center, Istanbul, Türkiye

Y. Karabulut
GITEC Consult, Cologne, Germany

G. Kayakutlu, M. Ö. Kayalica (eds.), *Decision Making Using AI in Energy and Sustainability*, Applied Innovation and Technology Management,
https://doi.org/10.1007/978-3-031-38387-8_4

4.1 Introduction

Investors, regulators, and other stakeholders are increasingly demanding companies to implement long-term strategies for managing their *Climate Change* (CC)-related risks. According to the 2021 report of the Intergovernmental Panel on Climate Change (IPCC) [1], it is unequivocal that human influence has warmed the atmosphere, ocean, and land, and human-induced CC is already affecting many weather and climate extremes in every region across the globe. A rapid and effective response is needed to transform business practices for a better future.

New business models connected to sustainable and circular supply chains require an in-depth analysis for identifying environmental-friendly strategies, complemented with Industry 4.0 technologies as a leverage [2]. Changing business processes from the ground presents many challenges due to regulatory restrictions, uncertainties, and dynamic changes in the work environment. Consequently, many businesses are more vulnerable to CC risks than ever. Prioritizing such risks and developing commensurate risk reduction actions can lead to more robust and flexible strategies for the long term. Developing appropriate strategies to mitigate and adapt to CC is crucial for creating resilient supply chains. While designing these strategies, the uncertainty and the lack of information are the main drawbacks for companies, policymakers, and decision-makers.

Multi-criteria decision-making (MCDM) approaches can help in this matter, as they have been proven useful for strategy generation and risk analysis [3]. MCDM models allow decision-maker preferences to be collected in a structure and integrated into the assessment process. Linguistic decision-making (LDM) is first revealed due to the need for defining uncertainty, vagueness, and incomplete information that probabilistic models cannot model. Zadeh introduced a fuzzy linguistic approach [4] to manage vagueness and impreciseness rising from linguistic descriptors. Yager [5] brings the need for LDM to the fore by emphasizing the facilitating influence on decision-making. Considering all the benefits of LDM and fuzzy sets, the risk assessment procedure is conducted in this chapter under a 2-Tuple Linguistic environment. The 2-Tuple Linguistic model was first introduced by Herrera and Martinez [6]. The 2-Tuple Linguistic model can successfully manage linguistic and numerical data in decision-making without losing or distorting the information content of the collected and processed linguistic data [7].

In this chapter, the essential objectives are:

- Determining the crucial CC risks for supply chains
- Providing an index for these risks
- Generating a linguistic-based framework to prioritize these risks to obtain an efficient roadmap to plan the strategies

By doing that, the chapter aims to show how identifying and ranking CC risks can lead to robust evaluation for future scenarios and definition of suitable responses to them. The proposed decision-making model employs LDM, combined with group decision-making (GDM) approaches. In GDM, the main idea is to generate a meaningful decision that is acceptable to all group members. For that purpose, consensus models and/or aggregation procedures can be followed depending on the situation. Generally, consensus models take longer to arrive at the ultimate decision since each group member should meet on the same idea. Aggregation procedures are quicker to reflect the group's vision into one meaningful value since they combine each member's point of view and propose an integrated solution that involves each member's idea.

In this chapter, the next section provides the theoretical basis of CC and supply chain management. Section 4.3 provides the suggested methodology. Section 4.4 provides the case study to see the plausibility of the recommended risk and strategy analysis. Section 4.5 provides the managerial implications, and finally, the conclusions and future perspectives are provided.

4.2 Theoretical Backgrounds

This section provides a rapid overview of the recent literature about CC and supply chain management, encompassing both academic and industrial publications.

4.2.1 Climate Change and Supply Chain Management in Academic Literature

Various reviews have been conducted in the academic literature to explore how to manage CC risks in supply chains. Ghadge et al. [8] highlighted the grave necessity to identify CC risks and develop options to deal with them, complemented with a research agenda for supply chain management with the help of a novel Supply Chain Risk Management framework. In another publication, Nimsai et al. [9] focused on the knowledge of sustainable supply chain management. They conducted a bibliometric literature review of sustainable supply chains and investigated internal processes and their environment respectively. Moshood et al. [10] also concentrated on sustainable supply chains by examining green and low-carbon manufacturing processes. They found that introducing environmental practices at the external level (downstream and upstream) is quite inadequate and that sustainable supply chain practices require enhanced organizational performance, reputation/risk management, customer pressure, and top management support. Few organizations are working together successfully with allies to expand the chain operation.

Apart from the reviews on CC and supply chain management, a mapping technique is used in this chapter to better understand the current approaches in the academia. For that purpose, "climate change" and "supply chain management" words are searched in the Scopus[1] database, and the results are analyzed according

[1] www.scopus.com

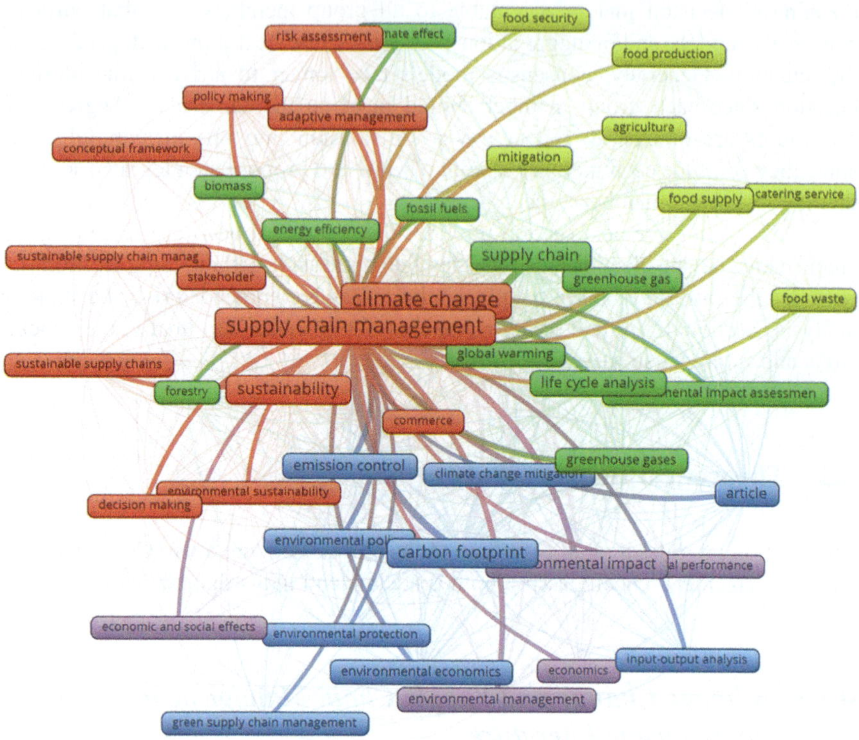

Fig. 4.1 Map visualization of keyword co-occurrences from Scopus Database

to the keyword co-occurrences. The VosViewer[2] application is used to obtain the mapping.

As Fig. 4.1 indicates, the VosViewer application divides the identified publications into five groups. The groups can be intersected in common areas such as supply chain management and CC (both nodes are the bigger ones in the figure), each group being concentrated on the different aspects of CC and supply chain management, as discussed next.

A quick investigation of the yellow group suggests that these studies are mostly concentrated on food supply chains, and they are the most recent studies about supply chain management and CC [11–14]. Current research accordingly emphasizes the impact of CC on the food industry, which might also be related to the COVID-19 pandemic.

The red group mainly concentrates on the conceptual frameworks to reach and design "sustainable" supply chains. This group predominantly employs MCDM

[2] https://www.vosviewer.com/

methods, including TOPSIS, SAW, linear integer modeling, as well as conceptual modeling, among others [15–18].

The green and blue groups are content-wise close to each other. The energy efficiency and consumption-related studies are primarily green [19–21]. On the other hand, the blue group highlights the emissions, and furthermore recommends mitigation policies to reduce the carbon footprint [22–24].

The last group is the purple one. The nodes in this group are smaller than the other groups, suggesting that the number of keyword occurrences is fewer. However, studies in this group frequently focus on the environmental impact analysis and socio-economic effects of CC in supply chains.

Figure 4.1 illustrates that the most prominent node is "carbon footprint," indicating that many academic studies concentrated on reducing carbon footprint with mitigation strategies. On the other hand, as accentuated in industrial reports, economic and social impact reduction is also a critical issue in the CC context. The following section reviews recent approaches and recommendations about CC and supply chain management in the industrial reports.

4.2.2 Climate Change and Supply Chain Management in Industrial Reports

The need for a new business model that can cut GHG emissions is the main challenge for the companies. Boston Consulting Group name the supply chain as a "game-changer" while fighting against the CC [2]. Concerning the immense effect of GHG emissions on global warming, going net-zero is accordingly the main goal for businesses by decarbonizing their supply chains. Nonetheless, according to the Carbon Disclosure Project [25], an international initiative to reduce business emissions, focusing on direct emissions and risks, will not be enough to diminish vulnerability to the CC. Accordingly, those business leaders who lower such risks throughout their supply chain collaboratively will gain a competitive advantage for today and be more resilient for the economic shifts of tomorrow. The whole value chain should be engaged to assess and progress performance.

The report by Slay and Dooley defined various CC risks to generate convenient strategies [26] and emphasized the COVID-19 pandemic situation and the importance of resilience in facing such challenges. The report not only highlighted the importance of a strong supplier relationship as a potent strategy for resilient supply chain management, but also the significance of collaborative planning and control. These publications emphasize that CC risk and strategy prioritization promise critical value for companies.

In their case study report in August 2020, McKinsey defined the climate as "the weak link" in the supply chain [27]. They have conducted nine case studies to investigate leading-edge CC impacts across all major geographies, sectors, and systems. The case studies were investigated from the perspectives of livability and workability, physical asset, infrastructure services, natural capital, and food systems. They primarily examined the effects of extreme weather on supply chains and claimed that supply chains are already damaged and the vulnerabilities and risks will further amplify in the future [28].

Other recent industrial reports mainly focus on the vulnerabilities related to CC and underline that today's supply chains are already damaged and vulnerable to significant changes. Most of them also mention COVID-19 as another source of vulnerability and risk in value chains. The most effective solution can be a collaboration between value chains and novel digital technologies. The following sections list detected CC risks and supply chain management strategies derived from the existing literature.

4.2.3 Climate Change and Supply Chain Risks

The CC risks in the supply chain and powerful strategies to create a climate-resilient supply chain are gathered from the reviewed academic and industrial literature. Table 4.1 gives these two-leveled risks. First, the most affected socio-economic areas identified by the McKinsey report are selected for the first level of CC risks [28]. Then, the second-level risks are defined with the help of the literature.

4.3 Suggested Methodology

The main objective of this chapter is to generate an index for CC risks in the supply chain and prioritize them to obtain a basis for strategical planning. For that purpose, a 2-Tuple Linguistic model-based framework is proposed.

Herrera and Martinez first unveiled the 2-Tuple Linguistic approach in 2000 [6]. This model helps to work with multi-granular information. It is suitable for GDM, where group members have a different experience about the same subject. As aforementioned, the 2-Tuple Linguistic model is generally used with various MCDM models to emphasize their ability to deal with linguistic data and diminish data loss during the translation phase.

The 2-Tuple Linguistic representation model represents the linguistic information using a 2-Tuple *(S, α)*. Here, *S* is a linguistic label and *α* is a numerical value representing the value of the symbolic translation. The function is defined as:

$$\Delta_S : [0, g] \rightarrow \overline{S}$$

$$\Delta_S(\beta) = (S_i, \alpha), \text{ with } \begin{cases} i = \text{round}(\beta) \\ \alpha = \beta - i \end{cases}$$

Table 4.1 Risks detected from the literature for the recommended methodology [2, 8, 9, 26, 28]

First-level socio-economic areas affected by CC	Second-level risks related to SC
R1 Livability and workability	**R1.1** Expansion of infectious diseases **R1.2** Changes in the quality of resources **R1.3** Changes in the labor effectiveness. (Exposure to extreme heat or cold)
R2 Food systems	**R2.1** Increased waste due to food systems' inappropriate supply chain structure. (Transportation issues) **R2.2** Capacity and scheduling issues due to the increased demand volatility **R2.3** Environmental risks due to the long-mile logistics with high GHG emissions
R3 Physical assets	**R3.1** Damage or destruction of facilities, energy networks, communication systems, and transportation due to climate change-related disasters **R3.2** Delays in delivery of products due to extreme/bad weather conditions **R3.3** The obligation to transfer facilities to resilient geographical regions
R4 Infrastructure services	**R4.1** Cost of improvement and embracing of new technologies and innovative solutions **R4.2** Destruction of critical service and utility infrastructure by climatic disasters **R4.3** Temporary functional loss due to landslides by rainfall
R5 Natural capital	**R5.1** Changes in energy demand for cooling and heating, transportation, and storage **R5.2** Cost of protecting vulnerable ecosystems **R5.3** Augmented cost due to environmental regulations

The linguistic term set S can be converted into 2-Tuple form by adding zero value as in the following relation:

$$S_i \in S \Rightarrow (S_i, 0) \tag{4.1}$$

The 2-Tuple Linguistic model is a symbolic computational model that modifies the fuzzy linguistic approach by including a parameter to the linguistic representation to increase the accuracy and the interpretability of the results [7]. The 2-Tuple Linguistic model enables us to deal with variables closer to human cognitive processes and augment the computations' accuracy.

The 2-Tuple Linguistic model provides the group members the opportunity to express their opinions comfortably by proposing linguistic scales compatible with their knowledge about the subject. The *Linguistic Hierarchies (LH)* can unify multi-granular data obtained by the experts. There is a set limitation that can be used in *LH* since the level $t + 1$ is generated from the following relation [7]:

$$S^{n(t)} \longrightarrow S^{n(2(n(t)-1))=n(t+1)} \tag{4.2}$$

where *n(t)* is the granularity of the linguistic term set.

The transformation function to translate a linguistic term set with granularity *n(t)* to a linguistic term set with granularity *n(t')* is as follows. Equation (4.3) defines the *n(t)* level can be transformed. Then, the TF function calculates the 2-Tuple form of the new level.

$$\mathrm{TF}_{t'}^{t} = \left(S_i^{n(t')}, \alpha^{n(t')}\right) = \Delta\left(\frac{\Delta^{-1}\left(\left(S_i^{n(t)}, \alpha^{n(t)}\right)\right) \times (n(t') - 1)}{n(t) - 1}\right) \tag{4.3}$$

The transformation function enables the normalization of multi-granulated information under one form. In GDM, another challenge is how to aggregate all of the opinions. Various aggregation operators are proposed in the literature [7] to combine different assessments of decision-makers, and they are widely used in the decision-making field.

The following aggregation operator is the *2-Tuple Weighted Harmonic Average Operator (TWHAO)*. It has been chosen to be used in this study. The TWHAO relation is computed as:

$$\vec{x} = \Delta\left(1 / \sum_{j=1}^{n} \frac{w}{\beta_i}\right), \beta_i = \Delta^{-1}(S_i, \alpha_i) \tag{4.4}$$

where w_i values are the weights of decision-makers and β_i values are the normalized assessments of each decision-maker.

4.4 Case Study

The ABC company is seeking new strategies to create a more durable and resilient supply chain management to consolidate its marketplace and gain a competitive advantage for its future strategies. Accordingly, we have offered this new methodology to the company. In this section, a case study is applied for ABC Company from Turkey to test the plausibility of the recommended strategy ranking. The detailed steps of the case study are elaborated below:

1. Forming a decision-making group:

   ABC company has gathered a decision-making group of three experts from the company to assess the risks detected from the industrial and academic literature. The decision-making group comprises two managers and one junior manager from the supply chain management department. Since their level of knowledge differs, we have applied different granulated linguistic sets to assess each

Table 4.2 First expert's assessment underS_i^9 for risk importance and the final aggregated values

First-level risks	Importance	Second-level risks	Importance	Utmost aggregated importance	Ranking
R1	(H,0)	**R1.1**	(L,0)	(VL,0.39)	15
		R1.2	(P,0)	(H,0.35)	7
		R1.3	(AH,0)	(H, −0.39)	9
R2	(AH,0)	**R2.1**	(H,0)	(H,0)	8
		R2.2	(P,0)	(P, −0.39)	1
		R2.3	(AM,0)	(AM,0.22)	14
R3	(P,0)	**R3.1**	(P,0)	(P, −0.43)	3
		R3.2	(VH,0)	(VH,0.22)	4
		R3.3	(AH,0)	(H, −0.39)	9
R4	(P,0)	**R4.1**	(M,0)	(M, −0.39)	12
		R4.2	(P,0)	(VH, −0.22)	5
		R4.3	(VH,0)	(VH, −0.22)	5
R5	(AH,0)	**R5.1**	(AM,0)	(M, −0.39)	12
		R5.2	(AH,0)	(AH,0.22)	11
		R5.3	(P,0)	(P, −0.39)	1

decision-maker. A linguistic set with the lower granularity S_i^5 *(Very Low (VL)-Low(L)-Medium(M)- High(H), Perfect(P))* is provided to the junior-level decision-maker. The other linguistic set with the high granularity S_i^9 *(Very Low (VL), Low(L), Medium Low (ML), Almost Medium (AM), Medium(M), Almost High (AH), High(H), Very High(VH), Perfect(P)*) is provided to the managers.

2. Assigning importance to each risk by decision-makers:

   Each decision-maker assessed identified risks using the provided linguistic sets. First LH [7] approach from the 2-Tuple Linguistic Model is applied with Eq. (4.3) to unify the multi-granular assessments obtained from experts. These assessments are unified under the linguistic set with the highest granularity (S_i^9), and afterward, the assessments are aggregated using Eq. (4.4). The aggregated relation matrix is provided in Table 4.2.

3. Obtaining the ultimate ranking by aggregated importance:

   The final aggregated ranking values are obtained by the TWHAO. The main reason to employ the harmonic average-based aggregation tool is its superiority for reflecting the assessments of decision-makers with smaller weights.

4.4.1 Results and Analysis

The final ranking shows that the "augmented cost due to the environmental regulations" and the "capacity and scheduling issues due to the increased demand volatility" are the most significant risks for ABC company. These are followed by "damage or destruction of facilities, energy networks, communication systems and transportation due to CC-related disasters."

At the socio-economical level, *Natural Capital, Food Systems,* and *Physical Assets* are the most vulnerable parts of the ABC company. Accordingly, ABC company will need a strategic roadmap for its future management operations. The suggested framework may provide valuable insights to the company generating its medium- and long-term strategies for CC resilience.

4.5 Managerial Implications

The case study demonstrates the applicability of the proposed method. This 2-Tuple-based assessment model can be convenient as a first step in understanding the underlying risks in the supply chain and generating CC adaptation strategies. It must be noted that these assessments may vary related to different prioritization of the socio-economic risks. To address this, a sensitivity analysis is carried out to observe how the ranking of risks changes under different circumstances.

In this analysis, five different scenarios are generated. Each scenario emphasizes one first-level socio-economic risk. This analysis may lead the ABC company to create various risk prioritizations according to the changing environment. In Case 1, livability and the workability risk are emphasized. In Case 2, food systems risk is emphasized. Figure 4.2 shows that the risk prioritizations may vary related to seniorities.

When all scenarios are examined together, some second-level risks are ranked as the most important in more than one scenario. Accordingly, R2.2 and R5.3 emerge from this analysis as being the most consistent risks for the ABC company to make its supply chain more resilient against the expected impacts of CC.

4.6 Concluding Remarks

This chapter aims to present a linguistic-based decision-making method for CC risk prioritization regarding today and future trends. A lingual approach is preferred to provide a comfortable environment for decision-makers to express themselves, enabling a better and more efficient decision-making process.

Uncertainties are unavoidable in business activities, as also proven by the COVID-19 pandemic. The extent of such uncertainties, however, can be managed. Linguistic-based processes are useful for reducing vagueness and creating such decision-making processes that are more similar to human cognitive processes. Carefully designed linguistic models can also be leveraged to close the experience gap between the decision-makers. Thanks to 2-Tuple Linguistic model integration, this methodology suggests separate multi-granular linguistic sets to each decision-maker, based on their experience on the subject.

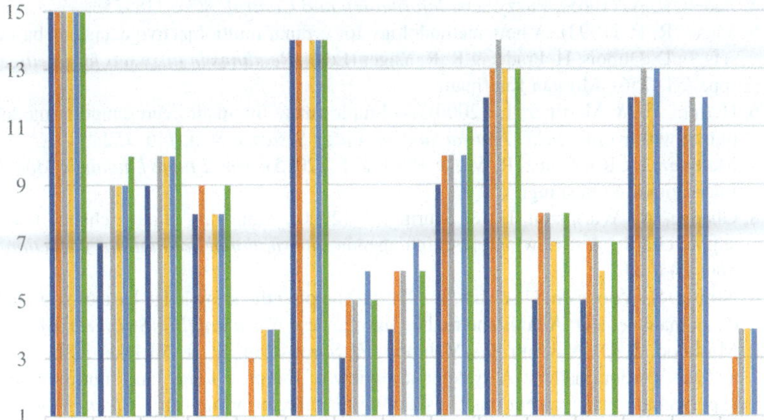

	R1.1	R1.2	R1.3	R2.1	R2.2	R2.3	R3.1	R3.2	R3.3	R4.1	R4.2	R4.3	R5.1	R5.2	R5.3
Case Study	15	7	9	8	1	14	3	4	9	12	5	5	12	11	1
Case 1 (R1)	15	1	1	9	3	14	5	6	10	13	8	7	12	11	3
Case 2 (R2)	15	9	10	1	1	1	5	6	10	14	8	7	13	12	4
Case 3 (R3)	15	9	10	8	4	14	1	1	1	13	7	6	12	11	4
Case 4 (R4)	15	9	10	8	4	14	6	7	10	1	1	1	13	12	4
Case 5 (R5)	15	10	11	9	4	14	5	6	11	13	8	7	1	1	1

Fig. 4.2 Sensitivity analysis

Risk assessments form the basis for strategy development. Therefore, this proposed model can be extended to the strategy determination and the selection processes. The comparative analysis of different methods may give a good insight into the most efficient way to develop such strategies. Prioritizing actions concerning the CC risk management strategy can also feed into the other fields of CC as a future research area.

Acknowledgments This research has received financial support from Galatasaray University Research Fund (Project No: FOA-2021-1059). The authors would like to thank the experts for their appreciation and support in the application.

References

1. Masson-Delmotte, V., et al. (2021). Summary for policymakers. In *Climate change 2021: The physical science basis. Contribution of Working Group I to the sixth assessment report of the intergovernmental panel on climate change*. Cambridge University Press.
2. Burchard, J., Frédeau, M., Hadfield, M., Herhold, P., & O'Brien, C. (2021). *Supply chains as a game-changer in the fight against climate change*. Boston Consulting Group.
3. Marle, F., & Gidel, T. (2012). A multi-criteria decision-making process for project risk management method selection. *International Journal of Multicriteria Decision Making, 2*(2), 189–223.

4. Zadeh, L. A. (1965). Fuzzy sets. *Information and Control, 8*(3), 338–353.
5. Yager, R. R. (1993). A new methodology for ordinal multiobjective decisions based on fuzzy sets. In D. Dubois, H. Prade, & R. R. Yager (Eds.), *Readings in fuzzy sets for intelligent systems* (pp. 751–756). Morgan Kaufmann.
6. Herrera, F., & Martinez, L. (2000). A 2-tuple fuzzy linguistic representation model for computing with words. *IEEE Transactions on Fuzzy Systems, 8*(6), 746–752.
7. Martínez, L., Rodriguez, R. M., & Herrera, F. (2015). *The 2-tuple Linguistic Model.* Springer International Publishing.
8. Ghadge, A., Wurtmann, H., & Seuring, S. (2020). Managing climate change risks in global supply chains: A review and research agenda. *International Journal of Production Research, 58*(1), 44–64.
9. Nimsai, S., Yoopetch, C., & Lai, P. (2020). Mapping the knowledge base of sustainable supply chain management: A bibliometric literature review. *Sustainability (Switzerland), 12*(18), 7348.
10. Moshood, T. D., Nawanir, G., Mahmud, F., Sorooshian, S., & Adeleke, A. Q. (2021). Green and low carbon matters: A systematic review of the past, today, and future on sustainability supply chain management practices among manufacturing industry. *Cleaner Engineering and Technology, 4*, 100144.
11. Wang, Y., Zhao, Y., Wang, Y., Ma, X., Bo, H., & Luo, J. (2021). Supply-demand risk assessment and multi-scenario simulation of regional water-energy-food nexus: A case study of the Beijing-Tianjin-Hebei region. *Resources, Conservation and Recycling, 174*, 105799.
12. Chen, W., et al. (2021). Environmental impacts of animal-based food supply chains with market characteristics. *Science of the Total Environment, 783*, 147077.
13. Chanda, S., Bhat, M., Shetty, K. G., & Jayachandran, K. (2021). Technology, policy, and market adaptation mechanisms for sustainable fresh produce industry: The case of tomato production in Florida, USA. *Sustainability (Switzerland), 13*(11), 1.
14. Pham, T. T. T., et al. (2021). Guidelines for co-creating climate adaptation plans for fisheries and aquaculture. *Climatic Change, 164*(3–4), 62.
15. Mageto, J. (2021). Big data analytics in sustainable supply chain management: A focus on manufacturing supply chains. *Sustainability (Switzerland), 13*(13), 7101.
16. Wheeler, J., Galán-Martín, Á., Mele, F. D., & Guillén-Gosálbez, G. (2021). Designing biomass supply chains within planetary boundaries. *AICHE Journal, 67*(4), e17131.
17. Siluthanyi, M., Yahia, Z., & Pradhan, A. (2020). *Supply chain management of milled grain products: A review and conceptual framework.* International Association for Management of Technology (IAMOT)At: Cairo, Egypt, pp. 604–617.
18. Sudarshan, S. V., Priyadarshuan, A., & Anbuudayasankar, S. P. (2019). Analysis of the barriers in implementing green supply chain management (GSCM) practices: A hybrid approach. *IOP Conference Series: Materials Science and Engineering, 577*(1), 012005.
19. Li, X., et al. (2022). Evaluating food supply chain emissions from Japanese household consumption. *Applied Energy, 306*, 118080.
20. Bressanin, J. M., et al. (2021). Advanced technologies for electricity production in the sugarcane value chain are a strategic option in a carbon reward policy context. *Energy Policy, 159*, 112637.
21. Negri, V., et al. (2021). Life cycle optimization of BECCS supply chains in the European Union. *Applied Energy, 298*, 117252.
22. Zhang, H. (2021). Regulations for carbon capture, utilization and storage: Comparative analysis of development in Europe, China and the Middle East. *Resources, Conservation and Recycling, 173*, 105722.
23. Morfeldt, J., Davidsson Kurland, S., & Johansson, D. J. A. (2021). Carbon footprint impacts of banning cars with internal combustion engines. *Transportation Research Part D: Transport and Environment, 95*, 102807.
24. Martinez, S., Delgado, M. D. M., Martinez Marin, R., & Alvarez, S. (2019). Organization environmental footprint through input-output analysis: A case study in the construction sector. *Journal of Industrial Ecology, 23*(4), 879–892.

25. CDP. (2021). *Transparency to Transformation: A chain reaction CDP global supply chain report 2020*. CDP.
26. Slay, C., & Dooley, K. (2020). *Improving supply chain resilience to manage climate change risks*. HSBC and The Sustainability Consortium.
27. Woetzel, J., et al. (2020). *Could climate become the weak link in your supply chain?* McKinsey Global Institute.
28. Woetzel, J., et al. (2020). *Climate risk and response: Physical hazards and socio-economic impacts*. McKinsey Global Institute.

Chapter 5
Creating a Net-Zero Carbon Emission Scenario Using OSeMOSYS for the Power Sector of Turkey

Atilla Kılınç and Gökhan Kirkil

Abstract The summer of 2021, when the effects of global climate change began to be felt more rapidly with wildfires, floods, and other environmental disasters, gave developing countries the opportunity to re-consider green energy investments. Therefore, creating new optimization models to minimize carbon emission rates in energy production has become mandatory, particularly for power generation. Based on the principle that the energy demand will increase in the long run, any step for reducing carbon emissions will save energy resources through energy efficiency networks. Its decision-making mechanism benefits from AI-centered platforms. The main objective of this chapter is to consider all new energy investments, the installed power production capacity, and the power generation projections in Turkey to create a zero-carbon emission scenario in the long run. As the Paris Agreement targets to reduce carbon emissions to zero by 2050, our scenario for Turkey will be based on the years between 2020 and 2050. It is estimated that Turkey will have an increasing electricity demand as a developing country. OSeMOSYS, an open-source optimization tool, is used to design our energy model. As the energy trilemma—security, affordability, and sustainability—take a principal role on the path to zero-net, the OSeMOSYS can be a powerful optimization tool for creating long-term scenarios for countries. This chapter presents Turkey's zero-carbon emission-based optimization model using a linear programming formulation.

Keywords Net-zero carbon · OSeMOSYS · Turkey · Energy modeling · AI

A. Kılınç (✉)
Graduate School, Istanbul Technical University, Istanbul, Türkiye
e-mail: kilinca20@itu.edu.tr

G. Kirkil
Department of Industrial Engineering, Kadir Has University, Istanbul, Türkiye
e-mail: gokhan.kirkil@khas.edu.tr

G. Kayakutlu, M. Ö. Kayalica (eds.), *Decision Making Using AI in Energy and Sustainability*, Applied Innovation and Technology Management,
https://doi.org/10.1007/978-3-031-38387-8_5

5.1 Introduction

The energy sector is affected by global conjunctures, geopolitics, production systems, and social structures [1]. Thanks to structural policy changes in Turkey's energy sector in the last two decades, it has risen among developing countries, increasing its infrastructure and power plant construction investments [2]. Along with these developments, when the awareness of air pollution started in the early 1990s, cleaner energy sources were politically supported [3]. Also, this situation has created pressure to change Turkey's power plant construction regulations. By this approach, Turkey has implemented natural gas and natural gas conversion power plants for the whole country in 20 years. In this respect, power consumption has constantly increased for the last 20 years, except for the economic crisis periods.

The natural disasters [4] that have occurred in the last 10 years have caused global concerns that have become even more evident after the Paris Agreement [5]. By signing the Paris Agreement in the last quarter of last year, Turkey took the first step toward changing the energy vision [6]. Power generation in Turkey strongly depends on the import of energy resources. Therefore, even the slightest possible profit situation will cause the investment strategy to change. There are more antecedent situations related to energy in Turkey. Turkey can be seen as a country that is not rich in energy resources, and there is a strong relationship between the foreign trade deficit and energy production [7]. Long-term power generation planning will be seen as a critical parameter for Turkey in terms of energy security in economic growth. The Ministry of Energy and Natural Resources in Turkey considered the diversification of power plants as a political maneuvering capability [8]. The new investments in renewable energy resources are subsidized before supporting the existing coal power plants [9].

The main objective of this chapter is to consider all new power investments, the installed power production capacity, and the electricity generation projections in Turkey to create a zero-carbon emission scenario in the long run. Since the main aim of the Paris Agreement is to reduce carbon emissions to zero by 2050, our scenario created for Turkey will be based on the years between 2020 and 2050.

The remaining part of this chapter is organized as follows. Creating a net-zero carbon emission scenario using OSeMOSYS as a topic could be seen strategically in recent optimization models in the literature. That is why the literature review is based on the topic, and the second section is reserved for reviewing several articles and reports on current power generation research in Turkey. Section three is based on the methodology to design the optimization model for power generation in Turkey. The OSeMOSYS optimization model will be briefed. Finally, the conclusion analyzes the results of the optimization model regarding the current facts in Turkey.

5.2 Literature Review

The Turkish electricity market was opened to competition with the electricity market law [10] adopted by the Turkish Grand National Assembly in 2001 [11]. This law, which was considered in parallel with the operation of the new economic program put forward under the Ministry of Economics, Kemal Derviş, was enacted in the context of the future projections of the energy market in Turkey [12]. Rapid population growth, technological developments, and increasing industrial investments are the main precursors for Turkey to cause a marginal increase in energy investments systematically [13].

In the last 50 years, energy consumption has continuously increased in Turkey except during periods of economic crisis [11]. The first step of the country's energy vision is to increase energy efficiency and reduce carbon emissions by using modern technologies in energy investments of a country that produces electricity by importing energy resources. Thanks to domestic and foreign investments, Turkey's installed power plant capacity has increased uninterruptedly for the last 20 years. As stated in the energy sector summary report of the Presidential Annual Program in 2022, the installed power capacity will reach 102,423 MWh in Turkey [14]. Thus, it can be said that Turkey has come to a position to catch up with the increase in energy demand if it can ensure the supply security in energy [15].

Turkey is also progressing rapidly in the investment of renewable energy resources. This policy is used to design the new maneuver strategy to fulfill the Paris Agreement conditions and reduce the current account deficit by creating resource diversity in power plants [16]. Although the increase in hydropower, solar, wind, and geothermal investments does not guarantee energy security, it causes a reduction in Turkey's foreign dependency. Therefore, an increase is essential for the energy sector's future projections with decreased costs [17]. The strong relationship between climate change and energy demand has been recognized recently. Although the location of Turkey is a peninsula surrounded by seas on three sides, the effects of climate change have been felt since 2007 [18]. Drought in agricultural areas and forest fires rise yearly, emerging the possibility of a severe water crisis [19]. Therefore, Turkey's energy sector has started directing its investments in solar power and wind farms [20].

Turkey is ahead of many European countries with the annual number of sunny days and wind averages [20]. It is not yet able to fully utilize the opportunities in this area. Reduction in investment costs and ease of access to energy sources will lead to Turkey's new renewable energy revolution. Since the early 2010s, when decarbonization began to be seen as a new planning strategy in the energy sector, multi-source optimization models have come to the fore [21]. Therefore, it is essential to construct data on different sources with an open application program to see the energy system compared to other programs and to use the advantages created by this difference.

5.3 Methodology

Energy market models and optimization design have become more strategic with the diversification of energy resources [22]. The reduction of carbon emissions, the widespread use of energy efficiency networks, and the inclusion of new technologies in energy production systems have led to the diversification of optimization programs in entire energy systems [23]. Kılınç et al. displayed the effects of the zero net transition in the South Atlantic Census Region with a new optimization model. In that article, the authors emphasized the critical impact of carbon emission reduction via transition in the kitchens [24]. When we started Creating a Net-Zero Carbon Emission Scenario Using OSeMOSYS for the Power Sector of Turkey, we focused on the potential effect of reducing carbon emission, which Kilinc et al. argue in the Zero-Net transition article. Today, industrialization processes have a historical background created without considering environmental factors. This approach has reinforced the dependence of all production models on fossil fuels. The awareness factor created by the environmental disasters experienced in recent years has reinforced the orientation toward long-term optimization programs like LEAP [25, 26].

OSeMOSYS modeling system is designed to develop both regional and multiregional energy strategies. OSeMOSYS system covers all sub-consumption and production systems of the energy sector and aims to achieve multimodal optimization using specific data sets [27]. Besides, OSeMOSYS is defined as a deterministic linear optimization model and minimizes the total discounted costs. Furthermore, it is an application that will ensure the consideration of future capacity expansions.

First, since this research aims to predict a possible impact, this study will not consider energy market conditions. This research will be based on OSeMOSYS, the energy market modeling platform. When a model is created using OSeMOSYS, all the necessary data is uploaded to the platform (such as generation units, capacity factors, and availability). The power production analysis is made to meet consumption projections. This specialty creates the power generation projection for the future [28]. This research applies a linear version of the Open-Source Energy Modelling System (OSeMOSYS) to power generation in Turkey. Turkey's power generation system is both fossil fuel and hydro-based, like many hybrid energy systems in Brazil [29]. Optimizing with OSeMOSYS will show the applicable future projections. In pursuing sustainable development, it is imperative to consider capacity-building measures, such as integrating nuclear power as a low-carbon technology. This alternative energy source offers significant potential to aid Turkey in achieving long-term emission reduction objectives, thereby warranting careful consideration as a critical component of our country's energy strategy [28].

OSeMOSYS is designed to inform national and multiregional energy strategy development. OSeMOSYS, which covers all sub-consumption and production systems of the energy sector, has been designed for multimodal optimization using different data sets [30]. At the same time, OSeMOSYS is defined as a deterministic linear optimization model that minimizes the total discounted costs. In addition, it is an application that will ensure that the expansions that will occur in the capacities of the power plants are included in the optimization. The first step to creating a model with OSeMOSYS is to define the datasets. Clusters are fixed for the model and each scenario. It is the item groups that represent the structure of the model. The clusters' eight parameters to be filled can be listed: fuel type, method of use and mode of operation, region, storage, technology, time period, and year [31].

OSeMOSYS is an open-source energy modeling system that does not require a financial investment. It is designed to compare specific analyses and modify or update the model. Therefore, some restrictions and rules must be followed to create a model. The model's objective is to estimate the cheapest way to satisfy the demand, which is why OSeMOSYS is more fundamental and easier to use than long-established energy systems models such as MARKAL, LEAP, etc. [32]. According to research, OSeMOSYS and other energy modeling systems have roughly a 1–5% difference in insignificant indicators [27].

5.4 Proposed Model

The article will be based on a reference scenario for working on the effects of different policies in this model called business as usual (BAU). In this scenario, estimated production values for the coming years until 2050 will be calculated according to the 2019 production reports. Since 0.3% is unreliable, 99.7% of Turkey's electricity production was selected, and optimization was set up by accepting it as 100%.

In the cost section, there will be only capital costs. Performance has six different parameters in this scenario. These are the input activity ratio, output activity ratio, capacity factor, operational life, capacity to activity unit, and availability. The demand side has two parts: specified annual demand and specified demand profile. Activity is based on two limits: total technology annual activity lower limit and total technology annual activity upper limit. The final part is the emission. The emission will show the activity ratio in this model. The model could be seen as a crowdsourcing analysis in this case [33].

Capital cost is fixed, one-time expenses for a power plant to build. Therefore, it is in million $/GW unit. Capital costs for each technology are shown below. This model assumes that every technology gets cheaper by 1% every year. The related data is based on 2019 records [14]. The data is summarized in Table 5.1.

The input activity ratio is the commodity usage rate for technologies below. It is calculated by dividing one by efficiency. Efficiency values and calculated input activity ratios are shown in Table 5.2. The capacity factor section considers renewable sources' efficiency values since they have no input resources.

In contrast to the input activity ratio, the output activity ratio gives a commodity's output rate. Transmission and distribution technologies have energy loss while transporting electricity from one place to another. According to the inputs below,

Table 5.1 Capital cost data for Turkey between 2019 and 2023 (Million $/GW)

	2019	2020	2021	2022	2023
BIO	2140	2118.6	2097.414	2076.439	2055.675
GPP	3915	3875.85	3837.091	3798.720	3760.733
HYD	1709	1691.91	1674.990	1658.240	1641.658
IGCC	3700	3663	3626.37	3590.106	3554.205
NGCC	1100	1089	1078.11	1067.328	1056.655
NUCL	3000	2970	2940.3	2910.897	2881.788
OIL	1200	1188	1176.12	1164.358	1152.715
PV_UT	994	984.06	974.219	964.477	954.832
WIND	1473	1458.27	1443.687	1429.250	1414.958

Table 5.2 Efficiency and input activity values

	Efficiency (0 to 1)	Input Act. ratio
BIO	0.3388	2.952
IGCC	0.6701	1.492
NGCC	0.2635	3.794
NUCL	0.30	3.03
OIL	0.2880	3.471

it is expected to lose 4.9% of the electricity production while transporting and distributing electricity. The input activity ratio accounts for the energy loss of the power plants.

- All electricity production technologies have a one output activity ratio.
- All transmission technologies have a 0.9848 output activity ratio.
- All distribution technologies have a 0.9676 output activity ratio.

The capacity factor is the ratio of annual capacity to the available capacity for each time slice. We assume that all the fossil fuel power plants are operatable at maximum capacity over the years. The renewable technologies' capacity factor is calculated by dividing actual production by theoretical production capacity over a year, as shown in Table 5.3 actual production. Available capacity data was obtained from the Transparency Platform in Turkey [17].

Table 5.3 Capacity factors for each timeslice

	ID	IN	SD	SN	WD	WN
BIO	1	1	1	1	1	1
GEO	0.70	0.70	0.70	0.70	0.70	0.70
HYD	0.37	0.37	0.32	0.32	0.42	0.42
IGCC	1	1	1	1	1	1
NGCC	1	1	1	1	1	1
NUCL	1	1	1	1	1	1
OIL	1	1	1	1	1	1
PV_UT	0.38	0	0.38	0	0.38	0
WIND	0.34	0.34	0.34	0.34	0.34	0.34

Table 5.4 Shares of fossil fuel-based technologies between 2019 and 2025 and yearly descending values

	Share between 2019 and 2025	Decrease in share 2025–2049	2050
COAL	0.3733	0.01253	0
OIL	0.0024	0.00009	0
GAS	0.1875	0.00721	0

Lower technology activity limits are the only difference between the BAU and zero-emission scenarios. While performing the required calculations and assumptions, the model is based on the global production statistics of renewable energy [16].

To calculate the values for the zero-emission scenario, the model assumes that the production percentages will stay the same for every technology until 2025. After that, fossil fuel-sourced production will decrease until zero in 2050, and renewable sources-based production will increase to satisfy demand. From 2025 to 2050, the coal, oil, and natural gas-based power plant production ratio will decline linearly from zero to 2050. The change in the shares in production over the years can be seen in the table below. Summarized data for fossil fuel-based production can be seen in Table 5.4.

After calculating each technology's growth rates and production capacities in Table 5.5, production capacity needs to be scaled up to meet projected demand. Therefore, it aimed to take the final percentages for the technologies and include

Table 5.5 Growth ratio statistics of renewable technologies

	Initial ratio	Lowering value	Global avg	Reached to global avg.	Fixed value
HYDRO	0.0949	0.01	0.0247	2027	0.0249
GEO	0.2732	0.03	0.0326	2030	0.0332
BIO	0.2716	0.03	0.0625	2027	0.0615
WIND	0.1729	0.02	0.2330	2025	0.0728
SOLAR	0.3787	0.03	0.3787	2030	0.0487
NUCL.	0.35	0.02	0.0109	2035	0.11

Table 5.6 Production values of technologies for 5 years intervals

	2019	2025	2030	2035	2040	2045	2050
COAL	428.6101	440.0355	417.1676	356.6894	270.3883	148.0576	0.0000
OIL	2.7787	3.3613	3.2092	2.7767	2.1564	1.2744	0.0000
GAS	215.2818	260.4157	248.6314	215.1266	167.0691	98.7345	0.0000
HYDRO	336.4963	417.5712	464.0116	495.7025	531.3103	573.4521	613.3388
GEO	33.8064	45.1180	69.2905	97.7559	134.6339	183.6656	246.0832
BIO	12.4285	17.2357	30.1187	45.7629	66.1427	93.3460	128.2592
WIND	82.4527	110.4258	171.7498	244.2449	338.2317	463.2567	622.5821
SOLAR	36.2595	53.6407	111.9029	184.3419	279.0969	405.9460	569.7173
NUCL.	0.0000	99.1717	210.7898	349.8172	531.7305	775.3118	1089.9318

them in the model until 2050. The production values in the PJ are summarized in Table 5.6.

All the needed values are entered into the OSeMOSYS optimization program, and it is time to execute both scenarios. OSeMOSYS uses GLPK (GNU Linear Programming Kit) to calculate and find the optimal solution according to inputs. Once we conducted both scenarios, we got results, which means GLPK could find an optimal solution. The Results section below compares the most important results from each scenario. Demand, production, emission, and capital cost plots are prepared and compared for both scenarios below.

Demand charts for both scenarios can be seen in Fig. 5.1. Demand values till 2050 are calculated according to the *Republic of Turkey Ministry of Energy and Natural Resources* projection for Turkey, and it's shown in graphics by OSeMOSYS.

The technologies used to produce electricity for each scenario are given in Fig. 5.2. Unlike the BAU scenario, the zero-emission scenario uses fossil fuel-

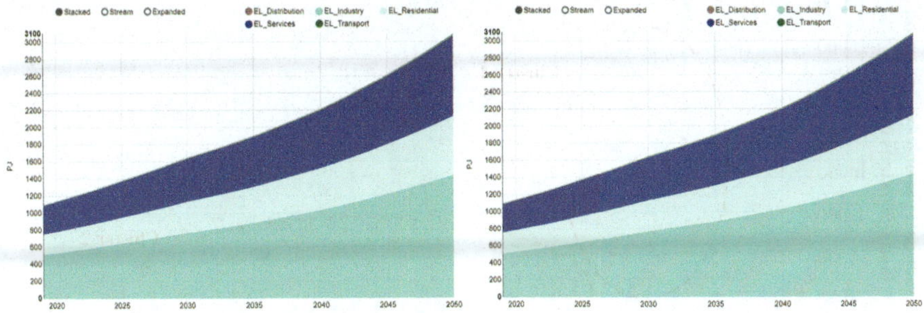

Fig. 5.1 Demand chart of both scenarios (left: BAU, right: Zero-emission scenario)

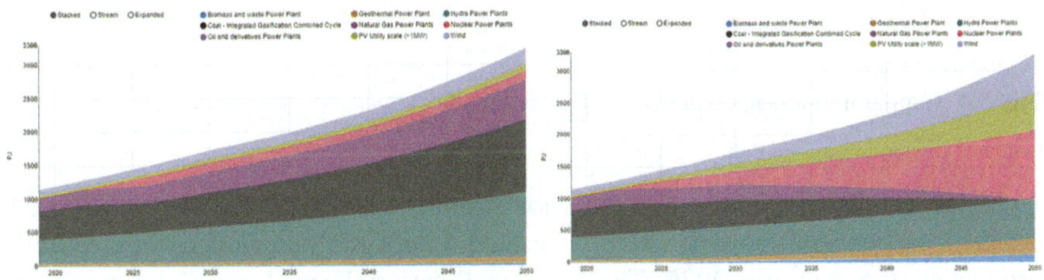

Fig. 5.2 Production chart of BAU scenario (left) and zero-emission scenario (right)

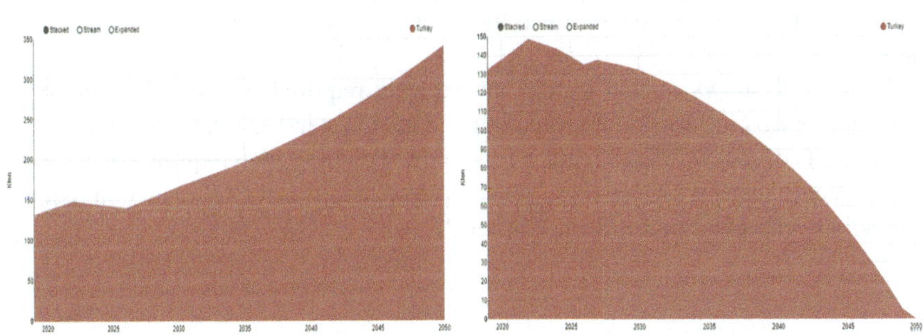

Fig. 5.3 Carbon emission chart of BAU scenario (right) and carbon emission chart of zero emission scenario (left)

based technologies less, and their share in meeting the demand also decreases over the years. Once the forecast reaches 2050, the net-zero scenario does not use fossil fuels and makes no emissions for electricity generation. By the Paris Agreement, the zero-emission scenario provides this model with a way to decrease the emission of the electricity production sector.

Figure 5.3 displays the expected emission values for Turkey from 2019 to 2050. The OSeMOSYS have made all the calculations. Calculations multiply the expected fuel use by the emission activity ratio. The zero-emission scenario's emission chart also can be seen in Fig. 5.3.

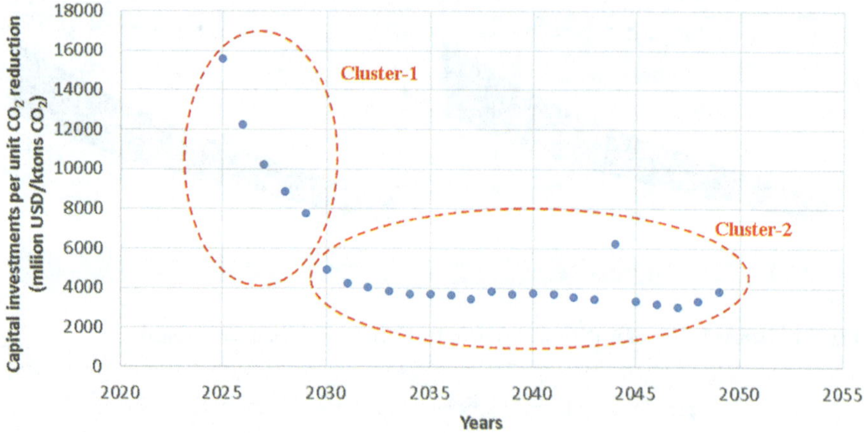

Fig. 5.4 Capital investments required for CO_2 reduction

Figure 5.3 shows that the zero-emission scenario shows that the model will reach net-zero carbon emissions in 2050. The model kept the percentages in electricity production the same till 2025. The BAU and zero-emission scenarios give us the same results for the first part of the simulation; after 2027, renewable technologies started to take a higher share in electricity production, which resulted in an exponential decrease in emissions over the years in the net-zero emission scenario until 2050.

Figure 5.4 shows annual capital investments required per unit CO_2 emission reduction. Two clusters are observed in the data. Cluster-1 is related with the initial investments of renewables to replace fossil power plants which require a high capital investment (Cluster-1). Once these are completed, additional CO_2 reduction requires a much lower capital investment (Cluster-2).

5.5 Conclusion

In this chapter, an energy system model is designed to forecast the amount of power generation in Turkey until 2050, called the business-as-usual scenario. An alternative scenario, the net-zero emission scenario, was developed in line with the ratios of existing energy production resources and a strategy to reduce carbon emissions. Intending to comply with the Paris Agreement, Turkey's carbon emission rates have to be matched to zero by 2050; it was deemed appropriate to use an open-source energy modeling system called OSeMOSYS in the model developed in the article. OSeMOSYS creates a wide area of use for these scenarios by making linear optimization calculations. As seen in this model, all energy source types and power plants were included in the same optimization. Thus, a long-term scenario strategy was realized using OSeMOSYS.

Turkey has been trying to create diversity in optimization models in recent years. The main reason for this is the decline in the impact of coal and the rapidly increasing hydro installed capacity with natural gas and many new investments. This chapter observes the contribution of new energy investments made in recent years to the system.

After creating an optimization model in OSeMOSYS, two working scenarios are developed for comparison. While the first is seen as the usual scenario, the other scenario pursues the goal of zero emissions. The same demand is met in both scenarios, and the same technologies are used to produce the needed electricity. The same demand is met in both scenarios, and technology shares were kept constant with 2019 data in the usual business scenario.

As the demand increases depending on the years in a business-as-usual scenario, fossil fuels are also growing. In this case, it causes carbon emissions to be 2573 times higher in 2050 compared to 2019. At the same time, the alternative scenario aims to reach zero emissions by 2050. Although the zero-emission scenario is desirable, it comes with a high cost. The zero-emissions scenario will cost 2.06 times more than the BAU scenario in 2050 and will be almost impossible with intermittent energy sources. The gap between capital investments in the two scenarios widens from mid-2030, as renewable generation technologies require more installed capacity. In addition, renewable energy sources are optimized for Turkey in this model. Finally, a model focused entirely on renewable energy sources in Turkey was obtained, and the results were shared. For environmental and ethical reasons, Turkey can start using a system like this model in the future.

References

1. The geopolitics of the energy transformation – World Energy Council | Turkey. https://www.worldenergy.org.tr/the-geopolitics-of-the-energy-transformation/. Accessed 15 Jan 2022.
2. I. Energy Agency. Turkey 2021 - Energy Policy Review. Accessed 15 Jan 2022. [Online]. Available: www.iea.org/t&c/
3. Turkey - Air pollution country fact sheet — European Environment Agency. https://www.eea.europa.eu/themes/air/country-fact-sheets/2021-country-fact-sheets/turkey. Accessed 15 Jan 2022.
4. IPCC report: 'Code red' for human driven global heating, warns UN chief | | UN News. https://news.un.org/en/story/2021/08/1097362. Accessed 15 Jan 2022.
5. The Paris Agreement | UNFCCC. https://unfccc.int/process-and-meetings/the-paris-agreement/the-paris-agreement. Accessed 4 Dec 2021.
6. Paris Anlaşmasının Onaylanmasının Uygun Bulunduğuna Dair Kanun. *T.C Resmi Gazete*, Ankara, pp. 1–22, 7 Oct 2021. Accessed 4 Dec 2021. [Online]. Available: https://www.resmigazete.gov.tr/eskiler/2021/10/20211007.pdf
7. Yilmaz-bozkus, R., Güvenliği, E., İşbirliği, E., Çıkarlar, U., & Merkezi, E. (2018). Main determinants of Turkey's foreign oil and natural gas strategy [Türkiye'nin Dış Petrol ve Doğal Gaz Stratejisinin Temel Belirleyici Faktörleri]. *Journal of Research in Economics, 3*(2), 114–132. https://doi.org/10.30784/epfad.426472
8. Turkey's International Energy Strategy / Rep. of Turkey Ministry of Foreign Affairs. https://www.mfa.gov.tr/turkeys-energy-strategy.en.mfa. Accessed 15 Jan 2022.

9. Ö. Üyesi Mustafa Emre Akbaş Marmara Üniversitesi İktisat Fakültesi İktisat Bölümü. (2019). Neoliberalizm, Küresel Enerji Sektörü Ve Seçilmiş Ülkeler Bazında Elektrik Sektörü Özelleştirmeleri. *Sosyal Bilimler Araştırma Dergisi, 8*(2), 21–43. Accessed 15 Jan 2022. [Online]. Available: https://dergipark.org.tr/tr/pub/ssrj/issue/45017/561356
10. 7835-7843 ENERJİ PİYASASI DÜZENLEME KURUMUNUN TEŞKİLAT VE GÖREVLERİ HAKKINDA KANUN (1)(2).
11. Yavuzdemir, M., & Gökgöz, F. (2015). Estimating gross annual electricity demand of Turkey. *International Business Research, 8*(4). https://doi.org/10.5539/ibr.v8n4p145
12. Türkiye Cumhuriyeti Katılım Öncesi Ekonomik Program Ankara Ekim 2001. Accessed 14 Feb 2022. [Online]. Available: http://ekutup.dpt.gov.tr/ab/kep/2001.pdf
13. Sirin, S. M., & Erdogan, F. H. (2013). R&D expenditures in liberalized electricity markets: The case of Turkey. *Renewable and Sustainable Energy Reviews, 24*, 491–498. https://doi.org/10.1016/J.RSER.2013.03.069
14. 2022 YILI CUMHURBAŞKANLIĞI YILLIK PROGRAMI ENERJİ SEKTÖRÜ ÖZET RAPORU.
15. TSKB. (2021, November). ENERJİ GÖRÜNÜMÜ, 2020. Istanbul. Accessed 14 Feb 2022. [Online]. Available: https://www.tskb.com.tr/i/assets/document/pdf/enerji-sektor-gorunumu-2020.pdf
16. Sirin, S. M., & Erdogan, F. H. (2013). R&D expenditures in liberalized electricity markets: The case of Turkey. *Renewable and Sustainable Energy Reviews, 24*, 491–498. https://doi.org/10.1016/j.rser.2013.03.069. Elsevier Ltd.
17. Madlener, R., & Stoverink, S. (2012). Power plant investments in the Turkish electricity sector: A real options approach taking into account market liberalization. *Applied Energy, 97*, 124–134. https://doi.org/10.1016/J.APENERGY.2011.11.050
18. As the climate bakes, Turkey faces a future without water. *Yale E360*. https://e360.yale.edu/features/as-the-climate-bakes-turkey-faces-a-future-without-water. Accessed 14 Feb 2022.
19. Fresh water resources – Turkey. *Climatechangepost.com*. https://www.climatechangepost.com/turkey/fresh-water-resources/. Accessed 14 Feb 2022.
20. Renewable energy statistics - Statistics Explained. https://ec.europa.eu/eurostat/statistics-explained/index.php?title=Renewable_energy_statistics#Share_of_renewable_energy_more_than_doubled_between_2004_and_2020. Accessed 14 Feb 2022.
21. Isik, M., Ari, I., & Sarica, K. (2021). Challenges in the CO2 emissions of the Turkish power sector: Evidence from a two-level decomposition approach. *Utilities Policy, 70*, 101227. https://doi.org/10.1016/J.JUP.2021.101227
22. Niet, T., Shivakumar, A., Gardumi, F., Usher, W., Williams, E., & Howells, M. (2021). Developing a community of practice around an open source energy modelling tool. *Energy Strategy Reviews, 35*, 100650. https://doi.org/10.1016/j.esr.2021.100650
23. Wang, F., et al. (2021). Technologies and perspectives for achieving carbon neutrality. *The Innovation, 2*(4), 100180. https://doi.org/10.1016/J.XINN.2021.100180
24. Kılınç, A., & Kayalica, M. Ö. (2022). Zero carbon energy transition in the kitchens. In *IFIP advances in information and communication technology* (Vol. 637 IFIP, pp. 44–62). https://doi.org/10.1007/978-3-030-96592-1_4/TABLES/6
25. Hernández, K. D., & Fajardo, O. A. (2021). Estimation of industrial emissions in a Latin American megacity under power matrix scenarios projected to the year 2050 implementing the LEAP model. *Journal of Cleaner Production, 303*, 126921. https://doi.org/10.1016/J.JCLEPRO.2021.126921
26. Massaga, D. J., Kirkil, G., & Celebi, E. (2019). A comparative study of energy models for Turkish electricity market using LEAP. In *International conference on the European energy market, EEM* (Vol. 2019-September). https://doi.org/10.1109/EEM.2019.8916283
27. Howells, M., et al. (2011). OSeMOSYS: The Open Source Energy Modeling System: An introduction to its ethos, structure and development. *Energy Policy, 39*(10), 5850–5870. https://doi.org/10.1016/J.ENPOL.2011.06.033

28. Kubulenso, S. (2019). *The OSeMOSYS teaching kit – an example of open educational resources to support sustainable development*. Accessed 3 Dec 2021. [Online]. Available: http://urn.kb.se/resolve?urn=urn:nbn:se:kth:diva-264273
29. Niet, T., et al. (2017). Hedging the risk of increased emissions in long term energy planning. *Energy Strategy Reviews, 16*, 1–12. https://doi.org/10.1016/J.ESR.2017.02.001
30. Gardumi, F., et al. (2018). From the development of an open-source energy modelling tool to its application and the creation of communities of practice: The example of OSeMOSYS. *Energy Strategy Reviews, 20*, 209–228. https://doi.org/10.1016/J.ESR.2018.03.005
31. Almulla, Y., et al. MoManI User Manual 201706-Rev04 | KTH-dESA 1 Model Management Infrastructure (MoManI) Training Manual. Accessed 14 Feb 2022. [Online]. Available: https://un-desa-modelling.github.io/
32. Yang, D., Liu, D., Huang, A., Lin, J., & Xu, L. (2021). Critical transformation pathways and socio-environmental benefits of energy substitution using a LEAP scenario modeling. *Renewable and Sustainable Energy Reviews, 135*, 110116. https://doi.org/10.1016/j.rser.2020.110116
33. Bazilian, M., et al. (2012). Open source software and crowdsourcing for energy analysis. *Energy Policy, 49*, 149–153. https://doi.org/10.1016/J.ENPOL.2012.06.032

Chapter 6
Prediction of Downward Surface Solar Radiation Using Particle Swarm Optimization and Neural Networks

Denizhan Güven, Eren Deliaslan, M. Berker Yurtseven, and Gülgün Kayakutlu

Abstract Concerning the impacts of climate change, renewable energy systems are becoming more prominent by providing more environmentally friendly energy. However, these systems are very dependent on meteorological parameters. In this study, it is aimed to forecast hourly downward surface solar radiation (DSSR), which is the main determinant of power output from PV systems, using Particle Swarm Optimization (PSO) and Long Short-Term Memory Recurrent Neural Networks (LSTM). The hourly data of 12 meteorological features in Karaman, Turkey, for the period between January 2020 and December 2021 were obtained from the European Centre for Medium-Range Weather Forecasts' (ECMWF) ERA5 reanalysis data sets. PSO is utilized to select the most significant features from this data. These selected variables are put into the LSTM model as inputs. The established hybrid forecasting model is examined for 1 hour, 2 hours, 4 hours, and 6 hours ahead. Furthermore, to measure the performance of the forecasting model, the Mean Squared Error (MSE) and Root Mean Squared Error (RMSE) are calculated. As a result, the lowest error rate is acquired for 1 hour ahead of time horizon with 8496 (W/m^2) MSE and 92 (W/m^2) RMSE. This study also exhibits that the proposed hybrid model may provide very accurate results to forecast hourly DSSR with selected variables in case it is not possible to measure the DSSR values directly.

Keywords Downward surface solar radiation · Particle swarm optimization · Long short-term memory

D. Güven (✉)
Eurasia Institute of Earth Sciences, Istanbul Technical University, Istanbul, Türkiye
e-mail: guvende@itu.edu.tr

E. Deliaslan · M. B. Yurtseven · G. Kayakutlu
Energy Institute, Istanbul Technical University, Istanbul, Türkiye

G. Kayakutlu, M. Ö. Kayalica (eds.), *Decision Making Using AI in Energy and Sustainability*, Applied Innovation and Technology Management,
https://doi.org/10.1007/978-3-031-38387-8_6

6.1 Introduction

Downward Surface Solar Radiation (DSSR) is one of the most significant elements for the surface energy balance. It has a great impact on hydrology, meteorology, agriculture, and energy generation [1, 2]. Despite the importance of DSSR, the observation stations to measure DSSR are not globally widespread since the measurement equipment of DSSR is costly and complex. Hence, most of the observation stations are not able to provide DSSR data [3].

There are four methods to acquire DSSR: (i) ground measurements, (ii) re-analysis, (iii) General Circulation Models (GCMs), and (iv) satellite observations [4]. Each method has different pros and cons. For instance, while ground measurements provide the most accurate data, they are not sufficient to derive the spatial dispersion of DSSR due to the limited number of observation stations. On the other hand, satellite observations may provide broad spatio-temporal coverage on a local, regional, or global spectrum [5].

In addition to these methods, machine learning techniques are gaining popularity to acquire DSSR. Despite the lack of explaining the physics, machine learning techniques have a number of advantages: (i) ascertaining the most influential variables in determining DSSR, (ii) presenting some mechanisms to measure the uncertainties, (iii) utilizing various types of data, (iv) catching the non-linear relations between variables, and (v) appraising the robustness of the model [6]. These advantages make machine learning techniques substantial to estimate DSSR using satellite observations or re-analysis data [7, 8].

As aforementioned, DSSR is very important for the energy sector, especially solar photovoltaic (PV) and solar thermal power plants. Unlike conventional power plants like coal or natural gas power plants, the electricity output of PV and solar thermal power plants is highly dependent on solar irradiation. Since solar irradiation is not constant, the electricity output from these systems varies in time. This causes frequency problems in the electricity grid. Thus, estimating the output of solar power plants in the future is very crucial for the sake of the electricity grid.

Considering climate change, the investment in solar power plants has been progressively increasing. By the end of 2021, the share of solar power in the global energy mix reached 3.7%. Furthermore, solar power has been the most dominant power plant type for new power plant investments for 17 years [9]. This situation makes the forecasting of potential electricity output from PV and solar thermal systems very important. Hence, to increase the benefit of PV and solar thermal systems and to avoid frequency problems in the electricity grid, predicting DSSR is very crucial.

Hence, this study aims to combine Particle Swarm Optimization (PSO) and Long Short-Term Memory (LSTM) methods to predict DSSR for different time horizons. The first phase is to detect the best variables among the 13 influencers of the DSSR utilizing the PSO algorithm. Following the first phase, an LSTM model is established to forecast the DSSR for 1 hour, 2 hours, 4 hours, and 6 hours ahead.

The purpose of this study is twofold: (i) determining the most important variables to predict DSSR and (ii) forecasting DSSR for 1 hour, 2 hours, 4 hours, and 6 hours ahead. More importantly, to the authors' knowledge, this is the first study combining PSO and LSTM methods to predict DSSR.

6.2 Literature Review

The number of studies for forecasting the surface solar radiation is increasing in the literature year by year (see Table 6.1). As mentioned in the previous section, there are four primary approaches to determine the DSSR. On the other hand, this literature review focuses only on studies with Neural Networks (NN).

Artificial Neural Network (ANN) models are one of the most common methods to forecast the DSSR. In this context, Chiteka and Enweeremadu [10] applied ANN to predict global horizontal irradiation for several locations in Zimbabwe. This study shows that ANN models are also capable of forecasting global horizontal irradiation (GHI) the day before and the day after successfully. In a similar study, Marzo et al. [11] established an ANN model to estimate daily global radiation in desert areas using daily minimum and maximum temperatures and extraterrestrial radiation. More recently, Pang et al. [12] used both Recurrent Neural Network (RNN) and ANN to evaluate the performance of deep learning methods in forecasting solar radiation. This study presents that the RNN model provides much higher prediction accuracy than the ANN.

In addition to ANN models, there are several studies using Convolutional Neural Networks (CNN) to forecast solar radiation. Using CNN, Cannizzaro et al. [13] predict short- and long-term GHI. They compared the model accuracy with that of the Random Forest and Long Short-Term Memory models. Jalali et al. [14] propose a new approach to promote the forecasting accuracy of GHI. They applied a swarm evolutionary optimization method, the so-called sine-cosine algorithm, and tested the model accuracies using three datasets from the United States. The result shows that the proposed method provides a noteworthy improvement compared to other models. Furthermore, Vakitbilir et al. [15] developed three hybrid models with CNN, LSTM, and Support Vector Regression (SVR) to forecast GHI. The accuracies of established models are compared using the dataset from Northern Cyprus. As a result, the CN-SVR model is found to be the best model.

With their high prediction accuracy, LSTM models are attracting more attention these days. Hence, the number of studies with LSTM in the literature is increasing persistently. In this regard, Qing and Niu [16] applied LSTM to forecast hourly day-ahead solar irradiance for Cape Verde, and compared the model accuracy with the persistence algorithm, linear regression, and neural networks with backpropagation algorithm (BPNN). The result shows that the LSTM model performs 18.34% better than BPNN in regards to root mean square error (RMSE). In a similar study, Srivastava and Lessmann [17] proposed an LSTM model to forecast day-ahead global horizontal irradiance in Europe and the United States. As a result, it

Table 6.1 Summary of literature survey

References	Location	Method	Note
[19]	India	FFNN, FFBN, RBNN, PNN, ELM, CFB, C1, C2	Hourly solar radiation
[20]	Iran and United States	ANN, GEP, WR	Daily global solar radiation
[10]	Zimbabwe	ANN	Global horizontal solar irradiance
[21]	China	ANN, MLP, GRNN, RBNN	Daily global solar radiation
[22]	Algeria	ANN	5 min solar radiation
[11]	Desert areas	ANN	Daily global solar radiation
[16]	Cape Verde	LSTM	Hourly day-ahead solar irradiance
[17]	Europe and US	LSTM	Day-ahead global horizontal irradiance
[23]	Iran	MLFFNN, ANFIS, ANFIS-GA, ANFIS-PSO, GMDH	Daily global solar radiation
[12]	United States	RNN and ANN	10 min, 30 min, 1 h solar radiation
[24]	Turkey	ANN, time series, empirical	Monthly average daily global radiation
[25]	Vietnam	LSTM	Global solar radiation
[26]	Turkey	ANN, SVM, k-NN, DL	Daily global solar radiation
[27]	Colombia	LSTM, GRU	Daily and weekly solar radiation
[13]	Italy	CNN, RF, LSTM	Global horizontal solar irradiance
[28]	Vietnam	RLMD-LSTM	Global solar radiation
[29]	United States	WPD–CNN–LSTM-MLP	Hourly solar radiation
[14]	United States	CNN-LSTM	Global horizontal irradiation
[15]	North Cyprus	CNN, LSTM, SVR	Global horizontal irradiation
[18]		LSTM	Global horizontal irradiation
[30]	United States	LSTM	Global horizontal irradiation

is shown that the LSTM algorithm surpasses a vast number of different methods. Moreover, Singla et al. [18] applied three different LSTM algorithms to predict the GHI. The results show that the LSTM model with the *adam* training algorithm performs better than the SGDM and RMSprop training algorithms.

Although there are some studies in the literature where hybrid LSTM models are used for forecasting, the numbers of these studies are quite inadequate. Furthermore, despite the number of studies surveyed, the downward surface radiation forecasting in Turkey has not been sufficiently analyzed. This study aims to fill this gap by choosing the best variables with the Particle Swarm Optimization (PSO) method and predicting the downward surface solar radiation with the Long Short-Term Memory (LSTM) method. Above all, to the authors' knowledge, this is the first study combining PSO and LSTM methods to forecast downward surface solar radiation.

6.3 Methodology and Data

6.3.1 Methodology

As aforementioned in the introduction section, this study combines Particle Swarm Optimization (PSO) and Long Short-Term Memory (LSTM) methods to predict DSSR. Firstly, a PSO algorithm is employed to detect the most effective variables. Then, these selected variables are utilized as input for the LSTM model to forecast the DSSR for 1 hour, 2 hours, 4 hours, and 6 hours ahead. In this context, the flowchart of the proposed model is shown in Fig. 6.1.

Particle Swarm Optimization (PSO) is an optimization method inspired by the social behavior of animals such as fish and birds. PSO has a similar evolutionary computation with Genetic Algorithm (GA). However, different from GA, PSO does not utilize any evolution operator [31].

The particles in PSO are located in the search space of the problem, and they estimate the objective function at their present positions. After that, each particle ascertains its action in the search space by consolidating the history of current and best positions with random perturbations (see Fig. 6.2).

Following the movements of all particles, the next iteration happens. At the end of the iterations, the swarm gets very close to the optimum solution of the fitness function (see Fig. 6.3). In this regard, the algorithm of PSO is shown below [32].

1. Create a particle population with random locations and velocities in the search space with D dimensions.
2. **Loop**.
3. Assess the fitness function in D variables for each particle.
4. Measure the fitness assessment of the particle with its previous best function (p_{best}). If the last fitness is better than p_{best}, set p_{best} as the present value.
5. Discover the most successful particle nearby.
6. Designate the index of the most successful particles to the variable g.
7. Modify the location and velocity of the particle as:

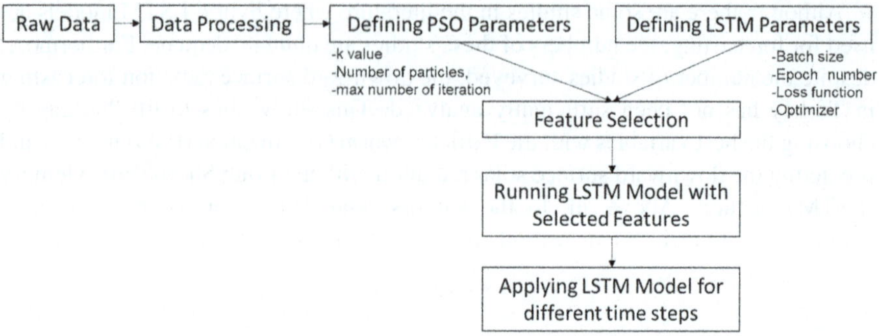

Fig. 6.1 Flowchart of the model

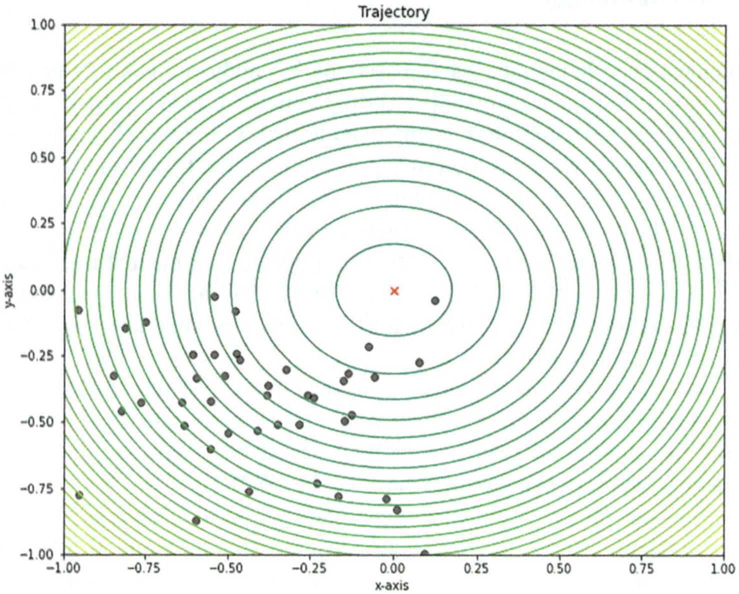

Fig. 6.2 An example of randomly defined particles

$$\begin{cases} \overrightarrow{v_i} \leftarrow \overrightarrow{v_i} + \vec{U}(0,\theta_1) \otimes \left(\overrightarrow{p_i} - \overrightarrow{x_i}\right) + \vec{U}(0,\theta_2) \otimes \left(\overrightarrow{p_g} - \overrightarrow{x_i}\right), \\ \overrightarrow{x_i} \leftarrow \overrightarrow{x_i} + \overrightarrow{v_i} \end{cases} \tag{6.1}$$

where $U(0,\theta_i)$ symbolizes a vector including uniformly distributed numbers in $[0,\varphi i]$ which is randomly created at each loop and for each particle, and $\otimes$ stands for the component-wise multiplication.

8. When the criterion is satisfied, exit loop.
9. **Stop loop**.

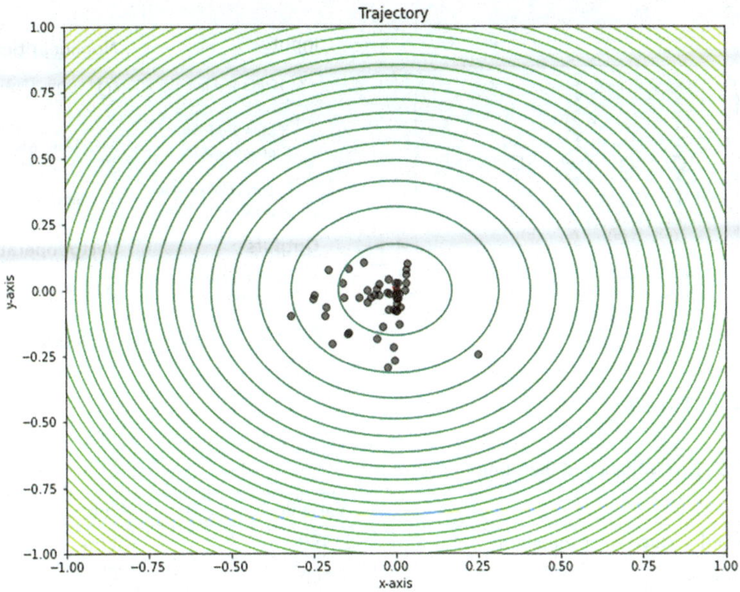

Fig. 6.3 New position of randomly defined particles

PSO algorithm is also a very useful tool for the feature selection process. In this process, the PSO algorithm seeks a subset of variables that optimizes the fitness function. In the fitness function, the error function should be detected first. In this procedure, k-nearest neighbors (k-NN) and multi-linear regression (MLR) methods were evaluated in this study. Although these methods show very similar performances, the k-NN method is selected since it works much quicker. Detecting the optimum input variable combination for forecasting models is very important since this process can remove unnecessary and/or insignificant information. Thus, computational costs can be lowered and overfitting issues can be improved [33].

Following the feature selection, a neural network with the new dataset is established. In this study, Long Short-Term Memory (LSTM), a special type of Recurrent Neural Network, was selected since it has a memory state that processes quicker, and has more accurate mapping capability [34]. The unique feedback mechanism of LSTM distinguishes LSTM from conventional RNNs. This feature provides to manage the whole sequence of data without processing each data item independently. An example for LSTM structure is shown in Fig. 6.4. More detailed information about LSTM architecture can be obtained from the study of Yu et al. [35].

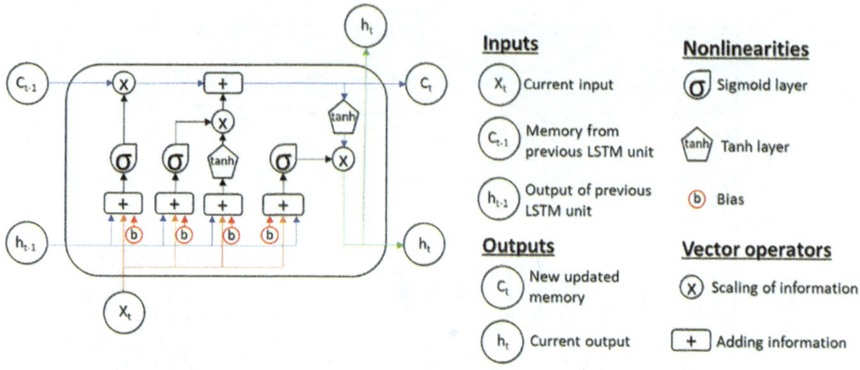

Fig. 6.4 LSTM cell

Table 6.2 List of data and their units

Data	Variable	Unit
Downward surface solar radiation	DSSR	W/m^2
Temperature	Temp	°C
Eastward wind speed	u10	m/s
Northward wind speed	v10	m/s
Ozone	O_3	Dobson unit
Precipitation	PREC	m
Surface pressure	PRES	hPa
High cloud cover	HCC	Dimensionless [0,1]
Medium cloud cover	MCC	Dimensionless [0,1]
Low cloud cover	LCC	Dimensionless [0,1]
Total cloud cover	TCC	Dimensionless [0,1]
PM_{10}	PM10	μg/m^3
$PM_{2.5}$	PM2.5	μg/m^3

6.3.2 *Data*

To set up the PSO and forecasting models, the hourly meteorology data covering the January 2020–December 2021 period was obtained from the European Centre for Medium-Range Weather Forecasts' (ECMWF) ERA5 reanalysis data sets. Moreover, hourly data of PM_{10} and $PM_{2.5}$ were gathered from Ministry of Environment, Urbanization and Climate Change [36]. The prediction model was trained and tested for Karaman, Turkey, which is one of the places with great solar energy potential. The data and its units are shown in Table 6.2.

Using exploratory data analysis before establishing the models assists the visualization of descriptive statistics. In this regard, the summary of data for the selected domain is given in Table 6.3.

Table 6.3 Brief statistical analysis of data

Variable	Min	q1	Mean	Median	q3	Max	S.D.	IQR
DSSR	0.0	0.0	218.94	9.86	417.44	1025.50	297.9	417.44
Temp	259.2	280	287.6	287.3	295.0	312.3	9.2	13.59
u10	−5.25	−0.44	0.25	0.19	0.86	7.08	1.24	1.39
v10	−9.89	−2.1	−0.28	−0.32	1.32	12.16	3.06	3.92
O_3	246	289.9	314.9	306.9	336.2	433.8	34.34	51.78
PREC	0.0	0.0	0.03	0.0	0.0	3.44	0.19	0.01
PRES	873.4	889	892.1	892.1	895.3	904.8	4.53	6.33
HCC	0.0	0.0	0.18	0.0	0.20	1.0	0.38	0.63
MCC	0.0	0.0	0.17	0.02	0.24	1.0	0.31	0.46
LCC	0.0	0.0	0.12	0.0	0.09	1.0	0.29	0.32
TCC	0.0	0.0	0.32	0.12	0.65	1.0	0.36	0.71
PM10	1.73	19.28	33.52	29.32	41.45	425.97	20.98	21.27
PM2.5	0.13	9.63	16.91	13.66	19.92	198.74	12.86	10.79

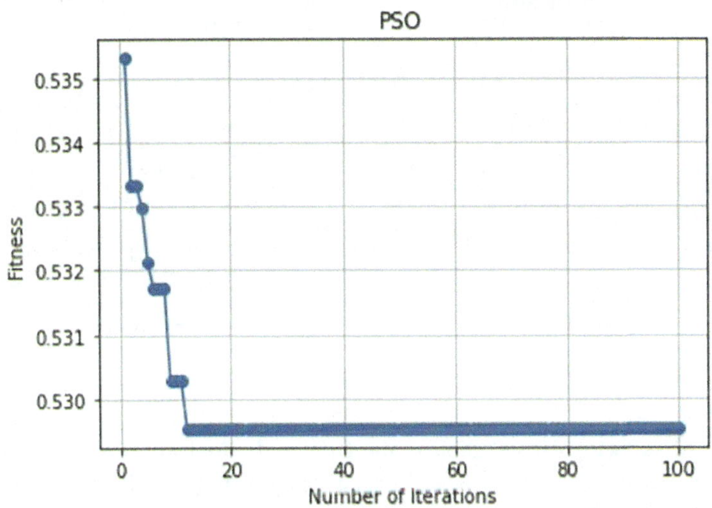

Fig. 6.5 Fitness function error depending on the iterations

6.4 Results and Discussion

Figure 6.5 shows the fitness function error depending on the iterations. As a result of the PSO, temperature, low cloud cover, eastward wind speed, and hour of the day were detected as the most important variables. These selected variables created a new data set to be input into the LSTM model.

In the LSTM model, the optimizer, activation function, and error function were chosen as *adam*, *ReLu (Rectified Linear unit)*, and *Mean Squared Error (MSE)*, respectively. The LSTM model was run for four different time horizons to estimate

Table 6.4 Performance metrics

Model	MSE	RMSE	R^2
1 hour-ahead	8496	92	0.904
2 hour-ahead	9537.8	97	0.89
4 hour-ahead	21554.6	146.81	0.76
6 hour-ahead	12731.55	112.8	0.86

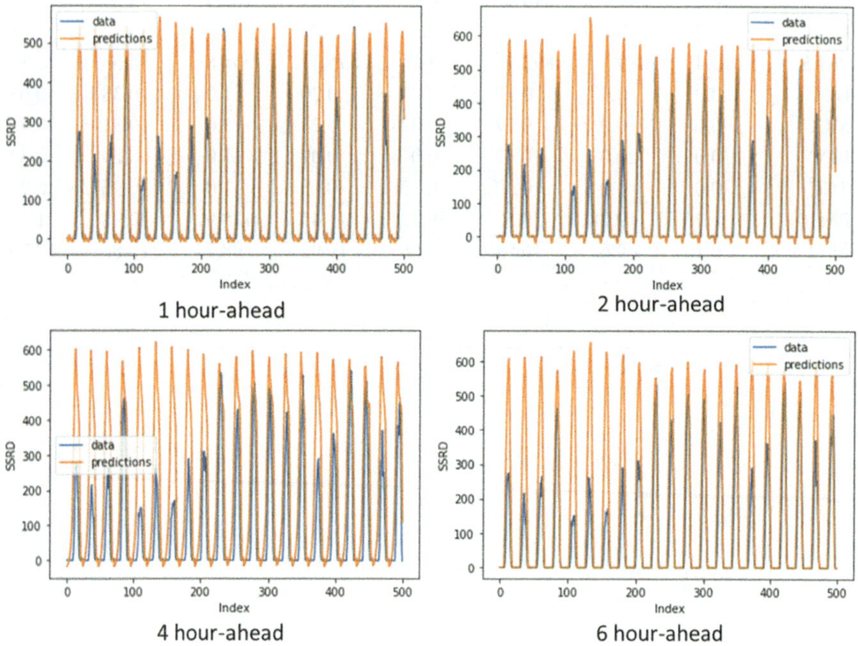

Fig. 6.6 Comparison of observed and predicted data

downward surface solar radiation in the case study domain. The performances of models were compared using Root Mean Squared Errors. While the lowest RMSE was obtained in the 1 hour-ahead model with a value of 92, the highest RMSE emerged in the 4 hour-ahead model with a value of 146.81. The performance metrics of all models are given in Table 6.4. Furthermore, the comparison of observed and predicted data for each time horizon is shown in Fig. 6.6 for a selected time period.

As a result of this chapter, it can be said that the most significant features in calculating downward surface solar radiation are temperature, low cloud cover, eastward wind speed, and the hour of the day. Considering global warming and climate change, the importance of solar energy systems as a clean and green energy generation technology has been growing more and more. On the other hand, the output of these systems is highly dependent on the shortwave solar radiation at the surface. Since downward surface solar radiation is exceedingly fluctuating because of the meteorological patterns, power generation from solar systems is not stable as distinct from dispatchable power plants, such as thermal power plants. Thus, to benefit from solar energy systems at their maximum level, predicting the downward surface solar radiation is very crucial and essential.

6.5 Conclusion

The aim of this study is twofold: (i) determining the most important variables to predict downward surface solar radiation and (ii) forecasting downward surface solar radiation for different time horizons. This study combines Particle Swarm Optimization and Long Short-Term Memory methods to predict DSSR. The first phase is to detect the best variables among the 13 influencers of the DSSR utilizing the PSO algorithm. Following the first phase, an LSTM model is established to forecast the DSSR for 1 hour, 2 hours, 4 hours and 6 hours ahead. As a result of the PSO algorithm, the most significant features in calculating downward surface solar radiation are detected as temperature, low cloud cover, eastward wind speed, and hour of the day. Following the detection of the most significant features, the prediction model was trained and tested for Karaman, Turkey, which is one of the places with great solar energy potential. The performances of models were compared using Root Mean Squared Errors. While the lowest RMSE was obtained in the 1 hour-ahead model with a value of 92, the highest RMSE emerged in the 4 hour-ahead model with a value of 146.81.

References

1. Lam, J. C., Wan, K. K. W., & Yang, L. (2008). Solar radiation modelling using ANNs for different climates in China. *Energy Conversion and Management, 49*, 1080–1090.
2. Dong, J., Olama, M. M., Kuruganti, T., Melin, A. M., Djouadi, S. M., Zhang, Y., & Xue, Y. (2020). Novel stochastic methods to predict short-term solar radiation and photovoltaic power. *Renewable Energy, 145*, 333–346.
3. Tymvios, F. S., Jacovides, C. P., Michaelides, S. C., & Scouteli, C. (2005). Comparative study of Ångström's and artificial neural networks' methodologies in estimating global solar radiation. *Solar Energy, 78*(6), 752–762.
4. Wei, Y., Zhang, X., Hou, N., Zhang, W., Jia, K., & Yao, Y. (2019). Estimation of surface downward shortwave radiation over China from AVHRR data based on four machine learning methods. *Solar Energy, 177*, 32–46. https://doi.org/10.1016/j.solener.2018.11.008
5. Hou, N., Zhang, X., Zhang, W., Wei, Y., Jia, K., Yao, Y., et al. (2020). Estimation of surface downward shortwave radiation over China from Himawari-8 AHI data based on random forest. *Remote Sensing, 12*(1), 181. https://doi.org/10.3390/rs12010181
6. Zhou, Q., Flores, A., Glenn, N. F., Walters, R., & Han, B. (2017). A machine learning approach to estimation of downward solar radiation from satellite-derived data products: An application over a semi-arid ecosystem in the US. *PLoS One, 12*(8), e0180239. https://doi.org/10.1371/journal.pone.0180239
7. Fan, J., Wu, L., Ma, X., Zhou, H., & Zhang, F. (2020). Hybrid support vector machines with heuristic algorithms for prediction of daily diffuse solar radiation in air-polluted regions. *Renewable Energy, 145*, 2034–2045. https://doi.org/10.1016/j.renene.2019.07.104

8. Chen, F., Zhou, Z., Lin, A., Niu, J., Qin, W., & Yang, Z. (2019). Evaluation of direct horizontal irradiance in China using a physically-based model and machine learning methods. *Energies, 12*(1), 150. https://doi.org/10.3390/en12010150
9. EMBER. Global electricity review 2022. https://ember-climate.org/app/uploads/2022/03/Report-GER22.pdf. Last accessed 19 Apr 2022.
10. Chiteka, K., & Enweremadu, C. C. (2016). Prediction of global horizontal solar irradiance in Zimbabwe using artificial neural networks. *Journal of Cleaner Production, 135*, 701–711.
11. Marzo, A., Trigo-Gonzalez, M., Alonso-Montesinos, J., Martínez-Durbán, M., López, G., Ferrada, P., et al. (2017). Daily global solar radiation estimation in desert areas using daily extreme temperatures and extraterrestrial radiation. *Renewable Energy, 113*, 303–311.
12. Pang, Z., Niu, F., & O'Neill, Z. (2020). Solar radiation prediction using recurrent neural network and artificial neural network: A case study with comparisons. *Renewable Energy, 156*, 279–289.
13. Cannizzaro, D., Aliberti, A., Bottaccioli, L., Macii, E., Acquaviva, A., & Patti, E. (2021). Solar radiation forecasting based on convolutional neural network and ensemble learning. *Expert Systems with Applications, 181*, 115167.
14. Jalali, S. M. J., Ahmadian, S., Kavousi-Fard, A., Khosravi, A., & Nahavandi, S. (2021). Automated deep CNN-LSTM architecture design for solar irradiance forecasting. *IEEE Transactions on Systems, Man, and Cybernetics: Systems, 52*(1), 54–65.
15. Vakitbilir, N., Hilal, A., & Direkoğlu, C. (2022). Hybrid deep learning models for multivariate forecasting of global horizontal irradiation. *Neural Computing and Applications, 34*, 1–22.
16. Qing, X., & Niu, Y. (2018). Hourly day-ahead solar irradiance prediction using weather forecasts by LSTM. *Energy, 148*, 461–468.
17. Srivastava, S., & Lessmann, S. (2018). A comparative study of LSTM neural networks in forecasting day-ahead global horizontal irradiance with satellite data. *Solar Energy, 162*, 232–247.
18. Singla, P., Duhan, M., & Saroha, S. (2022). Solar irradiation forecasting by long-short term memory using different training algorithms. In A. Khosla & M. Aggarwal (Eds.), *Renewable energy optimization, planning and control studies in infrastructure and control* (pp. 81–89). Springer.
19. Kashyap, Y., Bansal, A., & Sao, A. K. (2015). Solar radiation forecasting with multiple parameters neural networks. *Renewable and Sustainable Energy Reviews, 49*, 825–835.
20. Sharifi, S. S., Rezaverdinejad, V., & Nourani, V. (2016). Estimation of daily global solar radiation using wavelet regression, ANN, GEP and empirical models: A comparative study of selected temperature-based approaches. *Journal of Atmospheric and Solar-Terrestrial Physics, 149*, 131–145.
21. Wang, L., Kisi, O., Zounemat-Kermani, M., Salazar, G. A., Zhu, Z., & Gong, W. (2016). Solar radiation prediction using different techniques: Model evaluation and comparison. *Renewable and Sustainable Energy Reviews, 61*, 384–397.
22. Rezrazi, A., Hanini, S., & Laidi, M. (2016). An optimisation methodology of artificial neural network models for predicting solar radiation: A case study. *Theoretical and Applied Climatology, 123*(3–4), 769–783.
23. Khosravi, A., Nunes, R. O., Assad, M. E. H., & Machado, L. (2018). Comparison of artificial intelligence methods in estimation of daily global solar radiation. *Journal of Cleaner Production, 194*, 342–358.
24. Gürel, A. E., Ağbulut, Ü., & Biçen, Y. (2020). Assessment of machine learning, time series, response surface methodology and empirical models in prediction of global solar radiation. *Journal of Cleaner Production, 277*, 122353.
25. Huynh, A. N. L., Deo, R. C., An-Vo, D. A., Ali, M., Raj, N., & Abdulla, S. (2020). Near real-time global solar radiation forecasting at multiple time-step horizons using the long short-term memory network. *Energies, 13*(14), 3517.

26. Ağbulut, Ü., Gürel, A. E., & Biçen, Y. (2021). Prediction of daily global solar radiation using different machine learning algorithms: Evaluation and comparison. *Renewable and Sustainable Energy Reviews, 135*, 110114.
27. Narvaez, G., Giraldo, L. F., Bressan, M., & Pantoja, A. (2021). Machine learning for site-adaptation and solar radiation forecasting. *Renewable Energy, 167*, 333–342.
28. Huynh, A. N. L., Deo, R. C., Ali, M., Abdulla, S., & Raj, N. (2021). Novel short-term solar radiation hybrid model: Long short-term memory network integrated with robust local mean decomposition. *Applied Energy, 298*, 117193.
29. Huang, X., Li, Q., Tai, Y., Chen, Z., Zhang, J., Shi, J., et al. (2021). Hybrid deep neural model for hourly solar irradiance forecasting. *Renewable Energy, 171*, 1041–1060.
30. Bhattacharjee, A. D., & Chowdhury, A. R. (2022). Short-term solar irradiance forecasting using long short term memory variants. In *Proceedings of international conference on data science and applications* (pp. 227–243). Springer.
31. Eberhart, R. C., & Shi, Y. (1998). Comparison between genetic algorithms and particle swarm optimization. In *Lecture notes in computer science* (pp. 611–616). Springer.
32. Poli, R., Kennedy, J., & Blackwell, T. (2007). Particle swarm optimization. *Swarm Intelligence, 1*(1), 33–57.
33. Xue, B., Zhang, M., & Browne, W. N. (2014). Particle swarm optimisation for feature selection in classification: Novel initialisation and updating mechanisms. *Applied Soft Computing, 18*, 261–276.
34. Brito, R., Fong, S., Zhuang, Y., & Wu, Y. (2017). Generating neural networks with optimal features through particle swarm optimization. In *Proceedings of the international conference on big data and Internet of Thing - BDIOT2017* (pp. 96–101). ACM Press.
35. Yu, Y., Si, X., Hu, C., & Zhang, J. (2019). A review of recurrent neural networks: LSTM cells and network architectures. *Neural Computation, 31*(7), 1235–1270.
36. Ministry of Environment, Urbanization and Climate Change. (2022). *Air quality databank*. https://sim.csb.gov.tr/

26. Aghajani, [illegible] (2021). Prediction of hourly global solar radiation using different machine learning algorithms. [illegible] Energy [illegible]
27. [illegible] adaptation and solar radiation forecasting. [illegible]
28. [illegible]
[illegible]
29. [illegible]
30. [illegible] (2021). [illegible]
31. [illegible] (2018). [illegible]
32. [illegible]
33. [illegible]
34. [illegible]
35. [illegible]
36. [illegible]

Part III
Sustainability Energy Markets

Chapter 7
Electricity Demand Prediction: Case of Turkey

Ezgi Polat, Nezir Aydin, and Erkan Isikli

Abstract It is critical to generate accurate power demand forecasts for balancing supply and demand, as well as for the efficient operation and management of electricity generation. Accurate forecasts can also help to prevent energy wastage, reduce operating costs, and minimize environmental impacts. Several approaches for this task have been proposed in the related literature, including machine learning, statistical modeling, stochastic/fuzzy/gray modeling, and metaheuristics. The present study initially discusses the importance of accurately predicting electricity demand, providing an extensive overview of linear functions, neural network architectures, and regression models concerning their relevant concepts and applications, and then furnishes a case study of monthly electricity demand prediction in Turkey using data from 1998 to 2017. Considering several independent variables (IVs), namely average temperature, average precipitation, consumer price index, total industrial production index, total import, and total export, the performance of an Artificial Neural Network (ANN) model was compared to that of a Multiple Linear Regression (MLR) model and an Autoregressive Integrated Moving Average with Exogenous Variables (ARIMAX) model. The former has been shown to outperform the other two in terms of MAPE and R^2, indicating its capability to better explain the nonlinear relationship between the IVs and energy consumption. As a potential avenue for future research, to develop more comprehensive models that can generate reliable forecasts, the impacts of weather and climate change on energy demand

E. Polat (✉)
Department of Business Administration, Faculty of Economics, Administrative and Social Science, Işık University, Istanbul, Türkiye
e-mail: ezgi.polat@isikun.edu.tr

N. Aydin
Department of Industrial Engineering, Faculty of Mechanical Engineering, Yıldız Technical University, Istanbul, Türkiye

E. Isikli
Department of Industrial Engineering, Faculty of Management, Istanbul Technical University, Istanbul, Türkiye
e-mail: isiklie@itu.edu.tr

G. Kayakutlu, M. Ö. Kayalica (eds.), *Decision Making Using AI in Energy and Sustainability*, Applied Innovation and Technology Management,
https://doi.org/10.1007/978-3-031-38387-8_7

should be evaluated in different countries and regions adopting hybrid models that combine different methodologies such as metaheuristics and machine learning.

7.1 Introduction

Electricity is an essential type of energy that facilitates daily life and contributes to the sustainability in production and service industries. Due to technological advances, rapid population growth, and social development, demand for electricity has been increasing worldwide [1]. Sufficient power generation is thus required to meet global electricity demand, and the prediction of electricity demand with high accuracy is of great significance for balancing electricity supply and demand.

Electricity demand prediction models are generally developed for three different time periods: short-, mid-, and long-term. Short-term prediction (1 hour to 1 week) is adopted mainly for the scheduling operations of power systems. Mid-term prediction (1 month to 5 years) is important for the purchasing decisions of primary energy resources and the regulation of electricity tariffs. Long-term predictions (5–20 years) are used for planning resources, making investment decisions, and providing employment opportunities [2].

Several approaches have been proposed and applied in the related literature for high-accuracy electricity demand prediction. These may be classified into four main categories: machine learning, statistical modeling, stochastic/fuzzy/gray modeling, and metaheuristics.

Machine learning algorithms mainly seek a solution to regression, classification, or clustering problems. Since electricity demand should be treated as a continuous random variable, its prediction is classified as a regression-type problem. Artificial Neural Networks (ANN) and hybrid models with ANN have been primarily employed in the literature for solving this type of prediction problems. *Statistical methods* are traditional techniques that are based on specific assumptions. Much of the literature on electricity demand prediction through statistical methods seems to have been based on ARIMA and Multiple Linear Regression (MLR). Günay utilized MLR and ANN to predict the annual electricity demand of Turkey [3]. Bianco et al. developed an MLR model to forecast the annual electricity consumption of Italy [4]. Aydin modeled the annual primary energy consumption of Turkey using regression analysis [5]. Hamzaçebi et al. predicted Turkey's monthly electricity demand using four different ANN structures that considered seasonality and trend effect [6]. Son and Kim modeled the monthly electricity demand of the residential sector using a long short-term memory network [7]. Cui and Peng developed the short-term electric load forecasting model using an improved version of the Autoregressive Integrated Moving Average with Exogenous Variables (ARIMAX) [8]. Salkuti proposed a short-term forecasting model using the radial basis function neural networks [9].

Stochastic/fuzzy/gray methods usually deal with problems involving uncertainty and poor information that are often encountered during demand prediction. Thus, a

Table 7.1 Independent variables used for the prediction of electricity consumption [1, 4]

Type	Example
Economic/demographic variables	Gross domestic product (GDP), level of production, import and export level, consumer price index, population
Weather variables	Temperature, humidity, pressure, precipitation, solar radiation
Calendar variable	Time of day, day of the week, month, holidays, working hours, operating time of appliance

substantial portion of the literature is concerned with these methods. Metaheuristic algorithms are generally used with other techniques to increase the accuracy of prediction [10]. Laouafi et al. developed an adaptive two-stage methodology in which Holt–Winters exponential smoothing method was used in the first stage, and fuzzy-based methods were utilized in the next one [11]. Jiang et al. preferred Holt–Winters' exponential smoothing to estimate the monthly electricity consumption of a city in China. In their study, the fruit fly optimization algorithm was also used to determine the best parameters for the smoothing method [12]. Shepero et al. developed a residential electricity load forecasting model that was based on a Gaussian process [13].

Electricity demand are generally affected by a number of factors including weather change and time. Apart from these, social and economic factors may also contribute to the changes in electricity demand patterns [14]. In the related literature, economic/demographic, weather, and calendar variables have been essentially used to predict historical electricity load data [10]. Bianco et al. considered population, gross domestic product per capita, inflation, and average summer temperature as independent variables (IVs) to forecast Italy's annual electricity consumption [4]. Aydin predicted Turkey's electricity consumption using population and gross domestic product as well [5]. Chen et al. developed a model that links electricity consumption to month index, daylight time, and some weather variables [15]. Gonzalez-Romera et al. used historical electricity data to estimate monthly electrical energy demand in their study [16].

Table 7.1 provides an overview of the IVs that have been considered in electricity demand/consumption prediction problems.

After deciding on which methods and IVs should be used, performance metrics are required to evaluate how well the models fit to the data in hand. These metrics are utilized to measure prediction accuracy, compare the performances of alternative methods, and improve the available prediction model. Table 7.2 shows the most widely used performance metrics with their abbreviations and mathematical formulae.

Selecting the correct performance metrics is critical as they should be suitable for evaluating several model alternatives. MAE, which is based on the absolute difference between two continuous variables, is used in cases where continuous values are modeled because it is easily interpretable. MSE measures the average of the squared errors, and hence it is always a positive number. RMSE is basically the population standard deviation of prediction errors. It penalizes larger errors more heavily.

Table 7.2 Performance metrics used in prediction models

Performance metric	Abbreviation	Formula
Mean absolute error	MAE	$\frac{\sum_{t=1}^{n} \lvert e_t \rvert}{n}$
Mean squared error	MSE	$\frac{\sum_{t=1}^{n} e_t^2}{n}$
Root mean square error	RMSE	$\sqrt{\frac{\sum_{t=1}^{n} e_t^2}{n}}$
Mean absolute percentage error	MAPE	$\frac{100\%}{n} \sum_{t=1}^{n} \frac{\lvert e_t \rvert}{\lvert y_t \rvert}$
Coefficient of determination	R^2	$1 - \frac{\sum_{t=1}^{n} e_t^2}{\sum_{t=1}^{n} (y_t - \bar{y})}$

n: sample size
y_t: actual value
$\bar{y}$: arithmetic average of the actual values
e_t: error = actual value − predicted value

MAPE is very useful in regression problems as it is a good indicator of measuring relative error. The coefficient of determination, R^2, indicates the power of IVs in explaining the variance of the dependent variable(s) (DVs). It is more commonly used in regression models.

Based on MAPE and R^2, the present study evaluates the performance of three alternative models, namely Artificial Neural Networks, Multiple Linear Regression, and ARIMAX, in predicting the monthly electricity demand of Turkey with the consideration of several IVs. A brief overview of the methods employed, the description of data and variables, experimental results, and conclusions are provided in the next sections.

7.2 Methods

7.2.1 Artificial Neural Networks (ANNs)

ANN models were developed with an inspiration from the biological nervous system and the human brain. Neurons, the smallest units of artificial neural networks where information is processed, form the basis of ANN activities. Neurons are generally defined by their connections, adder, and activation functions [17]. Each connection has a unique weight. The connection weight w_{kj} that connects input j to neuron k is multiplied by the input data x_j, and these weighted inputs are summed up using an adder. This weighted linear combination u_k is formulated as follows:

Table 7.3 Activation functions used in ANN

Activation functions	Formula
Linear function	$\varphi(u) = u$
Rectified linear unit (ReLU)	$\varphi(u) = u^+ = \max(0, u)$
Sigmoid function	$\varphi(u) = \frac{1}{1+e^{-\beta.u}}$
Hyperbolic tangent function	$\varphi(u) = \frac{1-e^{-\beta.u}}{1+e^{-\beta.u}}$

$$u_k = \sum_{j=1}^{m} w_{kj}x_j + b_k \tag{7.1}$$

where b_k is the bias (threshold). The output of the neuron, denoted by a_k, is related to u_k via an activation function:

$$a_k = \varphi(u_k) \tag{7.2}$$

where $\varphi(\cdot)$ is the activation function that provides linear or nonlinear transformation in a weighted linear combination [18]. Table 7.3 presents the most widely used activation functions.

Activation functions can basically be divided into two groups as linear and non-linear functions. The linear function and the rectified linear unit are examples of linear functions, whereas the sigmoid and the hyperbolic tangent functions are examples of nonlinear functions. A linear function forwards the input to the next layer without changing it. If the input is negative, the ReLU function converts it to 0; if it is positive, the output of the activation function is simply determined as the input. The sigmoid and the hyperbolic tangent functions reduce the input to a range between 0 and 1, and a range between 1 and – 1, respectively. The advantage of these two functions is the fact that they are differentiable. In general, linear activation functions are preferred when working with a continuous output, while sigmoid and hyperbolic functions are more preferred in classification problems.

In general, neural network architectures are defined by three basically different classes: Single Layer Feedforward Networks (SLFN), Multilayer Feedforward Networks (MLFN), Recurrent Networks (RN). In feedforward networks, the information flows from the input layer to the output layer in the forward direction. The SLFN has an input layer and an output layer. The MLFN has one or more hidden layers apart from the input and output layers. The RN has at least one feedback loop unlike other feedforward networks [17]. As the number of hidden layers and the number of neurons in those layers increase, the network complexity and depth increase as well.

After a network is structured for a particular application, it becomes ready to be trained. The training process can basically be carried out in two different ways: If the inputs in hand is labeled, that is, the network consists of input and output pairs, we have supervised training. Conversely, if the inputs are not labeled, that is, the outputs of the network are unknown, what we have is called unsupervised learning.

Before any learning process, that is, before the data is introduced to the network, the data must be preprocessed if necessary. The preprocessing stage can include steps such as cleaning the data, dealing with the missing values, and normalizing the variables. After the necessary steps are followed, the data becomes available to the network.

Multilayer Feedforward Networks (MLFN)

The MLFN is composed of many interconnected layers and neurons (nodes) and tries to capture the nonlinear relationship between the input and the output [19].

The learning process for an MLFN can be summarized as follows [20]:

1. Assign connection weights randomly.
2. Apply an input vector, which is an element of the training data, to the network.
3. Feed forward the network through the layers to obtain an output.
4. Compare the calculated output with the targeted output and calculate the error.
5. Backpropagate the error through the network.
6. Revise the connection weights minimizing the error.
7. Repeat Steps 2–6 for each input vector. Stop when the error is satisfactorily small.

7.2.2 *Multiple Linear Regression (MLR)*

Linear regression analysis is a statistical method that models the linear relationship between one dependent variable (DV) and one or more independent variables (IVs). There are various regression applications in several fields ranging from engineering to social sciences [21]. In regression analysis, if the model contains only one IV, what we have is called simple linear regression; if the number of IVs is more than one, then the procedure is called multiple linear regression (MLR).

In general, MLR is formulated as follows:

$$y=\beta_0+\beta_1x_1+\ldots+\beta_nx_n+\varepsilon \tag{7.3}$$

where y denotes the DV, $\beta_0, \beta_1, \ldots, \beta_n$ denote the regression coefficients, $x_1, \ldots, x_n$ denote the IVs, and ε is the random error term.

Ordinary least squares (OLS) estimation is most often used to estimate the regression coefficients in Eq. (7.3). Using OLS, the regression coefficients are estimated by minimizing the sum of the squared errors, which are simply the squares of the difference between the predicted values and the actual values.

7.2.3 Autoregressive Integrated Moving Average Exogenous Variable Models (ARIMAX)

Autoregressive (AR) models describe the value of a time series at any period as a linear function of its past values along with an error term. An autoregressive model of order p, denoted as AR(p), can mathematically be expressed as:

$$y_t = \beta_0 + \beta_1 y_{t-1} + \beta_2 y_{t-2} + \ldots + \beta_p y_{t-p} + \varepsilon_t \tag{7.4}$$

where ε_t is error term and β_0, β_1, …, β_n denote the regression coefficients. An AR model is similar to an MLR model that treats the lagged values of y_t as IVs.

Moving Average (MA) models use the past values of the error term as IVs instead of using the past values of the time series itself. A moving average model of order q, denoted as $MA(q)$, can mathematically be expressed as:

$$y_t = \beta_0 + \varepsilon_t + \beta_1 \varepsilon_{t-1} + \beta_1 \varepsilon_{t-2} + \ldots + \beta_p \varepsilon_{t-q} \tag{7.5}$$

The Autoregressive Moving Average (ARMA) model is a combination of an AR model with an MA model. Hence, an ARMA model with p autoregressive and q moving average terms, denoted as ARMA(p, q), can be written as:

$$y_t = \beta_0 + \varepsilon_t + \sum_{i=1}^{p} \beta_i y_{t-i} + \sum_{i=1}^{q} \theta_i \varepsilon_{t-i} \tag{7.6}$$

The Autoregressive Integrated Moving Average (ARIMA) model is basically an ARMA model that is created by the inclusion of a step that considers the difference of the original time series. An ARIMA model that considers the dth difference of the original time series and incorporates p AR terms and q MA terms, denoted as ARIMA(p, d, q), can be written as:

$$y_t^d = \beta_0 + \sum_{i=1}^{p} \beta_i y_{t-i} + \varepsilon_t - \sum_{i=1}^{q} \theta_i \varepsilon_{t-i} \tag{7.7}$$

where d shows the degree of differencing.

The Autoregressive Integrated Moving Average with Exogenous Variables Model (ARIMAX) is basically an extended version of the ARIMA procedure. The only difference between the two is the incorporation of IVs, which are not the lagged values of the original times series or the error term, into the ARIMAX. An ARIMAX model can be mathematically be expressed as:

$$y_t = \phi x_t + \beta_0 + \sum_{i=1}^{p} \beta_i y_{t-i} + \varepsilon_t - \sum_{i=1}^{q} \theta_i \varepsilon_{t-i} \tag{7.8}$$

where x_t is called an exogenous variable and ϕ is its coefficient.

7.3 Case of Turkey

High-accuracy prediction of electricity demand for Turkey is of great importance since primary energy resources (e.g., natural gas, petroleum, coal) that are required to generate electricity are limited in Turkey, and thus most of these resources are imported from abroad. In addition, the need for electricity in Turkey is growing day by day with the increase in population and the advancements in technology and economy. This is why the present study considers the case of Turkey.

All of the applications in this study were carried out using Python programming language. Information on the data used and the methods applied are summarized below:

Data: Average temperature (IV_1) and average precipitation (IV_2) as the weather variables; consumer price index (IV_3), total industrial production index (IV_4), total import (IV_5), and total export (IV_6) as the economic variables; gross monthly electricity consumption as the dependent variable.

Time horizon: From January 1998 to January 2017 (228 months).

Time series plots of the DV and the IVs are provided in Fig. 7.1.

The stages of the methods employed are provided in detail below:

Method: Multilayer Feedforward Networks

1. *Normalization of data:* To ensure that all observations fall on the same scale, the study variables were normalized as shown in Eq. (7.9) to vary in between 0 and 1:

$$x_{\text{norm}} = \frac{x_0 - x_{\text{min}}}{x_{\text{max}} - x_{\text{min}}} \tag{7.9}$$

   where x_{max} and x_{min} are, respectively, the maximum and the minimum of the variable the observation x_0 belongs to.
2. *Data splitting:* The first 80% of the data was used for training and the rest was remained for testing the models.
3. *Training:* The training data was provided to the network for training. The training stage has the following features:

   Loss function: mean squared error
   Optimizer: adam
   Epochs: 100
   Batch size: 1

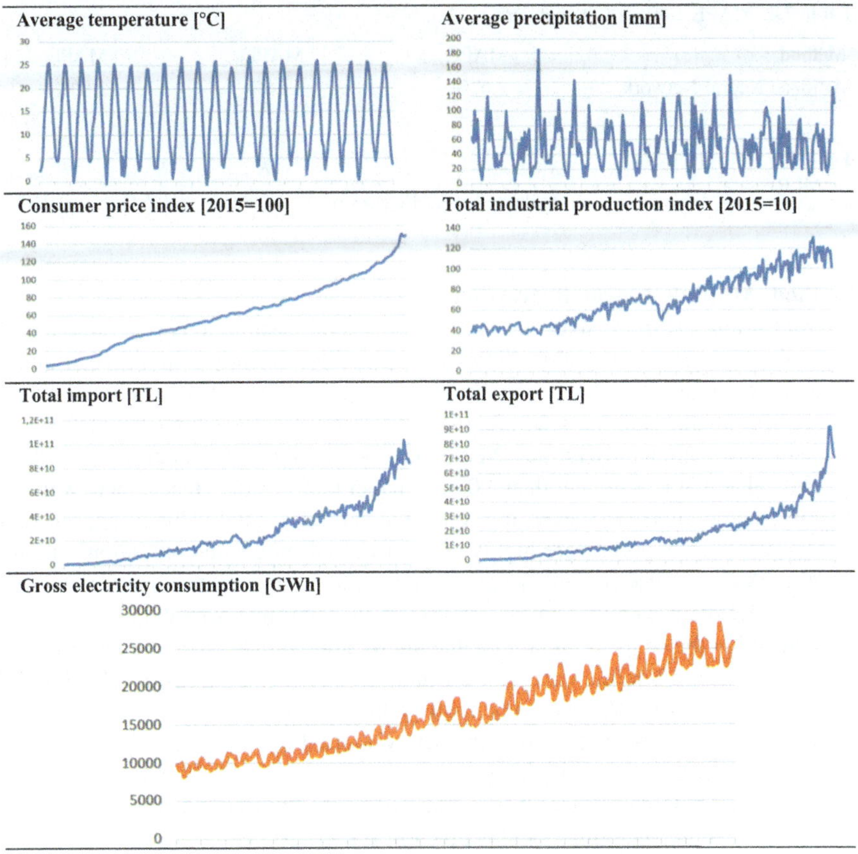

Fig. 7.1 Time series plots of the study variables for the period January 1998 through January 2017

4. *Test:* The test data was provided to the network, which had been trained. At this stage, the actual output of the test data was compared with the predicted output that was calculated in the network. If the error did not reach the desired level, the network would be retrained.

Within the scope of this study, the prediction model was tuned using various network structures. The performance of networks was measured with MAPE. The best network structure was constructed testing different activation functions, different numbers of hidden layers, and different numbers of nodes (neurons).

The best performing network consisted of two hidden layers, which had 15 neurons each. The Rectified Linear Unit function was used as an activation function in all hidden layers along with the output layer. Table 7.4 shows the MAPE values of the final network.

Table 7.4 MAPE values of the network

Method	$MAPE_{train}$	$MAPE_{test}$	$MAPE_{overall}$
Artificial Neural Network	2.67%	3.34%	2.80%

Table 7.5 R^2 values of the MLR

Method	R^2_{train}	R^2_{test}	$R^2_{overall}$
Multiple Linear Regression	0.921	0.531	0.940

Method: Multiple Linear Regression

1. *Data preprocessing:* Since MLR requires several assumptions (i.e., absence of outliers, absence of multicollinearity, normality, linearity, homoscedasticity, absence of autocorrelation) to be met for reliable parameter estimation and prediction, the data set was preprocessed to overcome potential violations to them before going any further. Neither the IVs nor the DV contained a standardized value that was greater than 3.00 or smaller than −3.00. Hence, there were no univariate outliers. However, the economic variables' total value of imports and total value of exports were highly correlated, and moderate correlations among other IVs were also present. To avoid multicollinearity and its consequences, we thus employed stepwise linear regression to determine which IVs should be included in the analysis and found that the weather variable *average temperature* (IV_1) and the economic variables *consumer price index* (IV_3) and *total industrial production index* (IV_4) should be used to predict the DV.
2. *Data splitting:* The first 80% of the data was used for training and the rest was remained for testing the model.
3. *Training:* MLR parameters were estimated using the training data through the OLS procedure. The full model is given in Eq. (7.10):

$$y = 2982.11 + 5.69IV_1 + +41.16IV_3 + 99.14IV_4 \tag{7.10}$$

4. *Test:* The DV was predicted using the parameter estimates given in Eq. (7.10) model and the actual values in the test data were compared with the predicted values produced by the training model. The *R*-squared values obtained at this stage are provided in Table 7.5.

Method: Autoregressive Integrated Moving Average Exogenous Variable Models

1. *Data splitting:* The first 80% of the data was used for training and the rest was remained for testing the model.
2. *Training:* Different combinations for the *p*, *d*, and *q* values were tried on the training data. ARIMAX(5,1,0) was chosen as the best model considering the autocorrelations, partial autocorrelations, and the stationarity of data along with the Akaike Information Criterion (AIC). Since the data is monthly, the seasonal order was set at 12.

Table 7.6 MAPE and R^2 values of the ARIMAX model

Method	R^2_{test}	MAPE $_{test}$
ARIMAX	0.5397	0.0482

Table 7.7 Comparison of the ANN and the MLR models (R^2)

Method	R^2_{train}	R^2_{test}	$R^2_{overall}$
Artificial Neural Network	0.710	0.980	0.982
Multiple Linear Regression	0.921	0.531	0.940

Table 7.8 Comparison of the ANN and the MLR models (MAPE)

Method	MAPE $_{train}$	MAPE $_{test}$	MAPE $_{overall}$
Artificial Neural Network	0.0267	0.0334	0.0280
Multiple Linear Regression	0.0574	0.0684	0.0596

Table 7.9 The predictive power of MAPE [22]

MAPE (%)	Evaluation
MAPE $\leq$ 10%	High accuracy forecasting
10% $<$ MAPE $\leq$ 20%	Good forecasting
20% $<$ MAPE $\leq$ 50%	Reasonable forecasting
MAPE $>$ 50%	Inaccurate forecasting

3. *Test:* The ARIMAX model that was trained was applied to the test data. At this stage, the actual values in the test data were compared with the predicted values produced by the model. As a result, the following MAPE and *R*-squared values were obtained (Table 7.6).

7.4 Comparison of Prediction Methods

Turkey's monthly electricity consumption from January 1998 to January 2017 was predicted using ANN, MLR, and ARIMAX models adopting the same data partitioning. The proposed ANN model can be said to outperform its alternatives since it produced higher R^2's and smaller MAPEs for both the training and test data as shown in Tables 7.7 and 7.8.

The scale proposed by Lewis [22] for the interpretation of model performance in terms of MAPE is given in Table 7.9. An MAPE smaller than 10% indicates "High Accuracy Forecasting" according to this scale. Since the MAPEs produced by the ANN and MLR models are both smaller than 10%, they can be said to forecast the DV with high accuracy; however, the relatively poor performance of the MLR model on the test data should not be overlooked. Figure 7.2 also illustrates that the predicted values obtained from the ANN model follow the actual values more closely compared to the MLR.

The ANN model outperformed the ARIMAX model as well, resulting in higher accuracy in terms of R^2 and MAPE as shown in Table 7.10.

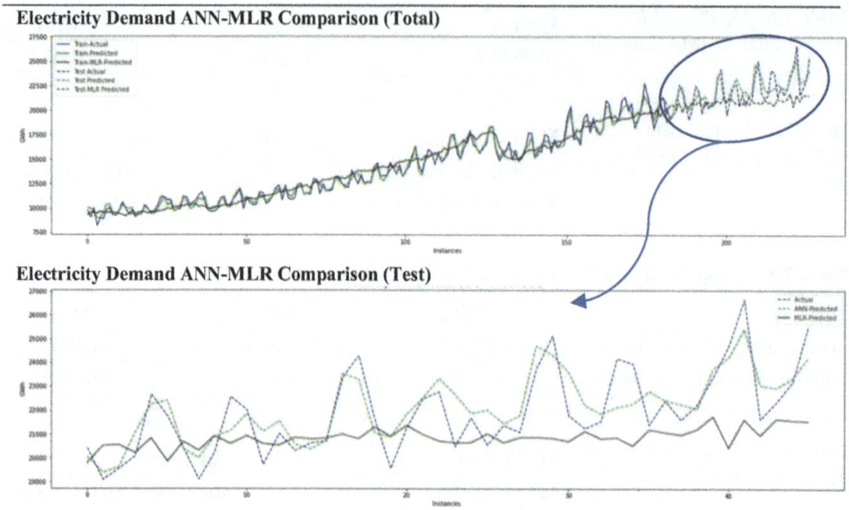

Fig. 7.2 Comparison of the ANN and MLR models (train-test)

Table 7.10 Comparison of the ANN model with the ARIMAX model

Method	MAPE $_{test}$	R^2 $_{test}$
Artificial Neural Network	3.34%	0.980
ARIMAX	4.82%	0.540

Based on these results, it can be inferred that even though the predicted values obtained from the ARIMAX model showed a similar trend with the actual test data, they did not follow each other closely.

7.5 Summary and Conclusion

The fact that electricity is becoming essential day by day and it is being generated with limited resources has necessitated the planning of electricity generation with higher accuracy. The prediction of electricity demand/consumption in the short-, medium-, and long-term is closely related to the planning of electricity generation. In the related literature, this prediction has been made using various approaches. In addition to the traditional statistical methods, machine learning algorithms, stochastic/fuzzy methods, and metaheuristics have been employed in predicting electricity demand. The review of the related literature revealed that the number of ANN-based applications is large. This is mainly due to the fact that ANN algorithms are easier to adapt to the nonlinear structure of electricity demand.

The scope of this study was determined as the monthly electricity demand prediction since the literature lacks applications based on monthly data. The case of Turkey was selected since Turkey depends on external sources for electricity generation and its technological dependency has grown in the past few decades along with its increasing population. In this study, economic, demographic, and weather variables were incorporated into a Multilayer Feedforward Network and its results were compared with those obtained from an MLR model and an ARIMAX model.

The application was carried out on a total of 228 data points, covering the period from January 1998 to January 2017. The first 80% of the data was reserved for training and the remaining 20% was used for testing the models. ANN, MLR, and ARIMAX models were trained with the same training data and they were then compared using the same test data. The ANN-based model outperformed the other two in terms of MAPE and R^2. The predicted values obtained from the ANN model were found to follow the actual data more closely than the other two models. This indicates that ANN can explain the non-linear relationship between the input and the output better than other statistical and traditional time series methods.

References

1. Polat, E. (2019). Türkiye'nin aylık elektrik tüketiminin yapay sinir ağlarıyla tahmini [Master's thesis, Fen Bilimleri Enstitüsü].
2. Abdel-Aal, R. E. (2008). Univariate modeling and forecasting of monthly energy demand time series using abductive and neural networks. *Computers & Industrial Engineering, 54*(4), 903–917.
3. Günay, M. E. (2016). Forecasting annual gross electricity demand by artificial neural networks using predicted values of socio-economic indicators and climatic conditions: Case of Turkey. *Energy Policy, 90*, 92–101.
4. Bianco, V., Manca, O., & Nardini, S. (2013). Linear regression models to forecast electricity consumption in Italy. *Energy Sources, Part B: Economics, Planning, and Policy, 8*(1), 86–93.
5. Aydin, G. (2014). Modeling of energy consumption based on economic and demographic factors: The case of Turkey with projections. *Renewable and Sustainable Energy Reviews, 35*, 382–389.
6. Hamzaçebi, C., Es, H. A., & Çakmak, R. (2019). Forecasting of Turkey's monthly electricity demand by seasonal artificial neural network. *Neural Computing and Applications, 31*(7), 2217–2231.
7. Son, H., & Kim, C. (2020). A deep learning approach to forecasting monthly demand for residential–sector electricity. *Sustainability, 12*(8), 3103.
8. Cui, H., & Peng, X. (2015). Short-term city electric load forecasting with considering temperature effects: An improved ARIMAX model. *Mathematical Problems in Engineering*, 2015.
9. Salkuti, S. R. (2018). Short-term electrical load forecasting using radial basis function neural networks considering weather factors. *Electrical Engineering, 100*(3), 1985–1995.
10. Verwiebe, P. A., Seim, S., Burges, S., Schulz, L., & Müller-Kirchenbauer, J. (2021). Modeling energy demand—A systematic literature review. *Energies, 14*(23), 7859.
11. Laouafi, A., Mordjaoui, M., Laouafi, F., & Boukelia, T. E. (2016). Daily peak electricity demand forecasting based on an adaptive hybrid two-stage methodology. *International Journal of Electrical Power & Energy Systems, 77*, 136–144.
12. Jiang, W., Wu, X., Gong, Y., Yu, W., & Zhong, X. (2020). Holt–Winters smoothing enhanced by fruit fly optimization algorithm to forecast monthly electricity consumption. *Energy, 193*, 116779.

13. Shepero, M., Van Der Meer, D., Munkhammar, J., & Widén, J. (2018). Residential probabilistic load forecasting: A method using Gaussian process designed for electric load data. *Applied Energy, 218*, 159–172.
14. Fallah, S. N., Deo, R. C., Shojafar, M., Conti, M., & Shamshirband, S. (2018). Computational intelligence approaches for energy load forecasting in smart energy management grids: State of the art, future challenges, and research directions. *Energies, 11*(3), 596.
15. Chen, J. F., Lo, S. K., & Do, Q. H. (2017). Forecasting monthly electricity demands: An application of neural networks trained by heuristic algorithms. *Information, 8*(1), 31.
16. Gonzalez-Romera, E., Jaramillo-Moran, M. A., & Carmona-Fernandez, D. (2006). Monthly electric energy demand forecasting based on trend extraction. *IEEE Transactions on Power Systems, 21*(4), 1946–1953.
17. Haykin, S., & Network, N. (2004). A comprehensive foundation. *Neural Networks, 2*(2004), 41.
18. Hu, Y. H., & Hwang, J. N. (Eds.). (2002). *Handbook of neural network signal processing*. CRC Press Taylor & Francis Group.
19. Gardner, M. W., & Dorling, S. R. (1998). Artificial neural networks (the multilayer perceptron)—A review of applications in the atmospheric sciences. *Atmospheric Environment, 32*(14–15), 2627–2636.
20. Bishop, C. M. (2006). Pattern recognition. *Machine Learning, 128*(9). 225–249.
21. Montgomery, D. C., Peck, E. A., & Vining, G. G. (2021). *Introduction to linear regression analysis*. Wiley.
22. Lewis, C. D. (1982). *Industrial and business forecasting methods*. Butterworths.

Chapter 8
The Impact of the Wind Energy Power Forecast Accuracy on the Price of Electricity

Uğur B. Kahveci and Burak Barutcu

Abstract The electricity market is a complex entity where many factors are taken into account to form the final price. Renewable energy systems, which are one of the many inputs, make it more difficult to forecast the price accurately due to their chaotic nature and the increasing amount of installed power. In this study, the effect of wind power forecasts on the electricity price has been investigated using autocorrelation and partial autocorrelation functions, as well as correlation coefficients. Afterwards, two different forecasting models were developed to forecast electricity price, named as standard model and improved model. The only difference between the models consisting of the Multi-Layer Perceptron network is that the wind power forecast error percentage and share of wind energy related productions to total energy production are also used for the training of the improved model. Results show that the accuracy of the wind energy production forecast is found to be related to the final electricity price, and there is a definite correlation between them. To carry out the study, a total of 8760 observations of one-year hourly data between January 1, 2020 and December 31, 2020 are used and provided from the Energy Exchange Istanbul (EXIST) platform.

Keywords Wind power forecast · Electricity price forecast · Multilayer perceptron

8.1 Introduction

According to data given by Turkey Wind Energy Association (TUREB), wind energy installed power amount exceeded 9.3 GW in 2020 and the share of wind energy electricity production in total electricity production is more than 10%. Wind energy market is growing continuously in Turkey. Figure 8.1 shows the development of wind energy in Turkey between years 2010 and 2020. With the policies and subsidies, the growth is expected to continue in the future.

U. B. Kahveci (✉) · B. Barutcu
Renewable Energy Division, Istanbul Technical University Energy Institute, Istanbul, Türkiye
e-mail: kahveciug@itu.edu.tr

G. Kayakutlu, M. Ö. Kayalica (eds.), *Decision Making Using AI in Energy and Sustainability*, Applied Innovation and Technology Management,
https://doi.org/10.1007/978-3-031-38387-8_8

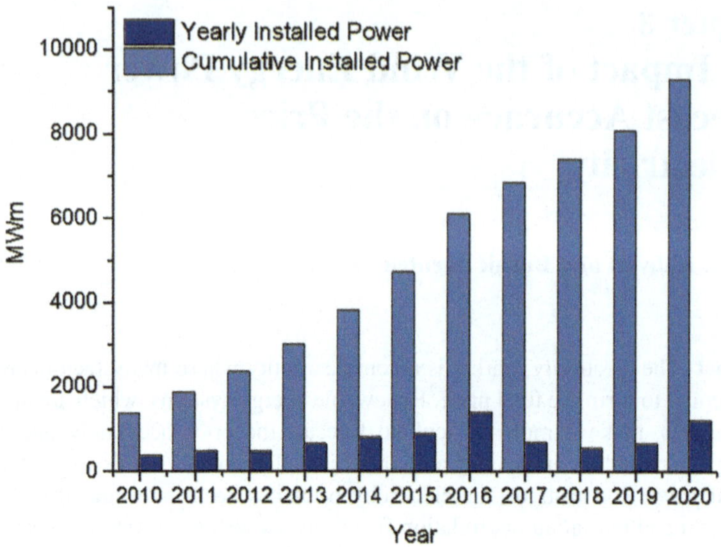

Fig. 8.1 Cumulative installed wind power versus yearly installed wind power in Turkey

Renewable energy sources, especially wind energy which is considered in this study, recently started to be an important player in the energy market. These are energy sources for which the amount of energy to be produced at any time cannot be determined precisely due to their nature. This uncertainty is not desirable for the energy market, where even seconds are crucial. The fact that wind energy is difficult to predict has an effect on the market to a certain degree. In order to fully understand what these effects are, a thorough analysis on the subject is required. This study aims to find out to what extent wind energy forecasts affect electricity price in the energy market.

Similar studies had been carried out in different countries. For instance, it has been demonstrated by Green and Vasilakos that the integration of renewable sources can increase the volatility in the electricity industry in Great Britain [1]. Such a change probably forces the industry to go for an optimal energy source combination in which gas-fired plants will be more in number. On the contrary, a different study in Denmark conducted by Mauritzen uses seasonal autoregressive moving average (SARMA) model and suggests that wind energy power generation decreases daily volatility of the local prices [2]. A more recent study in Germany by Ketterer utilizes a generalized autoregressive conditional heteroscedasticity (GARCH) model and demonstrates that wind power increases volatility of electricity prices [3].

The energy market differs for each country. As seen in the studies above, integration of wind energy increases the instability of the market in one country but the situation may be the opposite for another country. In order to make the most of renewable energy sources and make a better energy management, the effect of these sources on the market should be analyzed on a country basis. The motivation of this study is to carry out a similar analysis for the Turkish energy market.

The energy market is a phenomenon in which there are many players, and the final price is formed as a result of complex interactions. Knowing what the final price will be has become even more difficult, especially as renewable energy sources have become an important factor in the market [4]. The chaotic nature of renewable energy sources and the inability to produce energy at the desired rate can be cited among the reasons for this difficulty. Commercial sensitivity in the energy market of developed countries is lower than that of developing countries [5]. To reduce commercial sensitivity and ultimately increase the reliability of the energy market, highly accurate forecasting models should be developed and researched, as in developed countries, so that energy producers, consumers, distributors, and sellers can see the state of the market and plan their actions accordingly.

Building an efficient model to forecast electricity price can be a tedious task if available data is not reliable. There are many transparency policies currently in force that allow market players/analyzers to build models and to study the market dynamics. Energy Exchange Istanbul (EXIST) platform is one of the organizations that shares reliable information about the energy market within the framework of the transparency law in Turkey.

In this study, the correlation between the accuracy of wind power forecasts and the electricity price is investigated for the case of Turkey. For this purpose, a data preparation is utilized. Then, widely used statistical methods are applied to the dataset and the variables that are not related to the electricity price are separated. Afterward, the electricity price is forecasted with a multi-layered perceptron algorithm using the correlated variables, and it was examined whether these predictions improved with the addition of wind energy forecast error and share of wind energy data.

The chapter is organized as follows. An overview of studies that focus on the impact of wind energy forecasts on the electricity price is given in the literature review section. In the data section, the obtained dataset is explained. In Sect. 8.2, the proposed model is described in detail. Results and discussions of the study are given in Sect. 8.3 and the conclusion is presented in the last section.

8.1.1 Literature Review

There have been studies that focus on how renewable energy generations affect electricity price in different countries. The predictive accuracy of a collection of methodologies for day-ahead spot price forecasting in the Spanish electricity market is compared in the study conducted by Cruz et al. In this work, methods consist univariate, multivariate, linear, and nonlinear approaches. Results of the study suggest that if wind generation data is included in the forecasting process, the accuracy rises greatly [6].

Another study that is conducted in Ireland by Swinand and O'Mahoney investigated the effects of wind energy-related productions on system costs and prices. The main motivation of the study was to thoroughly assess the correlation between wind energy generation and price. Results show that 1% increase in the wind generation reduces SMP (System Marginal Price) by 0.06% [7].

Using a production cost model from the Independent System Operator – New England power system, Martinex-Anido et al., studied the impact of wind generation on energy prices. Different scenarios in terms of wind penetration, wind energy-related predictions, and wind curtailment are modeled in order to investigate the impact of wind power on electricity costs at various levels of wind penetration, visibility and, controllability. According to the results, over-forecasting increases the price and under-forecasting reduces the price [8].

Goodarzi et al. contributed to the general consideration of whether it would be good to incentivize system operators to increase the accuracy and market availability of renewable energy output predictions. The impact of wind and solar energy forecasting errors on imbalance volumes and subsequent spot electricity prices is explored using data from German electricity market. It was found out that wind forecast errors impact price more than solar forecast errors. Thus, policy incentives for system operators to increase the accuracy and availability of renewable energy predictions should be supported [9].

To have a brief idea as to how price changes when wind energy generations are higher in number than the other resources, Astaneh and Chen investigated the impact of high wind power penetration on electricity price change for the cases of Norway and Denmark. The main contribution of this chapter was to quantify patterns of electricity price volatility as wind power penetration rises. The study quantitatively explains a direct correlation between wind penetration and price volatility. The more wind penetration, higher the price variation [10].

Rubin and Babcock used a pre-developed equilibrium modeling approach to investigate how wind energy integration affects premiums in day-ahead power markets. In particular, premiums for various pricing mechanisms are compared as wind power capacity rises. Results suggests that the pricing of wind energy is critical. It was shown that in the case of imperfect competition, pricing wind energy in the market improves firms' potential to collect oversized day-ahead premiums, whereas market power is diminished as wind power capacity expands in the case of market-independent feed-in-tariff [11].

Wozabal et al. reviewed the hypothesis of the increase in renewable results in an increase in the electricity price variation by using a static market model. Results of the study suggest that increase in the intermittent renewable energy systems can increase but also decrease the price variation. The direction of the change is determined by whether the amount of increase is high or low. If the amount of increase in renewable energy systems and so the amount of energy produced by renewable energy sources are high, so is the price variation [12].

Using hybrid models to forecast electricity price are proven techniques in literature [13–16]. Some of these techniques use wind forecast data to train the models. Cerjan et al. developed a hybrid model to forecast electricity price, and wind forecast data is used as a training parameter for the forecasting algorithm. According to the results, the proposed method gave good results regardless of situations of low or high price volatility [17].

Another study that also uses renewable energy forecast as a training parameter is the study done by Bhatia et al. The effect of renewable energy sources on price forecasts is investigated and used for model training. The proposed embedded model is further improved by random forest and gradient boosting. Results show that the improved model captures the dynamics in price series without compensating the computational ease [18].

To the best of our knowledge, there have been no studies that investigate the impact of wind energy forecasts on the electricity price for the case of Turkey.

8.1.2 Data

Data is gathered from Energy Exchange Istanbul (EXIST) platform. Recent 1 year hourly data consists a total of 8760 rows of entries. Dataset consists of wind energy production forecasts, and actual generations both in MWh, forecast error percentage, share of wind energy production in total energy production, total energy production and total energy consumed both in MWh, market clearing price and system marginal price both in USD. To have a brief idea about the dataset used, a statistical summary of the dataset is given in Table 8.1.

It is important to mention that this dataset is raw, which means the table above has not yet been prepared for the study. Looking at values of Table 8.1, it is seen that there are three different variables that are related with the amount of wind energy production; these are forecasted and actual wind energy productions and total wind energy-related productions. It is important to point out that total wind energy-related productions include unlicensed productions whereas the other two do not and it contains data only from the registered wind turbines which can be monitored throughout Turkey. Also "Categorized Forecast Error" is a derived variable in which forecasts below −5% accuracy is shown as −1, above +5% accuracy shown as 1, and between −5% and +5% accuracy shown as zero which means no error.

8.2 Methodology

This study consists of three phases. First phase involves data preparation in which the dataset is analyzed and formatted accordingly. Every feature has been assigned a proper data type and all outliers have been removed from the dataset and any empty cells or cells with wrong data type have been replaced with the mean value of the corresponding columns.

Second phase is correlation investigation and feature selection. There are many different factors that affect electricity price, and it would be a tedious task to uncover all of them which is not what this study tries to accomplish. Rather, what we are trying to show whether if forecast accuracy affects the electricity price and determine

Table 8.1 Statistical summary of the dataset

Variable Name	*Count*	*Mean*	*STD*	*Min*	*25%*	*50%*	*75%*	*Max*
Total Energy Production [MWh]	8760	33040.26	5499.70	15246.64	29075.36	32932.61	37507.47	45210.03
Total Energy Consumption [MWh]	8760	33127.05	5519.15	15333.30	29150.82	32997.63	37594.96	49341.89
Total Wind Energy Related Productions [MWh]	8760	2781.93	1607.45	139.76	1365.73	2574.04	4075.26	6861.04
Forecasted Wind Energy Production [MWh]	8760	2244.99	1444.01	72.63	967.15	2005.67	3408.81	5827.80
Actual Wind Energy Production [MWh]	8760	2165.18	1252.65	91.17	1073.37	2002.35	3154.43	5585.79
Share of Wind Energy in Total Production [%]	8760	8.47	4.85	0.42	4.24	7.86	12.24	23.23
Forecast Error [%]	8760	1.96	25.60	-80.26	-14.28	3.71	16.82	99.79
Absolute Forecast Error [%]	8760	19.92	16.20	0.00	7.59	15.75	27.92	99.79
Categorized Forecast Error	8760	0.10	0.91	-1.00	-1.00	0.00	1.00	1.00
Amount of Forecast Error [MWh]	8760	79.80	479.21	-1691.18	-207.77	49.22	371.96	2161.90
Absolute Amount of Forecast Error [MWh]	8760	366.25	319.15	0.03	129.23	282.54	521.95	2161.90
System Marginal Price [$/MWh]	8760	39.17	15.41	0.00	30.28	41.69	48.58	272.42
Market Clearing Price [$/MWh]	8760	40.12	11.75	0.00	34.58	41.40	46.81	131.52

if this effect is positive or negative. For this reason, a correlation heatmap of all features is created using Pearson coefficient to separate the features that are going to be fed into the model and to determine which price will be used as a target variable. Correlation heatmap is given in Fig. 8.2.

Considering the rest of the correlation matrix, among prices, SMP seems to be correlated more with our input parameters. For this reason, SMP is used as a target variable for the analysis. Looking at the forecast performance parameters (forecast error, absolute forecast error, categorized forecast error, amount of forecast error and absolute amount of forecast error), forecast error and amount of forecast error are correlated with the SMP more than the other parameters. Since these two parameters are highly correlated with each other, and it is a bad practice to use dependent input parameters in forecast algorithms, one of them has to be chosen. Thus, forecast error is used for the rest of the study since amount of forecast error is derived from forecast error.

"Forecast Error (%)" and "Share of Wind Energy in Total Production (%)" are derived from the actual dataset as these three parameters are not provided by the EXIST platform. Derivation of forecasted error and share of wind energy-related production (shortly "share of WTs") in total energy production is given in Eq. 8.1 and Eq. 8.2 consecutively.

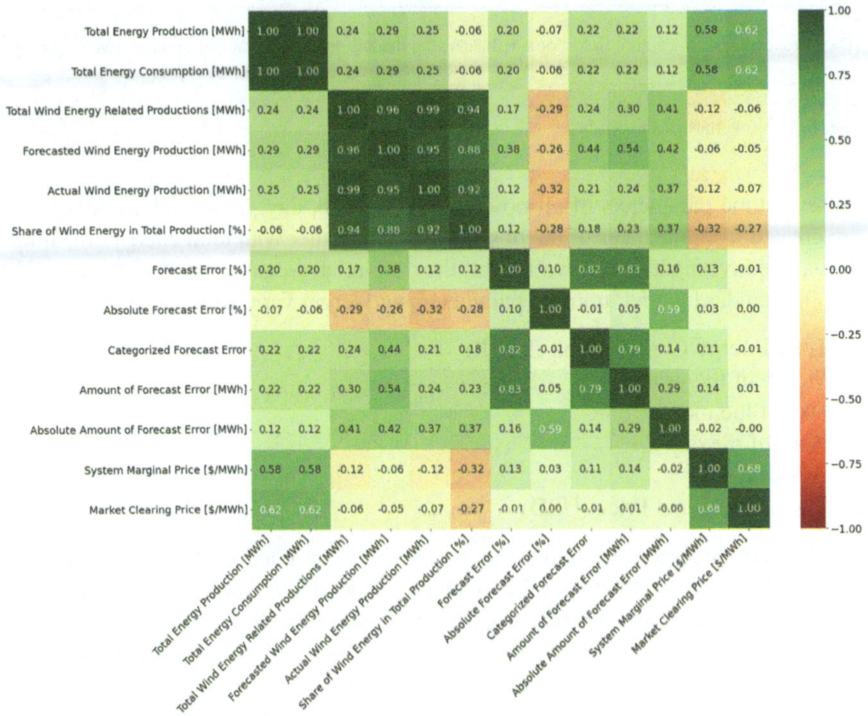

Fig. 8.2 Correlation matrix of input variables and proposed target variables

$$\text{Forecasted Error } (\%) = \frac{\text{PFA} - \text{APA}}{\text{AP}} \times 100 \quad (8.1)$$

$$\text{Share of WTs in Total Energy Production } (\%) = \frac{\text{APA}}{\text{TEP}} \times 100 \quad (8.2)$$

where PFA is production forecast amount, APA is actual production amount, and lastly TEP is total energy production. Since total energy production is derived from TEP, total energy consumption (TEC) is chosen to be a training parameter for this study.

Auto-correlation (ACF) and cross-correlation functions (CCFs) are calculated in time domain. Auto power (APSD) and cross power spectral densities (CPSD) are calculated in frequency domain for determining the relation of frequency components between forecast error and SMP series. This relation will show the effectiveness of forecast error on the final price.

The third and final phase of the study is model development. A flowchart of the study is given in Fig. 8.3. At this stage, selected features are fed into an MLP (Multi-Layer Perceptron) network to make electricity price forecasts. To this point, features are selected and the dataset is investigated using statistical methods. Outliers are removed from the dataset and blank cells are filled with interpolation. To

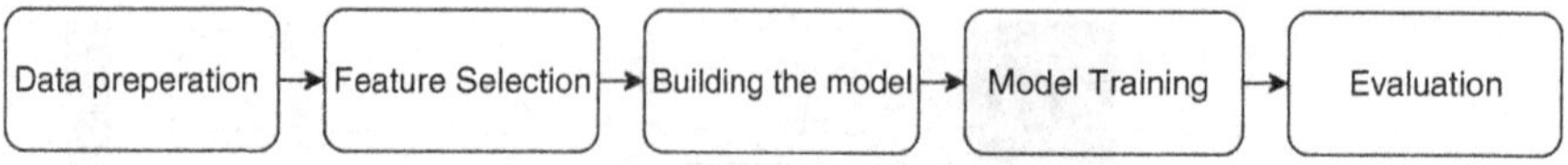

Fig. 8.3 Flowchart of the study

preserve time parameters of all observations, month, day, hour, and weekday data are created from the dates of the observations. Together with total energy consumption at the investigated observation, all the time-related data is fed into the network. For the standard model, training parameters and target variable were selected to be:

- Total Energy Consumption (MWh)
- Month of the year
- Day of the month
- Hour of the day
- Weekday
- System Marginal Price (USD)

Month of the year is a parameter that varies between 1 and 12. Day of the month is expressed as a number between 1 and length of the month. Hour of the day is also a number between 1 and 23 and weekday represents the day of the week. To use weekday parameter in the training process, an encoding is required in which corresponding values are represented with numbers. This is done by giving each day a numerical value between 1 being Monday and 7 being Sunday.

Neural networks created in this study consist of 3 hidden layers each having 100 neurons. Train and test set ratios are 0.8 and 0.2 of the whole dataset. Network topology is determined by trial-and-error technique. Batch normalization is applied between layers to standardize the inputs. This process makes the learning process stable and reduces the number of epochs needed to train the network [19]. Activation function at each layer is rectified linear unit (ReLU) except for the last hidden layer which has a linear activation function. In order for the model to have a good generalization, early stopping and dropout regularization techniques are used. Early stopping is a technique in which the model is evaluated on a holdout validation dataset after each epoch. If the performance of the model on the validation dataset starts to decrease (which can occur because of either loss beginning to increase or accuracy beginning to decrease) then the training process is stopped. This technique does not guarantee an increase in the generalization of the model, rather it helps training the model to the point where its generalization is maximized. Dropout technique, on the other hand, is a technique of network regularization in which randomly selected weights in the network are set to zero. Dropout approximates training a large number of neural networks with different topologies in parallel, as first proposed by Hinton et al. [20]. Some layer outputs are disregarded or "dropped out" randomly during training. The dropout approach is a successful method for making the model more robust, despite the fact that it is a noisy process. Model performance is evaluated using RMSE and R^2 evaluation metrics.

8.3 Results and Discussion

Autocorrelation function (ACF) is utilized to identify the periodicity of electricity price and the resulting graph is given in Fig. 8.4. The ACF plot quickly reaches zero for stationary time-series, while non-stationary time-series decreases rather slowly to the zero point.

Autocorrelation graph shows that the peak interval is 24 data points, which is in our case 24 hours. This corresponds to a diurnal cycle. For this reason, first daily cycle and weekly cycle of electricity price variations are investigated. Also, cross-correlation function (CCF) between electricity price and wind power forecast error is investigated. CCF is a measure of how similar two series are as a function of the displacement of one relative to the other. It is used commonly to investigate how two variables of interest correlate with respect to each other. CCF of wind energy forecast error and electricity price is given in Fig. 8.5.

The first thing to notice from the cross-correlation graph is that SMP and forecast error has a negative correlation with each other, simply because the points oscillate around zero. Another thing is that for a relation between two different series to be classified as "strong," the values should quickly become non-significant as the lag increases. Usually a correlation is significant when the absolute value is greater than the value calculated using Eq. 8.3.

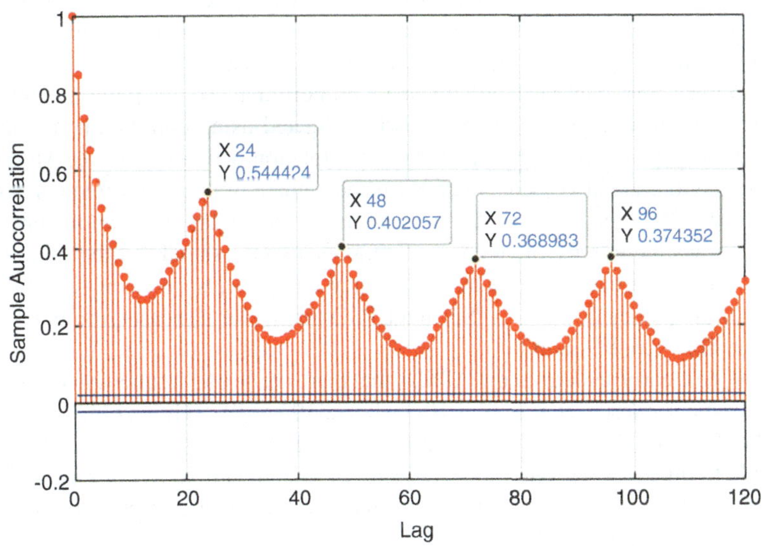

Fig. 8.4 Autocorrelation function graph of SMP with 120 lags

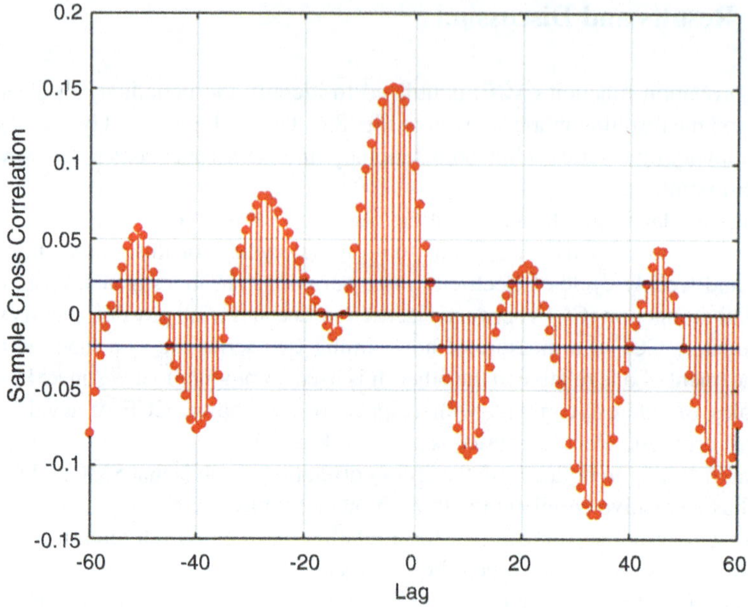

Fig. 8.5 Cross-correlation function graph of SMP and forecast error at lag 60

$$\frac{2}{\sqrt{n-|k|}} \tag{8.3}$$

where n is number of observations and k is the lag. Blue lines correspond to the confidence interval that is determined by Eq. (8.1) [21]. Considering that values do not fall between blue lines quickly, we can conclude that the relation between these two variables are rather weak and inversely proportional. To this point, the relationship between forecast error and SMP is investigated using correlation coefficients and in time domain. To further investigate the relation between the two, APSD and CPSD functions are also used, which shifts the domain of interest to frequency since spectral density functions are the Fourier transform of the correlation functions. APSD shows the energy of frequency components in a signal; on the other hand, CPSD shows the energy of shared frequency components between two different signals in frequency domain. APSD and CPSD graphs for forecast error and SMP is given in Figs. 8.6, 8.7, and 8.8.

It can be clearly seen from the APSDs that there are strong frequency components in forecast error series at 1.1529×10^{-5} Hz, 2.3193×10^{-5} Hz, 3.4722×10^{-5} Hz, and 4.6251×10^{-5} Hz frequencies and in SMP series at 1.1529×10^{-5} Hz and its multiple frequencies with first four frequencies shared with APSD of forecast error. These four shared frequency components between the forecast error and SMP series could be seen clearly in CPSD.

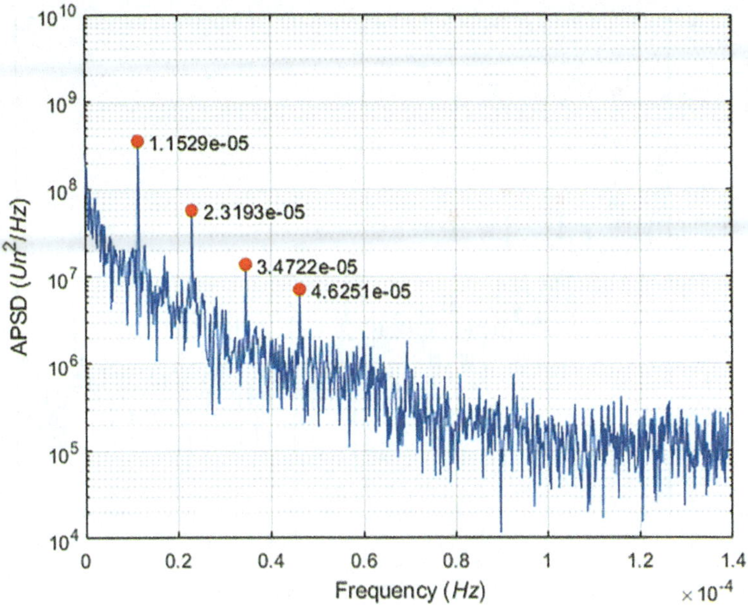

Fig. 8.6 APSD function graph of forecast error

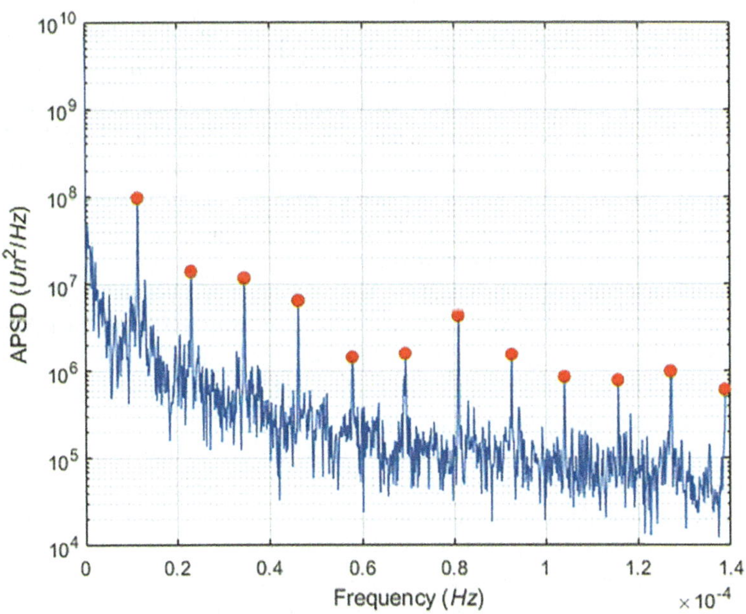

Fig. 8.7 APSD function graph of SMP

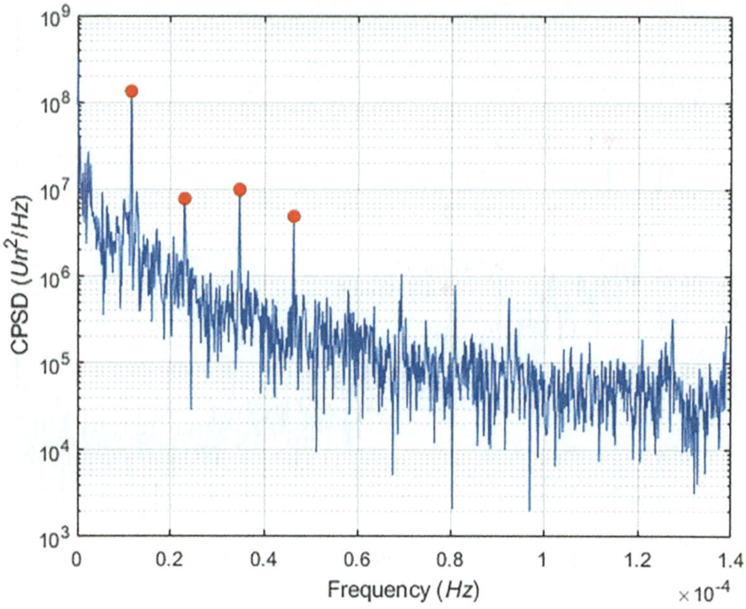

Fig. 8.8 CPSD function graph of forecast error and SMP

To investigate the dependent variable's daily cycle, values at each hour in a day are investigated. Intraday variations of forecast error and SMP are given in Fig. 8.9. It is seen that the price and forecast error reaches the lowest level on almost 5: 00 AM. Forecast error increases to its peak value from 5:00 AM to 13:00 PM. On the other hand, SMP also exhibits a similar behavior, reaches lowest value at 7:00 AM, and starts to increase till 20:00 PM to reach its peak and starts to decrease at the end of the day. It is worth mentioning that both series are sinusoidal which is in parallel with consumption cycle of the industry.

Daily mean values of both variables are calculated and plotted in Fig. 8.10 to investigate the situation on weekly basis. It is seen from the figure that there is a positive correlation between these two. When the forecasting error is low, so is the price, and vice versa. It is also seen that both variables exhibit the same behavior in terms of change. They start to increase at the early days of the week, oscillate through midweek, and start to decrease on Thursday to the end of the week. Coefficient of correlation is found to be 0.748 for these series.

To further investigate how wind power forecasting accuracy affects the final price, dataset is organized in such a way that weekdays and weekends are separated. Similar comparison between the price and the forecast error is also done for weekdays and weekends. Related comparison is given as a graph in Fig. 8.11. This is done simply because consumption habits differ between weekdays and weekends; therefore, this change in consumption results in a different price for each particular period.

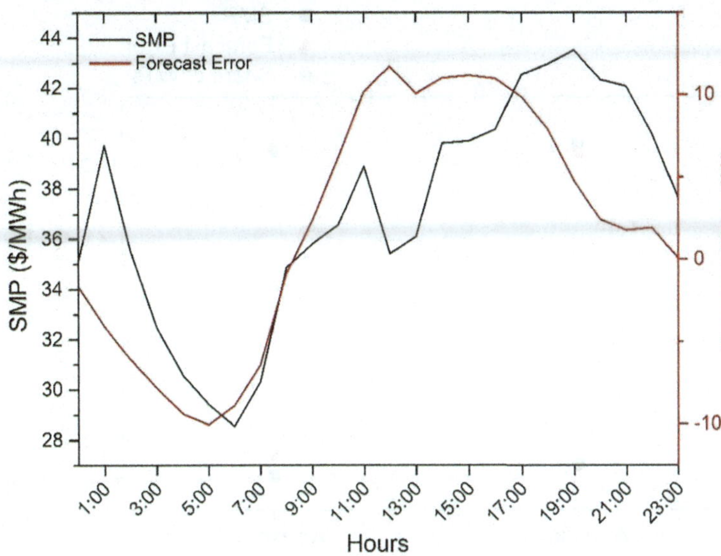

Fig. 8.9 Intraday variations of forecast error and SMP

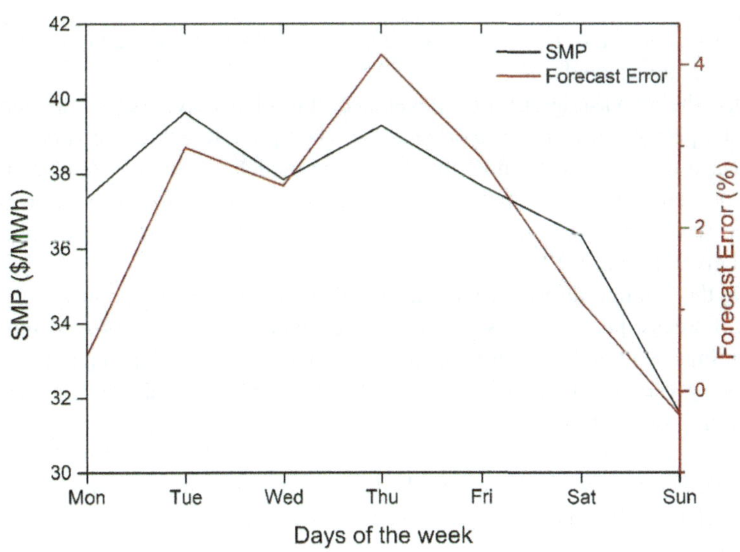

Fig. 8.10 Weekly variations of forecasting error and SMP

On weekdays both SMP and forecast error is relatively high as seen in Fig. 8.11, and on weekends both of them are relatively low. It is also clear that forecast error is low when share of wind energy-related production in total production is higher and

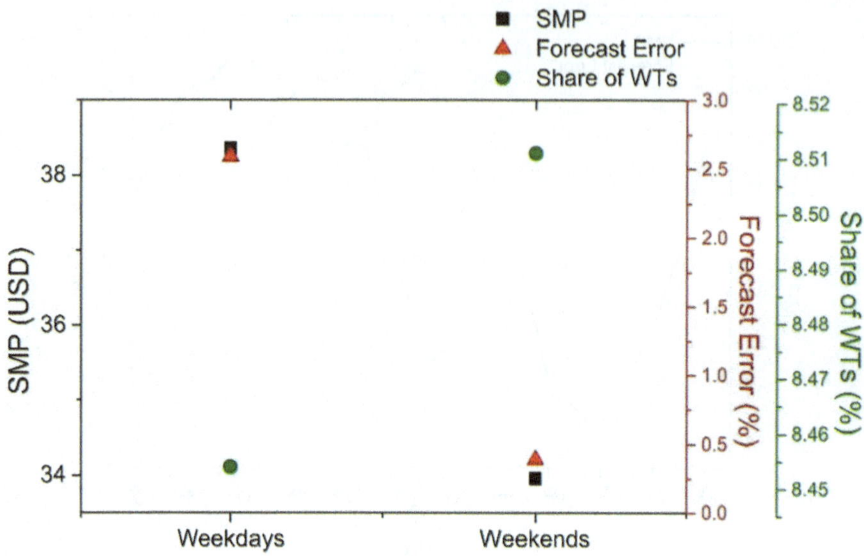

Fig. 8.11 Average SMP, forecast error, and share of wind energy-related production comparison for both weekdays and weekends

vice versa. Price and forecast error seem to exhibit similar behavior both on weekdays and weekends.

Lastly, the forecast algorithm is developed. It has been investigated that whether the model performance increase or decrease when price forecasts are supported by the forecast error series. Both of the developed models, including standard model and improved model, use multi-layer perceptron to make forecasts. Correlation heatmap which involves the final variables for the training of standard model and SMP is given in Fig. 8.12.

With the configuration mentioned above, both models are trained for 3500 epochs. Forecast performances of the standard model and the improved model is given in Figs. 8.13 and 8.14 (for the sake of visualization last 150 data points of the test set are given). R^2 and RMSE values of the fits are given in Table 8.2 and learning curve of the standard model is given in Fig. 8.15.

As it can be seen from the Table 8.2, improved model achieved higher R^2 and lower RMSE values which indicates a better fit. This shows that forecast error series are related to SMP, and this impact can be revealed to a certain extent by the multilayer perceptron.

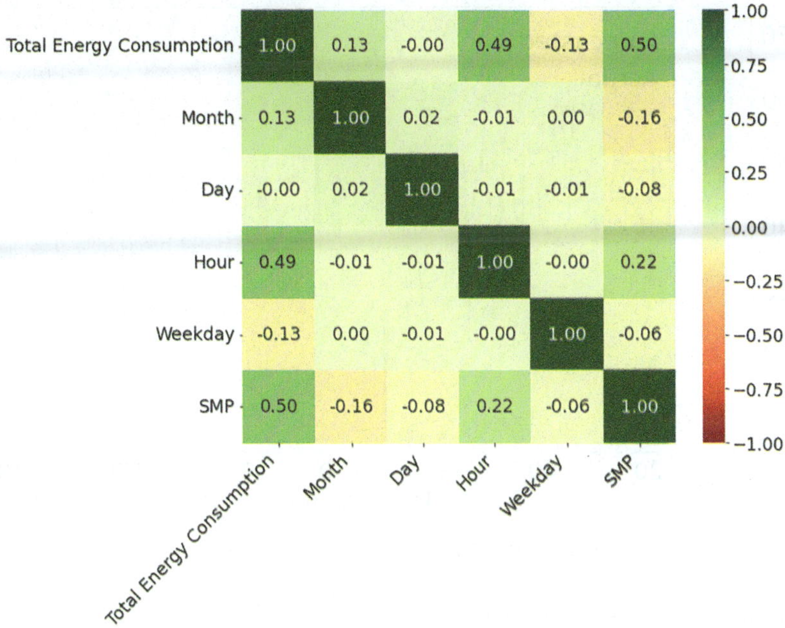

Fig. 8.12 Correlation heatmap of SMP and training variables for standard model

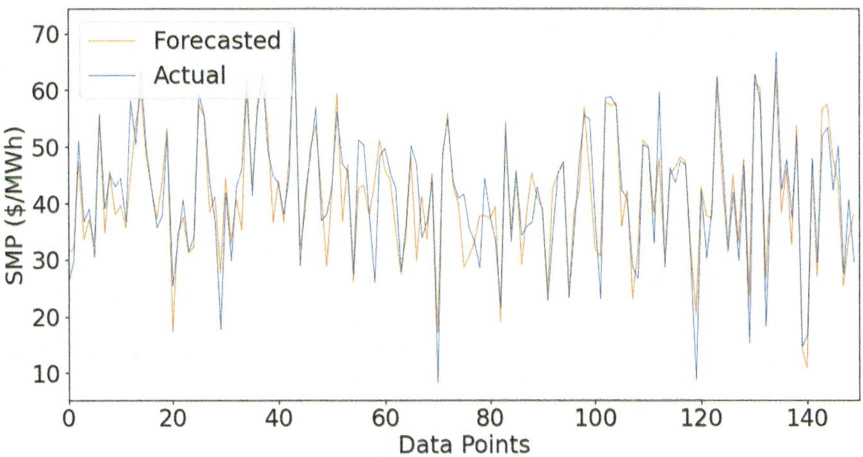

Fig. 8.13 Standard model forecast performance

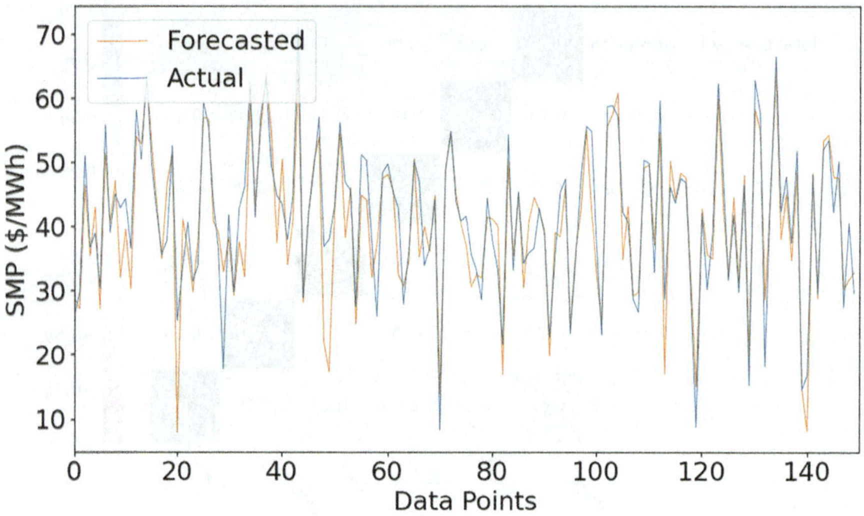

Fig. 8.14 Improved model forecast performance

Table 8.2 Forecasting performance comparison of standard and improved model

	Standard	Improved
RMSE	6.501	6.363
R^2	0.724	0.736

8.4 Conclusion

In this study, selected parameters such as total energy consumption and date information are used for determining the effect of forecast error on system marginal price. Both time domain and frequency domain's characteristic functions (as ACF, CCF, APSD, and CPSD) are used for finding the related frequency components between forecast error and SMP series.

Data is analyzed both for weekdays and weekends separately, and it is seen that they show same character (see Figs. 8.9, 8.10, and 8.11). Although the data is separated as weekdays and weekends, there seems to be no benefit of doing so in terms of revealing different characteristics. On the other hand, weekly and hourly analyses show similar forecast error and SMP behavior.

Two different models are developed, as standard and improved model. In standard model, training parameters are total energy consumption and date information,

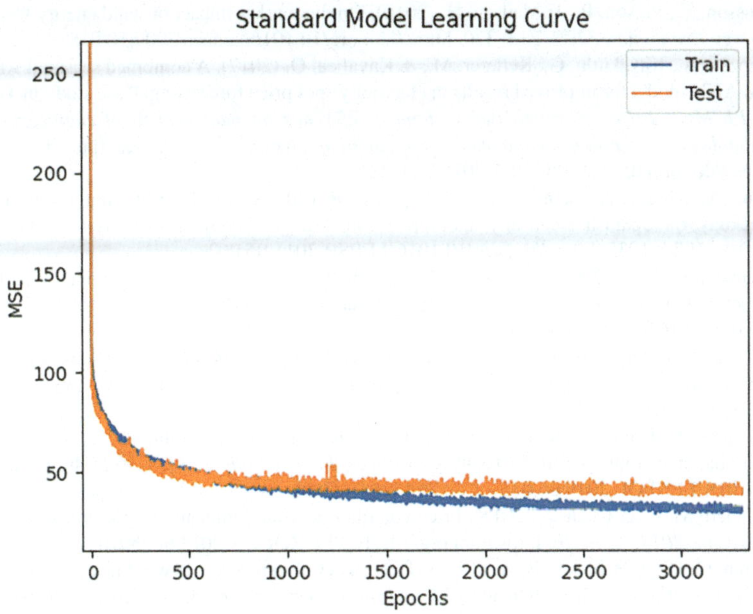

Fig. 8.15 Standard model learning curve

whereas in improved model forecast error and share of wind energy-related production in total energy production are also used in addition to standard model inputs. The improved model showed a higher success considering 0.736 R^2 and 6.363 RMSE, whereas standard model reached 0.724 R^2 and 6.501 RMSE. It is clear that forecast error and share of wind energy-related production have an impact on system marginal price. Although this impact is not too strong, it is significant enough to affect the price.

Acknowledgements The authors gratefully acknowledge the financial support of Istanbul Technical University Scientific Research Projects Commission (project code: MDK-2022-43753) for this research.

References

1. Green, R., & Vasilakos, N. (2010). Market behaviour with large amounts of intermittent generation. *Energy Policy, 38*(7), 3211–3220. https://doi.org/10.1016/j.enpol.2009.07.038
2. Mauritzen, J. (2012). What happens when it's windy in Denmark? An empirical analysis of wind power on price variability in the nordic electricity market. *SSRN Electronic Journal*. https://doi.org/10.2139/ssrn.2045489
3. Ketterer, J. C. (2014). The impact of wind power generation on the electricity price in Germany. *Energy Economics, 44*, 270–280. https://doi.org/10.1016/j.eneco.2014.04.003

4. Jónsson, T., Pinson, P., & Madsen, H. (2010). On the market impact of wind energy forecasts. *Energy Economics, 32*(2), 313–320. https://doi.org/10.1016/j.eneco.2009.10.018
5. Ozozen, A., Kayakutlu, G., Ketterer, M., & Kayalica, O. (2017). A combined seasonal ARIMA and ANN model for improved results in electricity spot price forecasting: Case study in Turkey. In *PICMET 2016 - Portland international conference on management of engineering and technology: Technology management for social innovation, proceedings* (pp. 2681–2690). https://doi.org/10.1109/PICMET.2016.7806831
6. Cruz, A., Muñoz, A., Zamora, J. L., & Espínola, R. (2011). The effect of wind generation and weekday on Spanish electricity spot price forecasting. *Electric Power Systems Research, 81*(10), 1924–1935. https://doi.org/10.1016/J.EPSR.2011.06.002
7. Swinand, G. P., & O'Mahoney, A. (2015). Estimating the impact of wind generation and wind forecast errors on energy prices and costs in Ireland. *Renewable Energy, 75*, 468–473. https://doi.org/10.1016/J.RENENE.2014.09.060
8. Brancucci Martinez-Anido, C., Brinkman, G., & Hodge, B. M. (2016). The impact of wind power on electricity prices. *Renewable Energy, 94*, 474–487. https://doi.org/10.1016/J.RENENE.2016.03.053
9. Goodarzi, S., Perera, H. N., & Bunn, D. (2019). The impact of renewable energy forecast errors on imbalance volumes and electricity spot prices. *Energy Policy, 134*, 110827. https://doi.org/10.1016/J.ENPOL.2019.06.035
10. Astaneh, M. F., & Chen, Z. (2013). Price volatility in wind dominant electricity markets. *IEEE EuroCon, 2013*, 770–776. https://doi.org/10.1109/EUROCON.2013.6625070
11. Rubin, O. D., & Babcock, B. A. (2013). The impact of expansion of wind power capacity and pricing methods on the efficiency of deregulated electricity markets. *Energy, 59*, 676–688. https://doi.org/10.1016/J.ENERGY.2013.07.020
12. Wozabal, D., Graf, C., & Hirschmann, D. (2016). The effect of intermittent renewables on the electricity price variance. *OR Spectrum, 38*(3), 687–709. https://doi.org/10.1007/S00291-015-0395-X
13. Qiao, W., & Yang, Z. (2020). Forecast the electricity price of U.S. using a wavelet transform-based hybrid model. *Energy, 193*, 116704. https://doi.org/10.1016/J.ENERGY.2019.116704
14. Cao, J., Li, Z., & Li, J. (2019). Financial time series forecasting model based on CEEMDAN and LSTM. *Physica A: Statistical Mechanics and its Applications, 519*, 127–139. https://doi.org/10.1016/J.PHYSA.2018.11.061
15. Zhang, J., Tan, Z., & Wei, Y. (2020). An adaptive hybrid model for short term electricity price forecasting. *Applied Energy, 258*, 114087. https://doi.org/10.1016/j.apenergy.2019.114087
16. Pourdaryaei, A., Mokhlis, H., Illias, H. A., Kaboli, S. H. A., Ahmad, S., & Ang, S. P. (2019). Hybrid ANN and artificial cooperative search algorithm to forecast short-term electricity price in de-regulated electricity market. *IEEE Access, 7*, 125369–125386. https://doi.org/10.1109/ACCESS.2019.2938842
17. Cerjan, M., Matijaš, M., & Delimar, M. (2014). Dynamic hybrid model for short-term electricity price forecasting. *Energies, 7*(5), 3304–3318. https://doi.org/10.3390/EN7053304
18. Bhatia, K., Mittal, R., Varanasi, J., & Tripathi, M. M. (2021). An ensemble approach for electricity price forecasting in markets with renewable energy resources. *Utilities Policy, 70*, 101185. https://doi.org/10.1016/j.jup.2021.101185
19. Santurkar, S., Tsipras, D., Ilyas, A., & Madry, A. (2018). How does batch normalization help optimization? *Advances in Neural Information Processing Systems, 2018*, 2483–2493.
20. Hinton, G. E., Srivastava, N., Krizhevsky, A., Sutskever, I., & Salakhutdinov, R. R. (2012). Improving neural networks by preventing co-adaptation of feature detectors. *arxiv.org*. Accessed 24 Feb 2022. [Online]. Available: https://arxiv.org/abs/1207.0580
21. Minitab Documentation. (2019). Interpret all statistics and graphs for Descriptive Statistics. *Minitab*. https://support.minitab.com/en-us/minitab/18/help-and-how-to/modeling-statistics/time-series/how-to/cross-correlation/interpret-the-results/all-statistics-and-graphs/. Accessed 9 Mar 2022.

Chapter 9
The Power of Combination Models in Energy Demand Forecasting

Erkan Isikli and Seyda Serdarasan

Abstract Accurate and reliable energy demand forecasting is essential for sustainable planning, operation, and control of energy systems in our ever-changing world, and it can assist planners and policymakers in making informed decisions to design more efficient power grids, identify cost-effective energy production methods, and optimize energy resource utilization. The related literature emphasizes the need for models that are accurate, fast, and not overly complex to make robust decisions based on forecasts. Time series forecasting offers potential benefits by reducing risks and costly infrastructure upgrades. Particularly, combining time series forecasts may aid in improving overall forecast accuracy in energy demand modeling. Using Turkey's daily household electricity consumption data from 2016 to 2021, this study thus investigates the potential benefits of combination models. Based on Schulze voting considering various performance metrics, the generated results indicate that combining forecasts from different traditional statistical models can result in better-performing models; however, it is not always the case that they outperform traditional models. In addition, static combination weights may not be suitable for wide forecast horizons. Potential avenues for future research include applying the investigated models to various time scales, such as hourly, weekly, monthly, or quarterly data for electricity consumption, to gain a deeper understanding of their performance and flexibility.

The original version of this chapter was revised. The correction to this chapter is available at https://doi.org/10.1007/978-3-031-38387-8_19

E. Isikli (✉) · S. Serdarasan
Industrial Engineering Department, Faculty of Management, Istanbul Technical University, Istanbul, Türkiye
e-mail: isiklie@itu.edu.tr

G. Kayakutlu, M. Ö. Kayalica (eds.), *Decision Making Using AI in Energy and Sustainability*, Applied Innovation and Technology Management, https://doi.org/10.1007/978-3-031-38387-8_9

9.1 Introduction

The world's energy needs are constantly changing. Our economies depend on energy for a variety of purposes such as heating and cooling buildings, manufacturing processes, and transportation. Accurate and reliable energy demand forecasting is the basis of sustainable planning, operation, and control of the systems that generate and use these energy sources. Therefore, there is a growing interest in developing models to accurately predict energy demand. A plethora of papers suggest that current approaches adopted for this purpose can be basically classified into two categories: causal modeling and time-series modeling. The former category attempts to forecast energy demand by adopting an analytical approach with the use of various factors such as economic activity, weather, and population, which are, in turn, influenced by technology, policy, and geopolitics [1, 2]. Models with more information or factors may be better at explaining the behavior of energy demand; however, since they may also require the prediction of these factors accurately, energy demand forecasting through causal models can become highly complex and tedious. On the other hand, time series forecasting has been a key mechanism for modeling energy consumption due to its relative flexibility, and it appears to be expanding its sphere of influence in the age of big data [3–5]. Research on energy demand forecasting adopting traditional time series techniques is quite comprehensive [1–3, 6, 7]; however, there is also a growing use of computationally intelligent techniques such as artificial neural networks, fuzzy logic, and genetic algorithms [8]. Considerable literature has recently grown up around the theme of hybrid or multi-stage models that adopt the combined use of multiple techniques to improve model performance in terms of accuracy, speed, and complexity [9–14]. Combining time series forecasts, which can also be considered as a hybrid modeling approach, is not a new concept, and compared to sticking to a single forecasting model, it has been shown to be superior regarding accuracy and speed as complexity increases [6, 7, 9, 14, 15]. Time series that are highly volatile with nonstationary means and variances or that are relatively shorter, for instance, may benefit from the combination of forecasts from alternative models whether they are computationally intelligent or not. However, research to date has mostly concentrated on combining only a limited number of time series models of the same type through a single forecast combination method. This chapter thus investigates how to properly combine some of the most prominent time series models to improve overall forecast accuracy in energy demand modeling considering several model selection criteria. Data on daily electricity consumption in Turkey, covering a period of 6 years, was used to generate predictions through various time series forecasting techniques. These base models were trained on the first 80% of the observations, and their predictions were then combined through different methods to attain an overall forecasting performance that is higher than theirs. Performance metrics such as MAPE, hypothesis testing based on the DM statistic, and graphical inspection were used to evaluate the performance of alternative models. The contribution of the present study lies in its extensive use of time series models for forecast generation and its consideration of various ways to combine them. In addition, based on a variety of performance metrics, Schulze voting was employed to compare the alternative models with each other to decide which of them performed better. Understanding how the past influences the future, planners and policymakers can evidently design more efficient power grids, identify the most cost-effective methods of energy production, and optimize the use of energy resources. The investigation in the current study thus contributes to such efforts flagging the importance of forecast combination in time series modeling.

9.2 Methodology

9.2.1 *Model Selection*

Deciding which time series model outperforms its contenders is not an easy task. In this subsection, we focus on the three most commonly used ways for model evaluation: performance metrics, hypothesis testing, and graphical inspection. The evaluation criteria to distinguish between alternative models should primarily be practical and efficient.

9.2.1.1 Performance Metrics

The most commonly used performance metrics that are given in Table 9.1 were written by adopting the in-sample notation whenever possible. However, we should note that the out-of-sample performance of a forecasting model, which depends on the number of observations used to generate forecasts and the forecast horizon, should essentially be used to distinguish between alternative models. Hence, these metrics should be modified accordingly when comparing the out-of-sample performances of the candidates. Most of the following metrics have adaptations for seasonal data, but here we provide only their generic versions. Note also that n denotes the in-sample size and h denotes the forecast horizon.

MAE, medAE, and RMSE are called scale-dependent metrics; MAPE, medAPE, maxAPE, sMAPE, and AMAPE are called percentage-error metrics; RelRMSE and TIC are called relative measures; MASE is called a scale-free error. The model with lower error measures (e.g., MAE, MAPE, maxAPE, RMSE) is preferred over the other alternatives. As the predicted values become closer to the actual values, all the error measures approach zero, indicating better forecasting accuracy. The same applies to FB and PBIAS; however, when FB is positive and PBIAS is negative, the model is said to underestimate the actual values [17]. MAPE is affected by the presence of relatively lower actual values [18]. TIC can take values within the interval $[0, 1]$. MASE, which compares the forecasting performance of a candidate model against the naive model, is particularly useful when the time series is intermittent.

9.2.1.2 Hypothesis Testing

To decide whether a model outperforms another, the DM test that was proposed by Diebold and Mariano [21] has been widely used [22]. The procedure compares the

Table 9.1 The most commonly used performance metrics

Name	Mathematical formula
Mean absolute error (MAE)	$\frac{1}{n}\sum_{i=1}^{n}\lvert y_i-\widehat{y}_i\rvert$
Median absolute error (medAE)	$\operatorname*{median}_{i=1,2,\ldots,n}\lvert y_i-\widehat{y}_i\rvert$
Root mean squared error (RMSE)	$\sqrt{\frac{\sum_{i=1}^{n}\left(y_i-\widehat{y}_i\right)^2}{n}}$
Mean absolute percentage error (MAPE)	$\frac{100}{n}\sum_{i=1}^{n}\left\lvert\frac{y_i-\widehat{y}_i}{y_i}\right\rvert$
Median absolute percentage error (medAPE) [16]	$100\cdot\operatorname*{median}_{i=1,2,\ldots,n}\left\lvert\frac{y_i-\widehat{y}_i}{y_i}\right\rvert$
Maximum absolute percentage error (maxAPE) [17]	$100\cdot\max_{i=1,2,\ldots,n}\left\lvert\frac{y_i-\widehat{y}_i}{y_i}\right\rvert$
Symmetric mean absolute percentage error (sMAPE) [10, 15]	$\frac{200}{n}\sum_{i=1}^{n}\frac{\lvert y_i-\widehat{y}_i\rvert}{\lvert y_i\rvert+\lvert\widehat{y}_i\rvert}$
Average mean absolute percentage error (AMAPE) [14]	$\frac{100}{n}\sum_{i=1}^{n}\left[\frac{\lvert y_i-\widehat{y}_i\rvert}{\frac{1}{n}\sum_{i=1}^{n}y_i}\right]$
Relative root mean squared error (RelRMSE) [10]	$\frac{\mathrm{RMSE}}{\frac{1}{n}\sum_{i=1}^{n}y_i}$
Theil's inequality coefficient (TIC) [18, 19]	$\frac{\sqrt{\frac{1}{h}\sum_{i=n+1}^{n+h}\left(y_i-\widehat{y}_i\right)^2}}{\sqrt{\frac{1}{h}\sum_{i=n+1}^{n+h}y_i^2}+\sqrt{\frac{1}{h}\sum_{i=n+1}^{n+h}\widehat{y}_i^2}}$
Mean absolute scaled error (MASE) [20]	$\frac{\sum_{i=n}^{n+h}\frac{\lvert\widehat{y}_i-y_i\rvert}{h}}{\sum_{i=2}^{n}\frac{\lvert y_i-y_{i-1}\rvert}{n-1}}$
Fractional bias (FB) [17]	$2\cdot\frac{\sum_{i=1}^{n}\left(y_i-\widehat{y}_i\right)}{\sum_{i=1}^{n}\left(y_i+\widehat{y}_i\right)}$
Percent bias (PBIAS) [17]	$\frac{\sum_{i=1}^{n}\left(\widehat{y}_i-y_i\right)}{\sum_{i=1}^{n}y_i}$
Nash–Sutcliffe (NS) criterion [9]	$1-\frac{\sum_{i=1}^{n}\left(y_i-\widehat{y}_i\right)^2}{\sum_{i=1}^{n}\left(y_i-\frac{1}{n}\sum_{i=1}^{n}y_i\right)^2}$
Correlation coefficient (r)	$\frac{\mathrm{Cov}\left(y_i,\widehat{y}_i\right)}{\sqrt{\mathrm{Var}(y_i)\mathrm{Var}\left(\widehat{y}_i\right)}}$

forecasting accuracy of two models using the following hypotheses based on the loss differential $\Delta_t = L(\varepsilon_t^{\mathrm{A}}) - L(\varepsilon_t^{\mathrm{B}})$:

$$H_0: \quad E[\Delta_t] = 0 \quad \text{for all } t$$
$$H_1: \quad E[\Delta_t] \neq 0 \quad \text{for all } t$$

where $L(\cdot)$ is the loss function that can be an $\mathcal{L}_1$ or an $\mathcal{L}_2$ norm and the DM statistics follows the standard normal distribution. The one-sided version of this test is also possible.

9.2.1.3 Graphical Inspections

Scatter plots of the actual values versus the predicted values or the residuals provide a simple way to comment on the model fit. Box plots and normal probability plots of the residuals generated by alternative models help differentiate between their performances. ACF and PACF plots of the residuals are particularly convenient to use in the ARIMA-type models to understand whether additional AR or MA terms are necessary or if the assumption of residual independence is violated.

9.2.2 Combining Time Series Forecasts

When working with a time series, it is a common practice to create alternative models to generate forecasts. The practitioner selects the best-performing one (s) among these models based on some predetermined criteria (see Table 9.1). However, these alternative models can also be combined to create some sort of synergy. Combining forecasts is not a recent methodology, with the first study dating back to the late 1960s [23]. Several methods currently exist in the related literature for combining forecasts produced by alternative time series models of the same type [24] or the different types [25]. Combined models were claimed to perform better than the worst-performing models and sometimes even outperform the best-performing model [26, 27] and they were especially recommended in the presence of outliers [28].

Several forecast combination methods have been proposed to achieve a lower overall forecast error:

1. *Mean combination*: The arithmetic, geometric, or harmonic mean of the predictions across all alternative models [25, 29]. The trimmed or winsorized means should be preferred when outliers exist in the data.
2. *Median combination*: The median of the predictions across all alternative models [25, 29].
3. *Weighted combination*: The weighted average of the predicted values is taken based on the procedure explained in Newbold and Granger [30]. The weights are

determined based on the covariances between the multi-step ahead forecast errors of the alternative models. The optimal combination weights, w_i, that minimize the variance of the forecast errors are calculated as follows:

$$\boldsymbol{w} = \begin{bmatrix} w_1 & w_2 & \cdots & w_k \end{bmatrix}^{\mathrm{T}} = \frac{\boldsymbol{S}^{-1}\mathbf{1}}{\mathbf{1}^{\mathrm{T}}\boldsymbol{S}^{-1}\mathbf{1}}$$

where $\sum_{i=1}^{k} w_i = 1$ and $w_i \in (0, 1)$ for $i = 1, 2, \ldots, k$, $\boldsymbol{S}$ is the covariance matrix estimated using the in-sample forecast errors, and $\mathbf{1}$ is a vector of 1's. Previously, Bates and Granger [31] had used the diagonal elements of $\boldsymbol{S}$ to determine the combination weights. More recently, Aiolfi and Timmermann [32] suggested ranking the alternative models first based on their in-sample performance, and then assigning each model a combination weight that is proportional to the multiplicative inverse of its rank:

$$w_i = \frac{\frac{1}{\mathrm{rank}_i}}{\sum_{j=1}^{k} \frac{1}{\mathrm{rank}_j}}$$

4. *Regression combination*: A constrained regression model is fitted to the time series with the predicted values from alternative models as explanatory variables [24, 29, 33]. The combining weights, which were estimated using the estimation data, must sum up to one to produce combined forecasts in the validation data. The unconstrained version is also possible, but it has as many disadvantages as its advantages. Elliott and Timmermann [34] suggested using a loss function that is based on the absolute values of the forecast errors rather than the squared errors.
5. *The aggregated forecast through exponential reweighting (AFTER)* was first proposed by Yang [35] to combine different forecasts which were later applied by Zou and Yang [36] to the case where forecasts were generated using the same type of models. It has been modified [37, 38] and applied to various cases [39] since its first introduction.

There are also other combination methods that are based on the eigenvectors of $\boldsymbol{S}$ and the Generalized Rolling Origin Evaluation Combination (GROEC) method that has been recently proposed as a weighted combination method in which the weights are determined through a loss function [40].

9.3 Results and Discussion

We conducted experiments with the time series of daily electricity consumption in Turkey between January 1, 2016 and December 31, 2021 (a total of 2192 observations) using the R open-source statistics software. Figure 9.1 depicts the original time series along with its three seasonal components: Season1 – weekly seasonality (frequency = 7), Season2 – Hijri annual seasonality (frequency = 354.37), and Season3 – Gregorian annual seasonality (frequency = 365.25). The series was split into a fitting sample (training set) of 1827 observations (5 years) and a forecasting sample (test set) of 365 observations (1 year). We first determined six base models and estimated them via **forecastHybrid** and **ATAforecasting** packages and then combined their results using **ForecastComb** and **AIafter** packages. All of the model orders were verified in several in-sample forecasting trials. Descriptions of all the models considered in this study are given in Tables 9.2 and 9.3. Once we estimated the model parameters using the training sample, we produced 365-step-ahead forecasts by the candidate models to evaluate their post-sample performances. Then, based on the performance metrics given in Table 9.4, we ranked all the models based on each metric and created rank ballots to employ Schulze voting [41]. The models were eventually ordered from the best-performing to the worst as follows: dynOLS, STLM, AFTER, ATA, MEDIAN, THETA, AMEAN, ARIMA, TBATS, WMEAN,

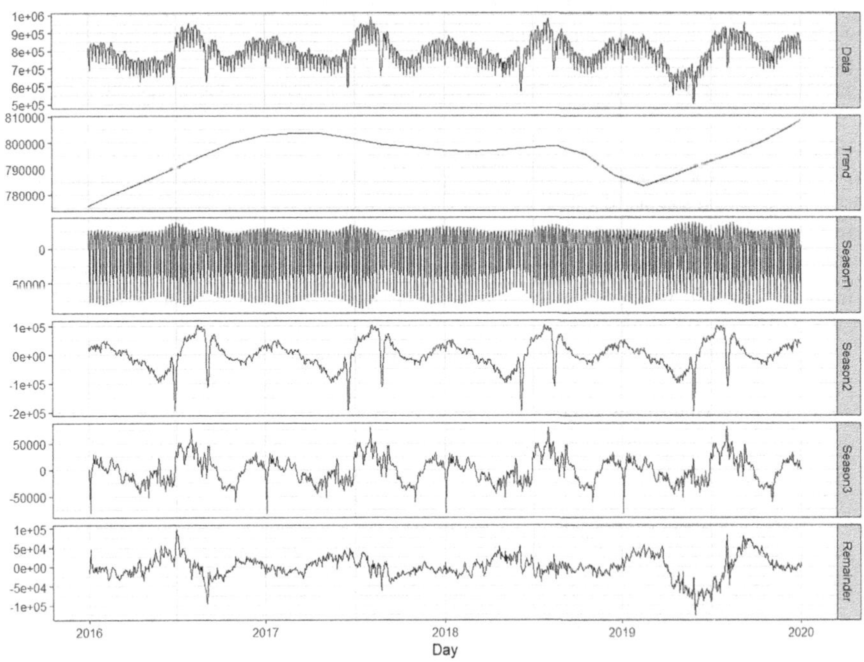

Fig. 9.1 Decomposition of Turkey electricity consumption time series

Table 9.2 Details on the base models

Abbreviation	Description and references
ARIMA	Non-seasonal Autoregressive Integrated Moving Average model with $p = 5$, $d = 1$, and $q = 2$.
NNETAR	Average of 20 networks, each of which is a Neural Network Autoregressive Model with 17 nodes in the hidden layer, 32 lagged inputs, and also the last observed values from the same season as inputs.
THETA	Theta model by Assimakopoulos and Nikolopoulos [42] with the smoothing parameter $\alpha = 0.7585$ and initial state $l = 725{,}128.24$.
STLM	STLM is a seasonal-trend decomposition procedure based on local regression smoothing (LOESS) that was developed by Cleveland et al. [43]. In the present study, a simple exponential smoothing with additive errors and multiple seasonality was constructed ($\alpha = 0.9686$, $l = 735{,}625.49$).
TBATS	TBATS is an exponential smoothing state space model developed by De Livera, Hyndman, and Snyder [44] for time series with multiple and complex seasonal patterns. In the abbreviation, B = Box-Cox transformation, A = ARMA errors, T = Trend, S = Seasonality, and the first T means that a trigonometric formulation was adopted to model seasonality. In the present study, the level parameter $\alpha = 0.0199175$ and the trend parameter $\beta = 0$. There are a total of six seasonal parameters, four AR coefficients, and one MA coefficient. The data was not Box-Cox transformed.
ATA	Proposed by Yapar et al. [45], the ATA method has been proclaimed to perform better than the other members of the family of exponential smoothing methods. The ATA model used in this study has additive errors, damped trend, and additive seasonality. Both the optimum level and trend parameters are equal to 55, and the optimum damped trend parameter equals 0.97.
AMEAN	Takes the arithmetic mean of the forecasts produced by the candidate models for each time period.
MEDIAN	Takes the median of the forecasts produced by the candidate models for each time period.
WEIGHTED	Takes the weighted average of the forecasts produced by the candidate models for each time period. The combination weights were determined based on the in-sample MASE.
BG	The weighted combination method proposed by Bates and Granger [31].
NG	The weighted combination method proposed by Newbold and Granger [30].
InvW	The weighted combination method proposed by Aiolfi and Timmermann [32].
OLS	The combination weights are simply the unconstrained beta parameters obtained by regressing forecasts from alternative models on the actual values.
LAD	The regression-based combination method proposed by Elliott and Timmermann [34].
dynOLS	The combination weights are found basically as in OLS, but they are allowed to vary by time.
AFTER	The Aggregated Forecast Through Exponential Reweighting method where the combination weights are non-linear.

Table 9.3 Combination weights

Abbreviation	ARIMA	THETA	NNETAR	STLM	TBATS	ATA
AMEAN	1/6	1/6	1/6	1/6	1/6	1/6
WEIGHTED	0.0855	0.0402	0.3257	0.2833	0.1925	0.0729
BG	0.0399	0.0109	0.4087	0.3841	0.1306	0.0258
InvW	0.1020	0.0680	0.4082	0.2041	0.1361	0.0816
LAD	−0.0150	−0.0249	0.7990	0.3897	−0.1738	0.0033
NG	−0.0108	−0.0061	0.7986	0.5427	−0.3288	0.0042
OLS	−0.0067	−0.0174	0.7951	0.5473	−0.3295	−0.0027
dynOLS	[−0.0323, 0.1359]	[−0.0576, 0.1259]	[−0.1240, 0.8544]	[0.4883, 0.9999]	[−0.7708, 0.1566]	[−0.0196, 0.1216]
AFTER	[0, 1/6]	[0, 1/6]	[0, 1]	[0, 1]	[0, 0.45)	[0, 1/6]

InvW, BG, NG, OLS, LAD, NNETAR. This ranking reveals several interesting results:

- dynOLS is the best-performing model both overall and among the combination models.
- AFTER is a promising alternative to dynOLS.
- STLM is the best-performing model among the base models and ATA is a promising alternative to it.
- MEDIAN, one of the simplest combination methods, ranked higher than four of the base models.
- dynOLS performed considerably better than both its static counterpart (OLS) and the other regression-based methods (BG, NG, and LAD).
- NNETAR is the worst-performing model both overall and among the base models.
- Regression-based methods, excluding dynOLS, performed worst among the combination models.

Next, we compared dynOLS with the second, third, and fourth best-performing models using the Diebold-Mariano test and graphical inspection. When dynOLS was tested against STLM, AFTER, and MEDIAN using a second-order loss-differential with $h = 1$, all of the one-sided null hypotheses were rejected. That is, dynOLS is significantly better performing than these models in terms of forecast accuracy. Finally, Fig. 9.2 depicts the actual testing dataset (blue line) and its forecast (red line) through the best base model and the three best combination models, and Fig. 9.3 illustrates how well the standardized residuals from each of these four models fit a proper normal distribution.

Table 9.4 Post-sample performances of the base and combination models

Model	MAE	medAE	RMSE	MASE	MASE* [a]	MAPE	AMAPE	sMAPE	medAPE	maxAPE	PBIAS	FB	TIC	NSC	Corr
ARIMA	92,124	72,505	114,072	2.19	1.34	9.88%	10.29%	10.47%	8.82%	39.78%	0.07	0.09	0.07	−0.69	0.17
ATA	61,381	38,128	87,877	1.46	0.89	6.56%	6.86%	6.84%	4.48%	31.73%	0.04	0.05	0.05	0.00	0.51
NNETAR	207,737	216,233	238,622	4.93	3.02	22.72%	23.20%	26.58%	34.11%	75.03%	0.22	0.25	0.15	−6.39	0.13
STLM	54,266	46,338	68,405	1.29	0.79	5.98%	6.06%	6.20%	5.31%	43.39%	0.04	0.04	0.04	0.39	0.77
TBATS	94,555	83,170	108,439	2.24	1.37	10.54%	10.56%	11.29%	9.99%	34.28%	0.10	0.10	0.06	−0.53	0.75
THETA	74,107	47,696	103,332	1.76	1.08	8.86%	8.28%	8.17%	4.90%	30.62%	−0.07	−0.06	0.06	−0.39	0.42
AMEAN	76,745	68,346	91,332	1.82	1.12	8.29%	8.57%	8.69%	8.50%	27.25%	0.07	0.08	0.05	−0.08	0.71
AFTER	54,753	47,607	68,885	1.30	0.80	6.03%	6.12%	6.03%	5.47%	43.39%	0.04	0.04	0.04	0.38	0.76
BG	116,910	118,398	133,424	2.78	1.70	12.75%	13.06%	13.82%	15.75%	46.79%	0.12	0.13	0.08	−1.31	0.57
dynOLS	45,905	34,526	61,069	1.09	0.67	5.01%	5.13%	5.01%	3.77%	45.49%	0.03	0.03	0.03	0.52	0.78
InvW	116,288	112,970	133,194	2.76	1.69	12.64%	12.99%	13.68%	14.96%	37.59%	0.12	0.13	0.08	−1.30	0.51
LAD	167,884	174,620	195,072	3.99	2.44	18.31%	18.75%	20.78%	24.62%	75.79%	0.17	0.20	0.12	−3.94	0.24
MEDIAN	67,450	55,409	83,814	1.60	0.98	7.25%	7.53%	7.58%	6.52%	26.56%	0.06	0.07	0.05	0.09	0.77
NG	158,768	167,404	187,118	3.77	2.31	17.29%	17.73%	19.53%	23.41%	80.27%	0.16	0.18	0.11	−3.54	0.24
OLS	159,612	168,088	188,214	3.79	2.32	17.38%	17.83%	19.65%	23.60%	81.29%	0.16	0.19	0.11	−3.60	0.23
WMEAN	106,733	104,093	121,634	2.53	1.55	11.62%	11.92%	12.48%	13.49%	37.60%	0.11	0.12	0.07	−0.92	0.62

[a] MASE was calculated by the formula given in Table 9.1, dividing the post-sample MAE of a model by the in-sample MAE of the naïve model. Since the forecast horizon is wide, it was later modified as MASE*, which replaces the value in the denominator with the out-sample MAE of the naïve model that accepts the last observation in the training set as the forecast value for all the observations in the test set. Nevertheless, MASE values should be approached carefully and be used to compare the base and combination models with each other instead of comparing them with the naïve model

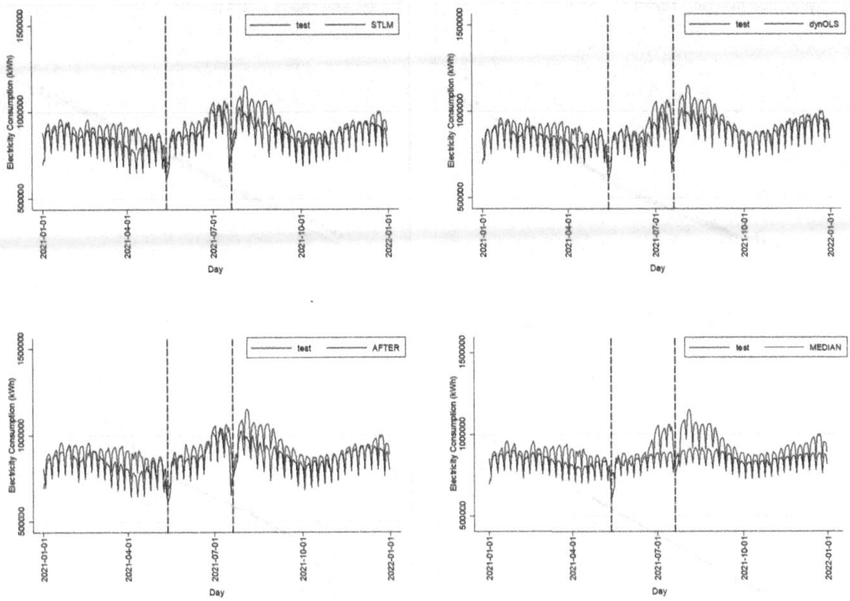

Fig. 9.2 Out-sample performances of the best base model and the three best combination models

9.4 Conclusion

Modern society is largely dependent on the supply of energy. To keep the lights on, the economy running, and the Internet humming, there is a clear need for better energy planning and management. Time series forecasting has the potential to improve the planning of sustainable energy systems by reducing the risk associated with long-term projections and the need for costly energy infrastructure upgrades. Recent developments in the field of computational intelligence have been leading researchers to modify traditional time series techniques or create hybrid models combining them with machine learning algorithms to improve prediction accuracy on complex data. Such approaches have the capability of improving overall forecast accuracy. Therefore, based on several performance metrics, the present study compares various ways to combine time series models using data on daily electricity consumption. In parallel with the related literature, our findings support combining the forecasts produced by different time series models. Even though it is not possible to claim that combination dominated the single (base) models considered, we realize that simple combination approaches (e.g., MEDIAN) can perform almost as well as their more sophisticated counterparts (e.g., AFTER). However, it is not recommended to work with static combination weights when the forecast horizon is wide.

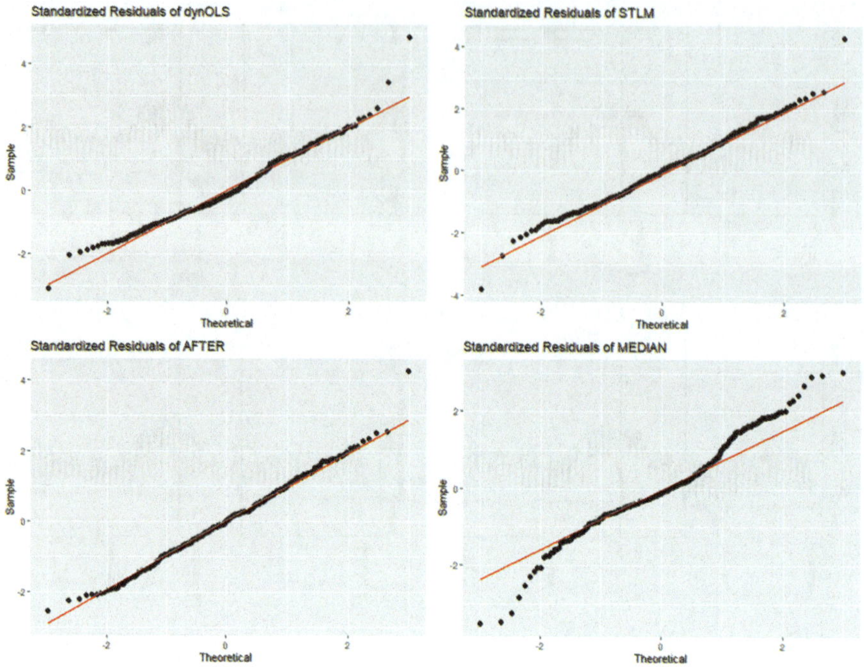

Fig. 9.3 Normal probability plots for the standardized residuals produced obtained from dynOLS, STLM, AFTER, and MEDIAN

As a possible avenue for future research, the investigated models can be applied to the cases of hourly, weekly, monthly, or quarterly data for electricity consumption to gain a better understanding of their performance and flexibility. Concerning sustainable energy production, transportation-related energy data or environmental data should also be investigated using the candidate models covered in this study accompanied by recently proposed methods such as Prophet.

References

1. Suganthi, L., & Samuel, A. A. (2012). Energy models for demand forecasting - A review. *Renewable and Sustainable Energy Reviews, 16.*
2. Ghalehkhondabi, I., Ardjmand, E., Weckman, G. R., & Young, W. A. (2017). An overview of energy demand forecasting methods published in 2005–2015. *Energy Systems, 8*, 411–447. https://doi.org/10.1007/S12667-016-0203-Y/TABLES/8
3. Verwiebe, P. A., Seim, S., Burges, S., et al. (2021). Modeling energy demand: A systematic literature review. *Energies (Basel), 14*, 7859. https://doi.org/10.3390/EN14237859
4. Makridakis, S., Spiliotis, E., Assimakopoulos, V., et al. (2021). The M5 uncertainty competition: Results, findings and conclusions. *International Journal of Forecasting*. https://doi.org/10.1016/J.IJFORECAST.2021.10.009
5. Makridakis, S., Spiliotis, E., & Assimakopoulos, V. (2022). M5 accuracy competition: Results, findings, and conclusions. *International Journal of Forecasting.* https://doi.org/10.1016/J.IJFORECAST.2021.11.013
6. Mosavi, A., Salimi, M., Ardabili, S. F., et al. (2019). State of the art of machine learning models in energy systems, a systematic review. *Energies, 12*, 1301. https://doi.org/10.3390/EN12071301
7. Debnath, K. B., & Mourshed, M. (2018). Forecasting methods in energy planning models. *Renewable and Sustainable Energy Reviews, 88*, 297–325. https://doi.org/10.1016/j.rser.2018.02.002
8. Roozbeh Nia, A., Awasthi, A., & Bhuiyan, N. (2021). Industry 4.0 and demand forecasting of the energy supply chain: A literature review. *Computers and Industrial Engineering, 154*, 10.1016/j.cie.2021.107128.
9. Anele, A. O., Hamam, Y., Abu-Mahfouz, A. M., & Todini, E. (2017). Overview, comparative assessment and recommendations of forecasting models for short-term water demand prediction. *Water (Switzerland), 9.* https://doi.org/10.3390/w9110887
10. da Silva, R. G., Ribeiro, M. H. D. M., Mariani, V. C., & Coelho, L. d. S. (2020). Forecasting Brazilian and American COVID-19 cases based on artificial intelligence coupled with climatic exogenous variables. *Chaos, Solitons and Fractals, 139.* https://doi.org/10.1016/j.chaos.2020.110027
11. Guo, Y., Xu, Y. P., Sun, M., & Xie, J. (2021). Multi-step-ahead forecast of reservoir water availability with improved quantum-based GWO coupled with the AI-based LSSVM model. *Journal of Hydrology, 597.* https://doi.org/10.1016/j.jhydrol.2020.125769
12. Ho, W. T., & Yu, F. W. (2021). Predicting chiller system performance using ARIMA-regression models. *Journal of Building Engineering, 33.* https://doi.org/10.1016/j.jobe.2020.101871
13. Ordóñez C, Sánchez Lasheras F, Roca-Pardiñas J, Juez FJ de C (2019) A hybrid ARIMA–SVM model for the study of the remaining useful life of aircraft engines. Journal of Computational and Applied Mathematics 346:. https://doi.org/10.1016/j.cam.2018.07.008.
14. Wu, L., & Shahidehpour, M. (2010). A hybrid model for day-ahead price forecasting. *IEEE Transactions on Power Systems, 25.* https://doi.org/10.1109/TPWRS.2009.2039948
15. Makridakis, S., Spiliotis, E., & Assimakopoulos, V. (2020). The M4 competition: 100,000 time series and 61 forecasting methods. *International Journal of Forecasting, 36.* https://doi.org/10.1016/j.ijforecast.2019.04.014
16. Tyralis, H., & Papacharalampous, G. (2017). Variable selection in time series forecasting using random forests. *Algorithms, 10.* https://doi.org/10.3390/a10040114
17. Ng, K. Y., & Awang, N. (2018). Multiple linear regression and regression with time series error models in forecasting PM10 concentrations in Peninsular Malaysia. *Environmental Monitoring and Assessment, 190.* https://doi.org/10.1007/s10661-017-6419-z
18. Fang, T., & Lahdelma, R. (2016). Evaluation of a multiple linear regression model and SARIMA model in forecasting heat demand for district heating system. *Applied Energy, 179.* https://doi.org/10.1016/j.apenergy.2016.06.133

19. Faghih, S. A. M., Gholipour, Y., & Kashani, H. (2021). Time series analysis framework for forecasting the construction labor costs. *KSCE Journal of Civil Engineering, 25*. https://doi.org/10.1007/s12205-021-1489-4
20. Makridakis, S., Spiliotis, E., & Assimakopoulos, V. (2018). Statistical and Machine Learning forecasting methods: Concerns and ways forward. *PLoS One, 13*. https://doi.org/10.1371/journal.pone.0194889
21. Diebold, F. X., & Mariano, R. S. (1995). Comparing predictive accuracy. *Journal of Business and Economic Statistics, 13*. https://doi.org/10.1080/07350015.1995.10524599
22. Halužan, M., Verbič, M., & Zorić, J. (2020). Performance of alternative electricity price forecasting methods: Findings from the Greek and Hungarian power exchanges. *Applied Energy, 277*. https://doi.org/10.1016/j.apenergy.2020.115599
23. Clemen, R. T. (1989). Combining forecasts: A review and annotated bibliography. *International Journal of Forecasting, 5*. https://doi.org/10.1016/0169-2070(89)90012-5
24. Tselentis, D. I., Vlahogianni, E. I., & Karlaftis, M. G. (2015). Improving short-term traffic forecasts: To combine models or not to combine? *IET Intelligent Transport Systems, 9*. https://doi.org/10.1049/iet-its.2013.0191
25. Milunovich, G. (2020). Forecasting Australia's real house price index: A comparison of time series and machine learning methods. *Journal of Forecasting, 39*. https://doi.org/10.1002/for.2678
26. Oh, C. O., & Morzuch, B. J. (2005). Evaluating time-series models to forecast the demand for tourism in Singapore: Comparing within-sample and postsample results. *Journal of Travel Research, 43*. https://doi.org/10.1177/0047287505274653
27. Song, H., & Li, G. (2008). Tourism demand modelling and forecasting-a review of recent research. *Tourism Management, 29*. https://doi.org/10.1016/j.tourman.2007.07.016
28. Beyaztas, U., Shang, H. L., & Yaseen, Z. M. (2021). A functional autoregressive model based on exogenous hydrometeorological variables for river flow prediction. *Journal of Hydrology, 598*. https://doi.org/10.1016/j.jhydrol.2021.126380
29. Andrawis, R. R., Atiya, A. F., & El-Shishiny, H. (2011). Combination of long term and short term forecasts, with application to tourism demand forecasting. *International Journal of Forecasting, 27*. https://doi.org/10.1016/j.ijforecast.2010.05.019
30. Newbold, P., & Granger, C. W. J. (1974). Experience with forecasting univariate time series and the combination of forecasts. *Journal of the Royal Statistical Society, Series A, 137*. https://doi.org/10.2307/2344546
31. Bates, J. M., & Granger, C. W. J. (1969). The combination of forecasts. *Journal of the Operational Research Society, 20*, 451–468. https://doi.org/10.1057/JORS.1969.103
32. Aiolfi, M., & Timmermann, A. (2006). Persistence in forecasting performance and conditional combination strategies. *Journal of Econometrics, 135*. https://doi.org/10.1016/j.jeconom.2005.07.015
33. Granger, C. W. J., & Ramanathan, R. (1984). Improved methods of combining forecasts. *Journal of Forecasting, 3*. https://doi.org/10.1002/for.3980030207
34. Elliott, G., & Timmermann, A. (2004). Optimal forecast combinations under general loss functions and forecast error distributions. *Journal of Econometrics, 122*. https://doi.org/10.1016/j.jeconom.2003.10.019
35. Yang, Y. (2004). Combining forecasting procedures: Some theoretical results. *Econometric Theory, 20*. https://doi.org/10.1017/S0266466604201086
36. Zou, H., & Yang, Y. (2004). Combining time series models for forecasting. *International Journal of Forecasting, 20*. https://doi.org/10.1016/S0169-2070(03)00004-9
37. Wei, X., & Yang, Y. (2012). Robust forecast combinations. *Journal of Econometrics, 166*. https://doi.org/10.1016/j.jeconom.2011.09.035
38. Qian, W., Rolling, C. A., Cheng, G., & Yang, Y. (2022). Combining forecasts for universally optimal performance. *International Journal of Forecasting, 38*. https://doi.org/10.1016/j.ijforecast.2021.05.004

39. Borthakur, P., & Goswami, B. (2020). Short term load forecasting: A hybrid approach using data mining methods. In *2020 international conference on emerging frontiers in electrical and electronic technologies, ICEFEET 2020*.
40. Fiorucci, J. A., & Louzada, F. (2020). GROEC: Combination method via generalized rolling origin evaluation. *International Journal of Forecasting, 36*. https://doi.org/10.1016/j.ijforecast.2019.04.013
41. Schulze, M. (2011). A new monotonic, clone-independent, reversal symmetric, and condorcet-consistent single-winner election method. *Social Choice and Welfare, 36*. https://doi.org/10.1007/s00355-010-0475-4
42. Assimakopoulos, V., & Nikolopoulos, K. (2000). The theta model: A decomposition approach to forecasting. *International Journal of Forecasting, 16*. https://doi.org/10.1016/S0169-2070(00)00066-2
43. Cleveland, R. B., Cleveland, W. S., McRae, J. E., & Terpenning, I. (1990). STL: A seasonal-trend decomposition procedure based on LOESS (with discussion). *Journal of Official Statistics, 6*.
44. de Livera, A. M., Hyndman, R. J., & Snyder, R. D. (2011). Forecasting time series with complex seasonal patterns using exponential smoothing. *Journal of the American Statistical Association, 106*. https://doi.org/10.1198/jasa.2011.tm09771
45. Yapar, G., Selamlar, H. T., Capar, S., & Yavuz, I. (2019). ATA method. *Hacettepe Journal of Mathematics and Statistics, 48*. https://doi.org/10.15672/hujms.461032

Part IV
Energy Efficiency

Chapter 10
Data-Driven State Classification for Energy Modeling of Machine Tools Using Power Signals and Part-Program Information

Nicla Frigerio, Paolo Albertelli, and Andrea Matta

Abstract There has been a growing interest in assessing the environmental impact and energy consumption of production facilities, resulting in the creation of various models for predicting energy usage of manufacturing equipment. These models are primarily based on empirical data gathered from field experiments, but the heterogeneity and complexity of the machines pose challenges in creating a model that can accurately represent specific machines or their components. Additionally, characterizing energy states can be costly in terms of time and experimentation, and experiments must often be repeated due to the changing nature of manufacturing equipment, such as changes in process conditions, degradation, or modifications made by operators. This study aims to address this issue by exploring a data-driven approach to classify energy models of machines by incorporating both field observations and prior knowledge. Machine learning algorithms are utilized for this purpose, using monitored signals like power requests and part-program information to guide the algorithms in a real-world application. The goal is to address the energy state classification problem as a first step toward developing autonomous energy modeling algorithms for machine tools.

Keywords Machine tools · Machine learning · Energy model

10.1 Introduction and Contribution

The manufacturing industry is a significant energy consumer, and the reduction of energy consumption is driven not only by ecological concerns but also by increasing electric energy costs. Monitoring manufacturing processes is critical for the development and sustainability of the industry, as it provides the necessary data for developing sustainable systems. With the growing awareness of energy consumption and the availability of new technologies for measurement and computation,

N. Frigerio (✉) · P. Albertelli · A. Matta
Department of Mechanical Engineering, Politecnico di Milano, Milan, Italy
e-mail: nicla.frigerio@polimi.it

G. Kayakutlu, M. Ö. Kayalica (eds.), *Decision Making Using AI in Energy and Sustainability*, Applied Innovation and Technology Management,
https://doi.org/10.1007/978-3-031-38387-8_10

innovative monitoring systems are being developed to extract useful information through the use of machine learning algorithms (MLAs).

Machine tools, specifically machining centers, have been the focus of energy reduction efforts, and new tools have been developed to monitor their operating conditions. Machining centers are equipped with software interfaces that allow for the acquisition of absorbed power signals, making it important for manufacturers to collect numerical data and assess the energy use of these machines. There is growing interest in developing data-driven machine energy models and mapping resource efficiency using acquired signals. The first step in achieving this goal is proper state identification, which is generally related to machine load and operating conditions.

This work focuses on the state identification problem and provides a data-driven approach to classify machine energy states. Information related to part-program execution are integrated with acquired power signals and machine learning algorithms (MLAs) are used to map machine behavior in a real case application. Indeed, in addition to energy-related data (e.g., electrical power, gas consumption), information about process condition or machine load are significant to represent machine-operative state, for example, the temperature, the pressure, and the torque. The use of load-related data or process data might be useful to identify machine states.

10.2 Energy Monitoring of Machine Tools and State Identification

The monitoring activity is composed by several tasks [11]. Among them, the following tasks are of interest for machine tools energy monitoring: visualization, performance evaluation, performance enhancement, correctness checking, and control. Visualizing energy data through proper dashboards enables the identification of energy flows and consumers. Also, it helps in identifying the evolution of signals over time. Performance evaluation enables the prediction of system KPIs so that performance enhancement can be pursued. Correctness checking compares actual signal with forecasts and control as a response to a certain deviation from predicted performance. Herewith we provide a description of machine energy model-related literature with particular detail on data-driven approaches.

10.2.1 Related Literature on Machine Energy Models

Recent literature includes systematic overviews discussing the development of energy-aware models of manufacturing systems (e.g., Mawson and Hughes [9]). In particular, energy modeling literature about chip removal machines has been widely explored in the last decades starting from the seminal works on the topic, e.g., Dahmus and Gutowski [3]. To provide some useful examples, the following review works collect recent literature on machine tool energy-related modeling approaches [4, 14, 15]. Analytical and empirical approaches are used to create machine tool energy models, and, specifically, empirical approaches are based on historical data or on data measured during specifically designed tests.

Recent empirical approaches use also the data acquired by monitoring systems during machining operations without an ad hoc shaped design of experiments and apply Machine Learning Algorithms to extract the machine model. Recent data-driven approaches that include MLAs are Hu et al. [5]; Abele et al. [1]; Vijayaraghavan and Dornfeld [12]; Avram and Xirouchakis [2]; Le et al. [8].

Hu et al. [5] propose an on-line energy efficiency monitoring method including an off-line identification of additional energetic losses. The model is built with the Least Square Method from historical data and represents machine power request. In addition, the proposed approach aims at providing an energy efficiency indicator, defined as the ratio of cutting energy to input energy, and a power efficiency indicator. This approach requires prior knowledge related to the machine and it also requires to measure many process parameters.

Using the Rete Algorithm, Vijayaraghavan and Dornfeld [12] describe an automated energy monitoring tool that obtains event-reasoning tables. The Rete algorithm is an efficient way to match patterns using a tree-structured sorting network to index and match time series patterns, in other words a similar sequence of power signal. Thus, the results are highly case-specific.

Avram and Xirouchakis [2] propose an analytical approach for the estimation of the energy consumption of a machine tool according to a given scheduling strategy. The consumed energy of spindle and feed axes is studied both in transient and in steady state. The energy required by auxiliary systems is also included in the model. The method requires information from CAM and CAD file and from FEM analysis, and then, it combines this information with the analytic formulation of the cutting force and motor torque to obtain the energy consumption prediction according to the GREEM software.

The use of neural network is exploited by Le et al. [8] to determine the operational state of the machine through a two-step method: first filtering and smoothing techniques are used to process input data, and secondly a neural network is used to classify data. The classification of energy consumption pattern is obtained dividing the power input signal into non-overlapping segments of a given length. Each segment serves as an input vector to the neural network, and when two segments contain the same pattern, they are classified in the same machine operational state. Therefore, the choice of the segment length is fundamental to proper assess the clustering and a formal method to choose such sequence length is missing.

The attempt to use energy consumption data to control and change operational states of subsystems is proposed by Abele et al. [1]. Their work is constituted by two modules: the EnergyMonitoring and the EnergyControl modules. Energy modes are defined (i.e., operational, ready, set-up, standby modes) and a global distribution system is used to process real acquired data and to perform a real-time monitoring. The power request data are clustered by supervised machine learning algorithms with self-organizing maps.

10.2.2 Challenges as in the Literature

Substantial endeavors have been undertaken to construct efficient online monitoring systems for machine tool energy; however, the challenge of creating a versatile and independent system with restricted information still persists. The approaches suggested in the scholarly literature are not yet practical for implementation in actual production environments. The assessment of performance and improvement requires the establishment of metrics that connect energy utilization with the operational state and workload of the machine, to accurately reflect efficiency measurements. To provide a comprehensive depiction of machine behavior, it is crucial to determine the operational state.

Hu et al. [5] developed a semi-empirical model for machine tool that requires prior knowledge about the finite state model of the machine and the machine's nominal parameters along with the power signal acquisition. The requirement of prior knowledge hampers the flexibility of the model's application. Abele et al. [1] utilized historical data to establish a correlation map of measured data, assuming a large amount of data had already been collected. Le et al. [8] proposed a Neural Network that is based on energy consumption data series comprising 60 records. The limited length of training tracks limits its further applications, as other production processes or products might need longer or shorter time windows to properly characterize the energetic state. Vijayaraghavan and Dornfeld [12] suggested the use of the Rete Algorithm, which requires a specific length of time series data and a large amount of input data, making it unsuitable for practical applications on a large scale where the variability of time series patterns might be high. Hu et al. [5] employed ad hoc created and static thresholds on spindle power to classify data.

The approaches proposed in the literature still lack generalization and require many prior pieces of information to be actually implemented. The approach proposed in this work focuses on state classification to enable performance evaluation and enhancement and aims to implement flexible learning algorithms using only acquired power signals and, eventually, other information collected from the machine controller, such as the family tool type.

10.3 Data-Driven Approach for State Classification

The proposed approach to solve the state classification problem is divided into three phases: (Phase 1) data pre-processing, (Phase 2) MLA training for state identification, (Phase 3) state identification on never-before-seen records.

10.3.1 Data Pre-processing

The goal of the first phase of data preparation and pre-processing (as described by Zhang et al. [13]) is to improve the quality of the data in terms of its completeness and consistency. This phase involves accessing, exploring, and evaluating the data, and enhancing it with additional information or transformations. The success of Machine Learning Algorithms (MLAs) is highly dependent on the quality of the data, making this phase crucial.

In this work, the term "data preparation and pre-processing" denotes all steps necessary to convert raw data into inputs for MLAs: data preparation, data reduction and partitioning, and feature engineering.

Data preparation includes data cleaning, conversion, integration, normalization, and transformation. In detail:

- Cleaning involves identifying missing values, eliminating outliers, and reducing noise (i.e., filtering).
- Conversion involves choosing the best available format.
- Integration involves combining data from heterogeneous data sources.
- Normalization involves scaling the data to a standard format for comparison.
- Transformation involves converting the data from one format or structure to another.

Data reduction involves reducing the amount of data stored in order to increase efficiency and reduce costs. The frequency of data acquisition from power meters can be adjusted to store only the minimum amount of data necessary for proper classification. Data partitioning is necessary only for the training phase of MLAs and involves dividing the data into training and validation sets.

Feature engineering is the process of transforming data into features that more effectively represent the underlying problem and simplify data mining. This phase involves determining the best way to present the data to the MLAs. Expert knowledge is essential for this phase.

10.3.2 MLAs for State Classification

MLAs are used to solve energetic state classification problem. Given x as an input record of data, the problem consists of assigning a category $c \in \mathcal{L}$ to the input. Thus, functional $\mathfrak{J} : \mathbb{R}^{|x|} \rightarrow \mathcal{L}$ is built during MLA training (Phase 2) and later used to classify never-before-seen records (Phase 3).

In this study, supervised machine learning algorithms (MLAs) are considered due to the known experimental conditions of data acquisition. While linear MLAs have desirable analytical and computational characteristics, they are limited to datasets that are linearly separable. Thus, non-linear MLAs are preferred, specifically the Neural Network (NN), Support Vector Machines (SVM), and K-Nearest Neighbor (k-NN) algorithms, which are more versatile with regard to input data requirements. A comprehensive comparison between NN, SVM, and k-NN can be found in Kotsiantis [6] and Kuhn and Johnson [7]. Typically, SVM and NN are recommended for use with multidimensional and continuous features, while k-NN is included for comparison due to its simplicity. These selected MLAs are capable of learning correlations through non-linear relationships among data, which is considered a crucial aspect in this field.

10.4 Real Case Application

The proposed approach to solve the state classification problem is applied to a real case as described in the remaining of the paper.

10.4.1 Case Description

A CNC machine tool JomaX 265 (i.e., a high-power Gantry Type Milling Centre) working in a production environment is considered. The machine has been equipped with voltage and current sensors so that machine power request can be acquired while the machine is executing machining operations on a set of products. In particular, power meters have a frequency of 5 Hz, collecting one data each 0.2 seconds. In particular, the spindle power P and the global machine power P_t are acquired. Additional information about actual machining operations are also available, e.g., spindle speed, tool type. Tools used along the acquisition are listed in Table 10.1 and classified into three families according to tool type: TF = 1 tools for drilling, TF = 2 threading tools, and TF = 3 milling cutters.

The acquired dataset $X = \{\boldsymbol{x}_i\}$ is composed by $N = 155{,}450$ samples corresponding to 8.5 hours of production. Record $\boldsymbol{x}_i = \{P_i, D_i, \mathrm{TF}_i\}$ is composed by the spindle power P_i, the difference power computed from the global power

Table 10.1 Classification of tools used during the acquisition period

Tool	Family *TF*	Description
PU6-8MDI	1	Drilling/Hole finishing
CE16-HSS-L	1	Drilling/Hole finishing
MA6	2	Threading
MA8	2	Threading
FR125-TOR-6INS-L	3	Milling
PU28-INS-L	1	Drilling/Hole finishing
PUNTA	1	Drilling/Hole finishing
FR80-SR-7INS-L-W	3	Milling

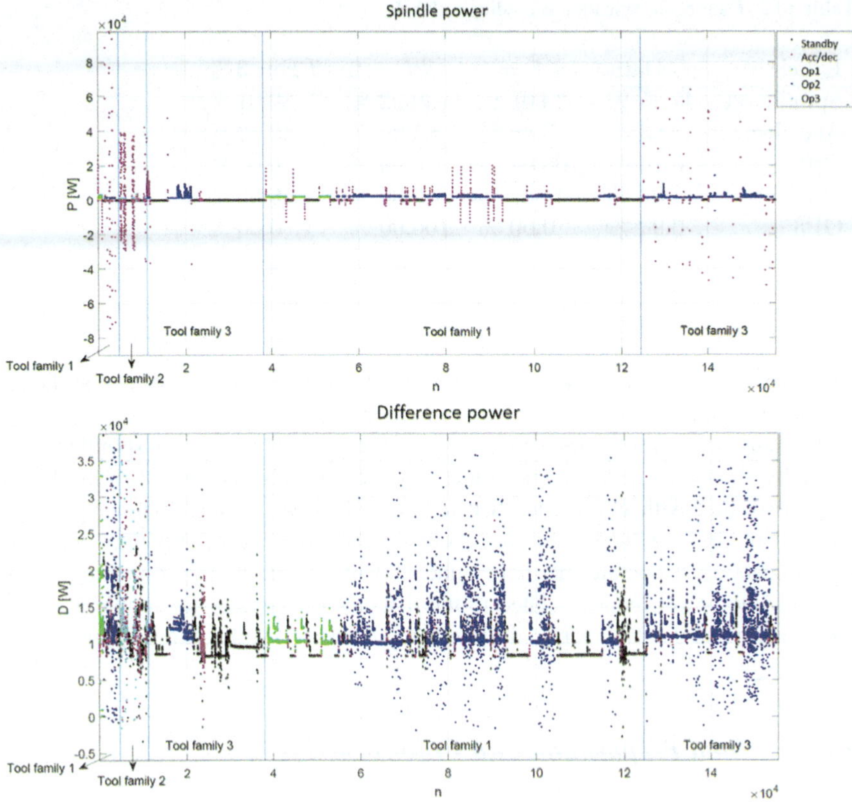

Fig. 10.1 Acquired signals P (top) and D (bottom) and associated labels

signal and the spindle power signal (i.e., $D_i = P_{t,\,i} - P_i$) and the tool family TF_i. If the tool is not loaded, TF_i indicates the family of the last tool loaded. Also, spindle speed s_i is acquired.

Label $\ell \in \mathcal{L}$ is associated to record x_i. Set $\mathcal{L}$ includes five states defined with prior knowledge: spindle off, spindle acceleration/deceleration, and three different machining operations op1, op2, and op3. Op1, op2, and op3 are considered different by an expert because of their geometric features and cutting parameters. It is assumed that for the application of interest, set $\mathcal{L}$ properly describes machine energy states.

Records are classified according to the labels as represented in Fig. 10.1, and Table 10.2 summarizes the percentage of records with the same label and the percentage of tool family type used. It's important to note that 47.45% of the records were obtained when the spindle was not in operation, which includes various activities such as executing part programs with no spindle speed, loading and unloading parts, brief machine stops and failures, as well as periods of machine idleness. This means that the power demand of the machine under the "spindle off"

Table 10.2 Dataset classification according to labels

Label	Data	$TF = 1$	$TF = 2$	$TF = 3$ %
Spindle off	47.45 %	30.08 %	39.92 %	30.01 %
Acc/dec	1.83 %	3.88 %	85.30 %	10.82 %
Op1	6.76 %	100.00 %	0.00 %	0.00 %
Op2	43.03 %	44.37 %	0.00 %	55.63 %
OP3	0.93 %	0.00 %	100.00 %	0.00 %

label may be quite diverse, although it is likely to be lower compared to other labels that involve cutting, spindle, and axis movements.

The range of power requests differs among the different labels. The spindle power generally shows a smaller range in the "spindle off" label compared to the larger range associated with acceleration/deceleration. The range of the difference power is wider than the spindle power range due to the presence of auxiliary machine components that operate partially independent of machining operations. For instance, the power request related to the machine chiller depends on the machine temperature and thermal inertia, despite being correlated with cutting operations.

10.4.2 Data Preparation and Pre-processing

In this work, Matlab functions are utilized to clean the data. Specifically, outliers are identified based on the deviation of the data from the median using the mean absolute deviation, and they are then replaced by using linear interpolation of neighboring data points (*filloutliers* function). Filtering is carried out by computing the moving average of the data over a fixed window length. This process reduces about 10% of the signal energy in the data through the use of the Fast Fourier Transform of the signal, which determines the frequency cutoff for the filtering (*smoothdata* function). Data cleaning transforms the acquired data matrix $\boldsymbol{X}$ into $\boldsymbol{X}^c$.

Min-Max Feature scaling is chosen to normalize data in the interval [0.1;0.9]. Normalized data matrix is denoted with $\boldsymbol{X}^n$ thus $x^n_{i,j}$, is computed as:

$$x^n_{i,j} = (0.9 - 0.1)\frac{x^c_{i,j} - \min\limits_i x^c_{i,j}}{\max\limits_i x^c_{i,j} - \min\limits_i x^c_{i,j}} + 0.1 \qquad (10.1)$$

where $x^c_{i,1} = P_i$, $x^c_{i,1} = D_i$, $x^c_{i,3} = \mathrm{TF}_i$ as after data cleaning.

Since the dataset is not replicated, we use data partitioning to obtain training and validation sets. Data matrix $\boldsymbol{X}^n$ is partitioned in three matrices selecting subsets of rows. Matrix $\boldsymbol{X}^n_{(k)} = \{\boldsymbol{x}^n_i\}$ with $k = 1, 2, 3$ and $i = k : 3 : N$. When matrix $\boldsymbol{X}^n_{(k)}$ is used as training set, remaining data are used as validation set so that three replications of the training can be performed for comparative purpose.

The dataset can be improved through feature engineering since the acquisition conditions are well understood. Upon examination of the data set, the label "acceleration/deceleration" does not accurately represent a process but rather a transition. Additionally, these records encompass a wide range of power values, which may contain outliers for peak power absorption. Given that the information about spindle speed is collected, it is also feasible to categorize records by computing the derivative of the spindle speed.

A set of $N = 111{,}228$ records can be used after data pre-processing operations and training matrices $X^n_{(k)}$, $k = 1, 2, 3$ are obtained as for validation matrices.

10.4.3 MLA Classification Performance

NN, SVM, and k-NN algorithms are compared in terms of the following classification performance indicators:

- Number of incorrect classified points N_{IC}
- Percentage of incorrect classified points P_{IC}
- Computation time obtained with Matlab executed on Intel(R) Core (TM) i5-3230M CPU @2.60 GHz

MLA parameters are collected in Table 10.3. Further, MLA training is replicated to demonstrate that training set is not significantly affecting MLA performance (Kruskal-Wallis test p-value $= 0.087$). SVM and k-NN are deterministic methods, while NN is stochastic, thus 10 replications of each NN training are performed varying in the initialization of weights and biases.

Comparing the three algorithms (Tables 10.4 and 10.5), it is evident that NN is the most time expensive MLA, while SVM and k-NN are trained in less than a minute. k-NN has the smallest N_{IC} and P_{IC}, followed by NN. Classified points as results of k-NN algorithm are in Fig. 10.2. More in details, the classification errors

Table 10.3 MLA parameters

	Hidden layers	Neurons	Max epoch
NN	2	20	250
	Kernel	Scale	Box constraint
SVM	Gaussian	Fine	2
	Distance	k	
k-NN	Euclidean	1	

Table 10.4 Comparison of MLAs performance (mean and 95%CI over three training sets are provided). Each training set results in the classification of 74'152 records

MLA	Training time [s]	*NIC*	*PIC* [%]
NN	1123 ± 0.46	613 ± 0.46	0.83 ± 0.00
SVM	37.3 ± 0.06	1199.333333 ± 0.18	1.62 ± 0.00
k-NN	6.3 ± 0.04	522 ± 0.14	0.70 ± 0.00

Table 10.5 Incorrect classified points according to state labels. A single training is reported as example

MLA	Standby	Acc/dec	Op1	Op2	Op3
NN	22	–	174	390	1
SVM	40	–	759	393	0
k-NN	22	–	177	317	3

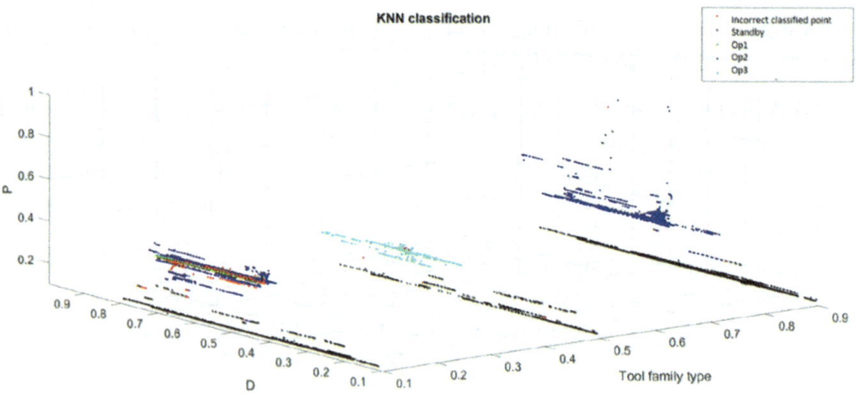

Fig. 10.2 Classified points by k-NN using training set $X^f_{(1)}$

interest mainly the distinction between op2 and op3 with family TF = 1 which is actually equally used for both operations (Table 10.5). NN and k-NN algorithms have a similar behavior with around 60% of incorrect classified points are in Op2.

10.4.4 Sensitivity Analysis

A sensitivity analysis was conducted to determine the effect of using additional information beyond the main signal, spindle power P, in classifying the machine state. Furthermore, the original data was classified to evaluate the significance of the data pre-processing phase. Four cases are compared varying the input record: *Case A*: $\mathbf{x} = \{P, D, \mathrm{TF}\}$ as in Sect. 10.4.3, *Case B*: $\mathbf{x} = \{P, \mathrm{TF}\}$, *Case C*: $\mathbf{x} = \{P, D\}$, *Case D*: $\mathbf{x} = \{P\}$, and *Case E*: $\mathbf{x} = \{P, D, \mathrm{TF}\}$ as acquired. Classification performance are reported in Table 10.6. A single training set is used for this analysis.

First of all, the comparison among *Case A* and *Case E* clearly shows the impact of data pre-processing phase. The P IC increases up to 4.9% and the computation time increases as well.

The training of MLAs is faster when less information is used as input, but it results in a decrease in classification performance. For *Case B* and *Case D*, the P_{IC} is around 2% for all MLAs, showing that difference power D is a relevant information. Similarly, when both *P* and *D* are used (*Case C*), MLAs perform similarly to *Case A* where all information is utilized.

The tool information in this analysis does not seem to be contributing to the classification. Nevertheless, it should be noted that the tool families included in this study may not be representative of all machining operations. The processes included in this analysis might have a small MRR (material removal rate) and might not represent very different processes in terms of power request. Also, tool families

Table 10.6 Comparison of MLA performance varying input record. Totally classified records: 74'152 for Cases A, B, C, and D; 103'633 for Case E

MLA	Case	Training time [s]	NIC	PIC [%]
NN	A	1145 ± 0.07	625 ± 31	0.84 ± 0.04
SVM	A	34.9	1192	1.61
k-NN	A	8.6	519	0.70
NN	B	854 ±154	1653 ± 154	2.23 ± 0.25
SVM	B	38.3	1634	2.20
k-NN	B	4.5	1572	2.12
NN	C	998±112	765 ± 112	1.03 ± 0.21
SVM	C	42.1	1224	1.65
k-NN	C	5.7	414	0.56
NN	D	643 ± 169	1629 ± 169	2.20 ± 0.21
SVM	D	46.5	1735	2.34
k-NN	D	5.4	1617	2.18
NN	E	2432 ± 137	5089 ±484	6.86 ± 0.47
SVM	E	287	4506	6.08
k-NN	E	12	5088	6.86

TF = 1 are used both in states $\ell = \text{op1}$ and $\ell = \text{op2}$ indicating that some tools might be more flexible than others. Further investigation is needed to understand how tool type information can aid in the machine state classification. This could be achieved by including a larger set of tools, such as roughing tools and critical operations, and varying the material removal rate of operations.

10.5 Conclusive Remarks

State identification is an important step toward achieving other results in energy assessment. Identifying a reference baseline consumption is crucial for accurately mapping machine efficiency. Knowledge of resource energy efficiency is key for sustainable management of machine tools and can be used to evaluate alternative process conditions. This work is a first attempt at applying supervised learning algorithms to identify the energetic state of a machine tool during machining operations. The approach is flexible and adaptable to different machine tools and operating conditions, and the creation of reference models for classifying machine energetic states is of high interest to practitioners.

Future research will focus on developing AI-based methods to help machine tool builders and users better understand how machining centers work. This represents a future challenge for sustainable production automation. Including information related to part-program execution may be useful in identifying machine states, and analyzing the sequence of events is also a challenge in creating machine models that include state transitions. The ultimate goal is to develop smart monitoring systems that provide an alarm when an unexpectedly low machine tool efficiency is detected.

Acknowledgment The authors would like to thank Laura Morè for her significant contribution as a Master's student [Morè, L. [10]].

References

1. Abele, E., Panten, N., & Menz, B. (2015). Data collection for energy monitoring purposes and energy control of production machines. *Procedia CIRP, 29*, 299–304. The 22nd CIRP Conference on Life Cycle Engineering.
2. Avram, O. I., & Xirouchakis, P. (2011). Evaluating the use phase energy requirements of a machine tool system. *Journal of Cleaner Production, 19*(6), 699–711.
3. Dahmus, J. B., & Gutowski, T. (2004). An environmental analysis of machining. In *Manufacturing engineering and materials handling engineering ASME* (pp. 643–652).
4. Fysikopoulos, A., Papacharalampopoulos, A., Pastras, G., Stavropoulos, P., & Chryssolouris, G. (2013). Energy efficiency of manufacturing processes: A critical review. *Procedia Forty Sixth CIRP Conference on Manufacturing Systems 2013, 7*, 628–633.
5. Hu, S., Liu, F., He, Y., & Hu, T. (2012). An on-line approach for energy efficiency monitoring of machine tools. *Journal of Cleaner Production, 27*, 133–140.

6. Kotsiantis, S. B. (2007). Supervised machine learning: A review of classification techniques. *Informatica, 31*, 249–268.
7. Kuhn, M., & Johnson, K. (2018). *Applied predictive modeling*. Springer.
8. Le, C. V., Pang, C. K., Gan, O. P., Chee, X. M., Zhang, D. H., Luo, M., Chan, H. L., & Lewis, F. L. (2013). Classification of energy consumption patterns for energy audit and machine scheduling in industrial manufacturing systems. *Transactions of the Institute of Measurement and Control, 35*(5), 583–592.
9. Mawson, V. J., & Hughes, B. R. (2019). The development of modelling tools to improve energy efficiency in manufacturing processes and systems. *Journal of Manufacturing Systems, 51*, 95–105.
10. Morè, L. (2019). *On line monitoring of energy consumption: machine learning for state identification* [Master's thesis]. Politecnico di Milano, Milano, Italy.
11. Schroeder, B. (1995). On-line monitoring: A tutorial. *Computer, 28*(6), 72–78.
12. Vijayaraghavan, A., & Dornfeld, D. (2010). Automated energy monitoring of machine tools. *CIRP Annals, 59*(1), 21–24.
13. Zhang, C., Cao, L., & Romagnoli, A. (2018). On the feature engineering of building energy data mining. *Sustainable Cities and Society, 39*, 508–518.
14. Zhang, Y. (2015). Review of recent advances on energy efficiency of machine tools for sustainability. *Proceedings of the Institution of Mechanical Engineers, Part B: Journal of Engineering Manufacture, 229*(12), 2095–2108.
15. Zhou, L., Li, J., Li, F., Meng, Q., Li, J., & Xu, X. (2016). Energy consumption model and energy efficiency of machine tools: A comprehensive literature review. *Journal of Cleaner Production, 112*, 3721–3734.

Chapter 11
Energy Efficiency Optimization Application in Food Production Using IIOT Based Machine Learning

Mete Omerali and Tolga Kaya

Abstract The energy sources of the universe are limited and draining with increasing speed. One of the biggest consumption of energy is in the manufacturing industry with 35% of the overall consumption. Collecting and analyzing data is becoming a crucial topic with the Internet of Things (IOT). Using the science of data is becoming a hotter topic in manufacturing industries as well. The study aimed to apply theoretical analytical methods in real-life cases and demonstrate potential returns on IOT investments for fully automated systems. This study not only analyzed and predicted the energy consumption but also provided a prescriptive receipt to minimize energy spending while keeping high-quality production standards. For the use case, the data gathered from real-production of a multinational food manufacturer and machine learning algorithms were applied. The study provides an energy efficiency of 5–6% for a single machine which consumes 35–45% of the total energy spending of the entire factory.

Keywords Internet of Things · Green factory · Artificial intelligence · Energy efficiency · Digitalization

11.1 Introduction

In today's world, manufacturing becomes smarter with the help of IOT applications. IOT systems can be used in various areas such as smart cities, smart products, and smart manufacturing. The usage of IOT systems in manufacturing is called the Industrial Internet of Things (IIOT). IIOT systems, increase the cooperation between machinery and people using analytic, to bring intelligence to manufacturing [28]. As IIOT devices can communicate, using a various number of protocols, [6] and the communication with machines becomes much simpler and common. This communication opens doors to advance data analytic capabilities which can be used for

M. Omerali (✉) · T. Kaya
Istanbul Technical University, Department of Management Engineering, Istanbul, Türkiye
e-mail: omerali18@itu.edu.tr

G. Kayakutlu, M. Ö. Kayalica (eds.), *Decision Making Using AI in Energy and Sustainability*, Applied Innovation and Technology Management,
https://doi.org/10.1007/978-3-031-38387-8_11

optimize energy efficiency. The benefits of using IIOT systems and their return on investment (ROI) are difficult to calculate for most companies as the industry lacks solid calculations and examples. In this study, two main objectives are identified. Firstly, whether IIOT systems can bring value in terms of reducing energy costs to cover their ROI is evaluated. Secondly, the IOT-based machine learning use case is demonstrated as a guide, to reduce energy spending for manufacturing companies to achieve green production.

11.1.1 Challenges in Production

After the introduction of IOT technologies, various new challenges are added, on top of thousands of existing challenges the manufacturing/production companies had. Globalization and the increased competition reshape the economy and political situation. Current manufacturing organizations are required to adapt to the changes to survive. The change has already shaped some of the industries faster like healthcare and marketing. But manufacturing companies are somehow behind in the technology game due to not correctly recognizing the potential [10, 16, 17]. To compete and survive, all manufacturing companies are required to move from traditional manufacturing to so-called "smart manufacturing". Smart manufacturing becomes the main objective of governments in most countries [14]. On the bright side, only in recent years, the developments in the IT domain make such a move possible [3].

Moreover, the manufacturing companies are aiming for standardization in their production. Irrelevant of the type of production or end product, all companies should aim to produce the same quality each time to avoid big variances between products. The quality is not about building the best product, but keeping the minimum variance between products [4]. The systematic approaches are key in standardization and continuous improvements over performance criteria such as sustainability, safety, environmental-friendly, and so forth, with the main aim of producing efficient and accurate products for the end customers [16, 18].

Among all other challenges, perhaps the most important one is on competitive pricing battle. Due to the increase in energy prices, companies aim to reduce their consumption and waste. Reduction in the end product's cost cannot be achieved solely by advanced manufacturing techniques. Therefore, the need of reducing energy waste becomes critical. As of today, almost 35% consumption of all electricity is used by manufacturing companies followed by additional 20% of carbon emission [21]. To avoid an energy crisis and apply a green and sustainable environment, energy management in manufacturing should be targeted as the highest priority factor [29].

11.1.2 Need of Analytic in Manufacturing

To find high consumption areas to survive under high market pressure, manufacturing companies require applications of advance analytic [2]. In modern manufacturing, the amount of data gathered with IIOT platforms are in extensive amounts, not only in terms of observations but also in terms of numbers of parameters collected, which cannot be analyzed by humans with traditional approaches [15]. On top of that, current manufacturing companies are not utilizing the data of other critical systems, such as computer-aided manufacturing, manufacturing execution systems, product lifecycle management systems, enterprise resource planning systems, etc. [12]. Most manufacturing companies are using simple, event-based, alert triggering mechanisms, which are far behind dedicated machine learning (ML) based analytic solutions [17]. The companies are required to implement IOT solutions with advanced ML capabilities to turn the data into knowledge to form high quality, real-time, intelligent, predictive, or preventive decision-making solutions [18]. Implementing such solutions require high effort and budget. But over time, the realization of smart manufacturing improves the cost saving activities, as well as, opening doors to sustainable and standardized production [18]. To select the best IOT/IIOT solutions, the analytic types need to be understood in detail.

11.1.3 Type of Analytic

The science of analytic aims at turning raw data into useful information. To do that, the machine learning system requires an extensive amount of data, to learn and test the end model. These vast amounts of data and their applications are named as Big-Data. The big data term refers to a set of actions and structures that cannot be fulfilled, with the conventional approaches or systems, due to specificity in size and other characteristics [16]. Analytical studies can be grouped into two types, those are: historical data analytic, in which the data is already collected, or real-time flowing data, which is called real-time analytic [14]. In the decision-making process, four types of analytics can be used, those are descriptive-analytic, diagnostic analytic, predictive analytic, and prescriptive analytic [3, 23, 28]. Figure 11.1 shows what each type of analytic does, as well as, what kind of complexity exists and what value is gained when moving from one to another.

As the more advanced analytic technique is used, the decision and insight gathered from the data become clearer. Therefore, human input and human-based mistakes reduce. In Fig. 11.2, the amount of decision support linkage between data and required human action is shown.

The remainder of this paper is structured as follows. Section 11.2 is devoted to the literature review. Section 11.3 shows the methodology, Sect. 11.4 explains the problem statement, and Sect. 11.5 demonstrates real-life production case. Finally,

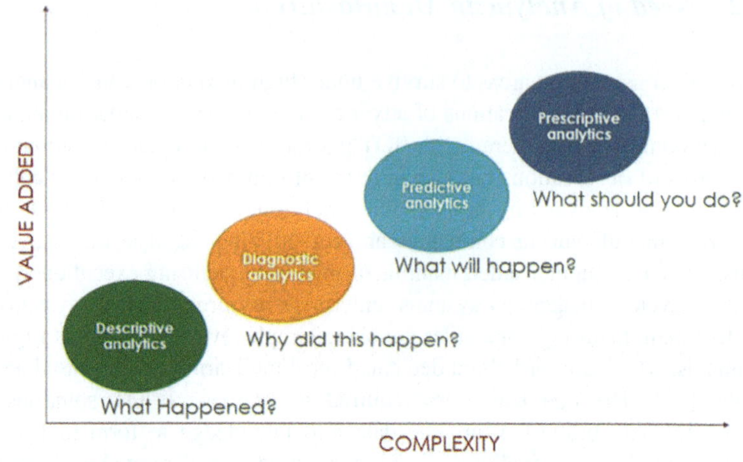

Fig. 11.1 Analytic types

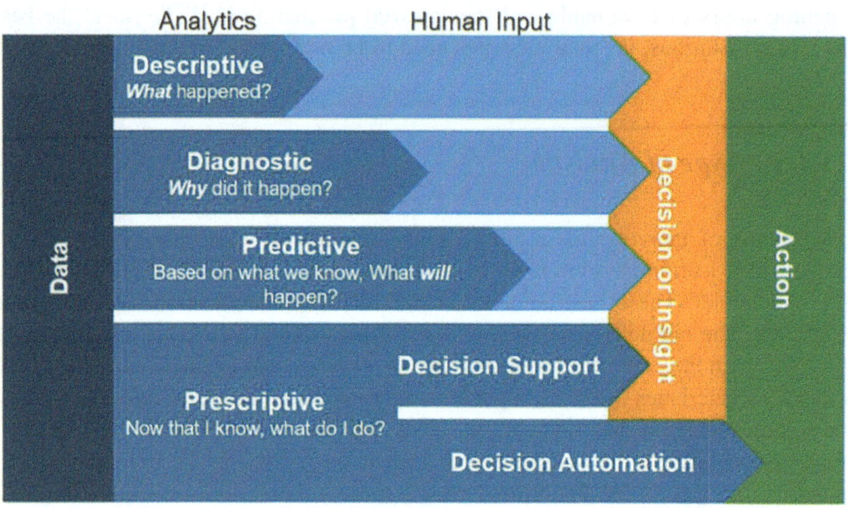

Fig. 11.2 Human input in analytic

Sect. 11.6 discusses the findings, includes concluding remarks and future work suggestions.

11.2 Literature Review

In the literature review, the aim is to find examples and methods used in the manufacturing industry. Where the data analytic studies were conducted, and the results were gathered concerning energy management and other similar cases. Most of the use-cases are related to equipment health, predictive maintenance, and fault or defect detection on industries such as semiconductors [1, 20]. Most of the papers claim that analytic studies are challenging since the knowledge of individuals is separated between the ones who know only about manufacturing and others who know only about analytics. Both sides have challenges in understanding the information provided by the other [14, 16, 20]. Other highly suggested methodologies depend on certain configurations such as collecting data from the edge but applying analytic on the cloud ([8, 14, 28]; Zuo et al. 2018). Using cloud-based systems could be interesting for some organizations, but it can be a game-breaker for others especially in industries like aerospace, defense, healthcare, etc. where the information is classified and pushing it to the cloud can be challenging. On top of that, some manufacturing plants are lacking trustable internet stability, which can affect the work of manufacturing companies as the stoppage time due to inaccessibility will not be tolerated.

Studies aiming at energy management with or without conducting analytical studies are aiming to reduce energy to save money for manufacturing companies. As well as supporting the environment for sustainable future production. However, none of which to our knowledge provides details of estimated savings in terms of percentages or actual dollars spent. The comprehensive study, done by Zuo et al. 2018, claims that all pre-production, production, and post-production activities should focus on energy consumption (Zuo et al. 2018). The configuration of machine parameters is the main source of energy spending, and how it should be monitored and controlled [27] is also the main idea and aim of this chapter. Proactive configuration control is crucial, as the reactive approaches cannot fulfill comprehensive management over high consumption [21]. Apart from the need of making savings on production costs, ecological awareness and regulations such as ISO-50001 are forcing organizations to reshape the production facilities. Although, only a few companies are using energy consultants and advanced methodologies to optimize the processes [25]. To maximize productivity, while reducing cost and minimizing resource usage, real-time predictive or prescriptive analytic with fully integrated end-to-end and autonomous IIOT platform with built-in ML capability is required [11, 15, 24]. If the analysis cannot be done on a real-time basis, the success is at high risk [17]. Recommendations of applying visual techniques on collected data are important in understanding what happened and up to some point, the reason why it has happened that way can be clarified. However, without an extensive ML study, it is certainly impossible to understand all correlations and even further, predict the future implications in manufacturing [5]. The study of [7] offers a parameter reduction mechanism, using the partial least squares (PLS) technique to increase the ratio of observations vs the number of parameters in the data [7]. However, in the manufacturing domain, the number of observations can be huge due to the extensive amount of data collected from production [23]. Therefore, parameter reduction was not required but used to generate simpler models in this study.

In order to make an analytical study, certain actions need to be taken. Firstly, the problem needs to be identified. Secondly, the relevant data need to be collected and certain data cleansing operations should be applied if/when needed. Parameter engineering in few cases is required and implemented to satisfy assumptions of various models. The observation unit needs to be clarified. Afterward, if the analysis will be done manually, the required steps should be defined and documented. Finally, the solution needs to be implemented and the model should be generated to be used [14, 16, 20].

11.3 Methodology

1. Multiple Linear Regression: For n independent predictors, the corresponding slope coefficient is multiplied. The linear combination of the slope of the curve (intercept) and error rate all together defines the quantitative response variable [13].

$$\gamma = \alpha + \beta_1 x_1 + \beta_2 x_2 + \cdots + \beta_n x_n + \epsilon_i \tag{11.1}$$

2. Ridge Regression: Ridge Regression is a parameter shrinkage method that is similar to least squares. By shrinkage penalty, the aim is to minimize the residual sum of squares. Due to the sum of all squared values of coefficients cannot be bigger than the predefined penalty, some parameters' coefficients get closer to zero and can be removed from the model [13].

$$\min_{\beta} \sum_{i=1}^{N} \left(y_i - \beta_0 - \sum \beta_j x_{ij} \right)^2 \text{ st.} \sum_{j=1}^{p} (\beta_j)^2 \leq s \tag{11.2}$$

3. Lasso: Lasso is also a parameter shrinkage method that is similar to ridge regression with an additional benefit. Due to the sum of all absolute values of coefficients cannot be bigger than the penalty value, some parameters shrink exactly to zero and are removed from the model [13].

$$\min_{\beta} \sum_{i=1}^{N} \left(y_i - \beta_0 - \sum \beta_j x_{ij} \right)^2 \text{ st.} \sum_{j=1}^{p} | \beta_j | \leq s \tag{11.3}$$

4. XGBoost: The idea behind XGBoost is to create m number of models, in which every new model fits to the residual of the previous model. As every model generates lower residuals, the end model fits more accurately [26].

$$F_{m+1}(x) = F_m(x) + y_m h_m(x) \tag{11.4}$$

5. Random Forest: The aim is to generate multiple bootstrapped training data sets, which finally averages total B trees predictions into the final prediction [13].

$$\hat{f}(x) = \frac{1}{B}\sum_{b=1}^{B} \hat{f}^{*b}(x) \tag{11.5}$$

6. Neural Network: The neural network aims to convert inputs to outputs using multiple layers and activation function, which consists of a certain weight and threshold values. The model constantly adjusts the weight of each neuron to predict the most accurate training result. In the end, all attached nodes with a certain weight and value make the prediction [19].

$$y(x) = f\left(\sum_{i=1}^{n} w_i x_i\right) \tag{11.6}$$

7. Decision/Regression Tree: Aims to create homogeneous segmentation defined by splitting rules. Each region gets the mean or mode of the segment. The goal of the equation is to find boxes to minimize residual sum of squares [13].

$$\sum_{j=1}^{j} \sum_{i \epsilon R_j} \left(y_i - \hat{y}_{R_J}\right)^2 \tag{11.7}$$

8. Support Vector Machines (SVM): Aims to fit a hyperplane (one dimension less than the total dimensions), which separates the data by enlarging the feature space, in multiple sides. Support vector machines are generally combined with a non-linear kernel. Kernel quantifies how similar the two observations are [13].

$$f(x) = \beta_0 + \sum_{i \epsilon S} \alpha_i K(x, x_i) \tag{11.8}$$

9. Residuals: The residuals show the difference between the expected outcome and the observed prediction value of a specific observation.

$$\hat{\epsilon}_i = y_i - \hat{y}_i \tag{11.9}$$

10. Residual Sum of Square (RSS): The aim of the regression is to minimize total residuals' sum of square in order to fit the model most accurately. The residuals' sum of square calculates squares of the total error [13].

$$\text{RSS} = \sum_{i=0}^{n} \left(y_i - \widehat{\beta_0} - \widehat{\beta_i} x_i\right)^2 \tag{11.10}$$

11. Pearson Correlation Coefficient: Pearson correlation coefficient shows the level of relation between two variables. Based on their correlation being positive or negative, they get a value between −1 to +1 ([9].

$$r = \frac{n(\sum xy) - (\sum x)(\sum y)}{\sqrt{\left[n\sum x^2 - (\sum x)^2\right]\left[n\sum y^2 - (\sum y)^2\right]}} \tag{11.11}$$

12. Root Mean Square Error: RMSE, calculates the rooted average error squares in order to calculate both positive and negative errors' total [9].

$$\text{RMSE} = \sqrt{\frac{\sum_{i=1}^{n}(u)^2}{n}} \tag{11.12}$$

13. Mean Squared Prediction Error: Similar to RMSE, the MSPE calculates regular (non-square rooted) error rate of the prediction vs the actual target values [9].

$$\text{MSPE}(L) = E\left[(g(x_i) - \hat{g}(x_i))^2\right] \tag{11.13}$$

14. Homoscedasticity: Homoscedasticity is the state of having same variance among all residuals, irrelevant of their value [9].

$$\text{var}(y_i) = \text{var}(e_i) = \theta^2 \tag{11.14}$$

15. Coefficient of Determination (R-Squared): R^2 shows the proportion of variance explained in the dependent variable by the independent variable(s) [22].

$$R^2 = 1 - \left(\frac{\text{Residual sum of squared}}{\text{Total sum of squares}}\right) \tag{11.15}$$

16. Variance Inflation Factor (VIF): VIF helps detect multicollinearity problem in regression models. For every predictor, VIF is calculated separately and parameters with VIF value higher than 5 or 10 are generally removed from the regression analysis [22].

$$\text{VIF} = \frac{1}{1 - R_i^2} \tag{11.16}$$

17. Coefficient Weight Calculation: In the weighted ensemble algorithm, the selected top models are used as independent variables to predict the dependent variable. The coefficients in the regression are normalized to generate a weight for the corresponding parameter.

$$\text{Weight}_x = \frac{\text{Coefficient}_x}{\sum_{i=1}^{m} \text{Coefficient}_i} \tag{11.17}$$

18. Weighted Average Ensemble: In order to calculate the weighted average ensemble model, the linear combination of each parameter is multiplied by normalized coefficient and its prediction value and sum all together.

$$\hat{y} = (W_1 \times P_1) + (W_2 \times P_2) + \cdots + (W_n \times P_n) \tag{11.18}$$

11.4 Problem Statement

The general problem in manufacturing is mainly making sure of keeping the high-quality standards as defined in previous sections. However, while keeping the high standards the cost of making such production increases. To balance the cost-reducing energy spending is essential. In this study, the main problem is no different. The multinational food production company producing a broad range of products is required to reduce energy spending while maintaining standardization and high quality. Throughout food production, various parameters can make an impact on quality and energy consumption. Perhaps, in various manufacturing types, quality implications can be taken as secondary concerns. However, when it comes to food production due to its exceptional impact such as the risk of toxication, the minimum standards have to be achieved with great concern. Consequently, food should be produced under a certain minimum heating level and required duration. Formerly, only the allowed limits were defined and all low-quality outputted sections of the data were left out from the study. Within the data, in which the output is in acceptable ranges among multiple parameters, the configurable ones were identified. All of the configurable heating variables are related to a particular subsection of the extruder machine. Our problem is then reduced to find the most energy-friendly heating target temperatures based on each type of product to be produced. As one can guess, different heating values have changing effects on total energy consumption. Hence, the main aim is then to suggest to the operators a set of machine configurations based on the upcoming production plan to minimize the cost of energy spending.

11.5 Industrial Case Study

11.5.1 Overview

To test energy reduction in real production, a supervised analytic study is applied in a multinational food production company. The data was collected during a period of 2 months. The collection of data was done in real-time, and each minute of production was saved to the database. The data unit consists of input parameters of the machine and the configuration of current calculated energy spending of the given product mix. The data is collected by a programmable logic controller (PLC) of the extruder machine using the IOT system. An extruder machine takes the inputs of wheat and corn as well as other inputs such as water, food chemicals, sweeteners, etc. Since different product groups have different sets of ingredients which could affect the results depending on what is produced, the data is grouped by product types as well. For simplicity, a certain group of product mixes is chosen for this study. However, as the type of data is the same and only the values of observations differ, the same study can be reapplied on different product groups to generate new models. The combination of all ingredients is called the mix. As a first step, the mix is sent to the mixer to create a homogeneous mixture by adding syrup, water, and oil and it is kneaded for a certain period. Ingredients are not added at a fixed time and from a fixed location. To keep the know-how secure, those details will not be explained explicitly. The mix kneading operation changes the heat of the mix, which could end up in a degree of around 80–100 °C. After this stage, since certain machine sections have different predefined heating value targets for the target product cooking operating begins. Since the temperature of the kneaded mix, additional ingredients and target heat of the section is not at the same temperature a gap occurs. If the target value is higher than the current temperature, the machine is set to the cooling mode for that specific subsection. By pushing cold water, the subsection is cooled down. Whereas, if the target temperature is lower than the set value; steam is pushed to increase the heat in the subsection. There are multiple subsections within the machine and each section works the same way toward the corresponding current and target values. The problem of variable heating and cooling values is the main reason for high energy consumption. By setting different set values, the production can still achieve the same quality with less variability on the heating and cooling. As a consequence, the energy consumption of the machine is reduced.

The aim of the analytic study has three main areas;

- Identify parameters affecting energy consumption in order for operators to fine-tune those to save energy during production
- Predict the energy consumption based on target set values to observe potential consumption rates and update values if the consumption is above expectations
- Prescript future heating set values for maximum energy saving for a given product in real-time to update machine configuration based on given production recipes.

The data collected from the machine has multiple quantitative parameters those are;

- The current temperature of all sections
- The flow rate of steam-water
- The flow rate of cold water
- The flow rate of oil and syrup
- The torque of the engine

Moreover, there are dummy Boolean qualitative values that show if heating or cooling is in progress for the subsection of the extruder. The data does not contain any value where the quality of the production wasn't satisfactory. Lastly, as the study was taking place, there was no mechanism such as a watt-meter to compute the energy consumption directly. The factors affecting the consumption are: the current torque of the engine, the flow rate of the water steam, etc. To identify how much energy is consumed, an additional energy column was added to the collected data. This column contains the weighted total of certain energy consumption fields which are calculated based on their effects on energy spending defined by the machine manufacturer.

11.5.2 Data Operations

There are numbers of machine learning algorithms that all learn from data and generate a model which can be used to make predictions. Moreover, those predictions can be tested on the known target value (response parameter), of the validation set. Different machine learning algorithms have different pros and cons. Selecting one or using an ensemble method cannot be decided beforehand due to the inability to know the outcomes with a given dataset. For that reason, in this study, multiple algorithms were tested and their models' fit to the data is analyzed. The most accurate ones are explained in detail in the upcoming sections.

In data science, as stated above, there are no exact correct answers or best models to be used. If one chooses a very complex model for a simple real-life case or vice versa, one can end up in a bias-variance trade-off dilemma. To avoid such a problem, different methods can be tested. But as the machine learning algorithms learn from the data, using the same data for testing could end up in an over-fitting problem. The over-fitting problem occurs when a model fits too closely to limited data. Therefore, cannot provide accurate information for similar but not yet observed part of the new data (test data). To overcome such problems, the data needs to be separated beforehand into two parts randomly. In the case of this study, the total available data of more than 26,000 observations were separated into training dataset which contains 80% of all data (around 21,000 observations) and test data as 20% (almost 5000 observations). There are multiple ways to separate the data and the method used in this study is called hold-out cross-validation, which exactly creates two sets of data without any overlapping. Data is randomized before being separated to 80% for training and 20% for the test. To be able to avoid bias/variance tradeoff, all models were trained on the training dataset and tested on the test dataset. Such operation increases the chances of creating a model that does not only work on the data at hand but also on the data which will be used in the future.

11.5.2.1 Linear Regression

To start the analytical study, firstly, the Eq. 11.1 of regression is used. Based on Eq. 11.1, it is easy to generate Eq. 11.9 which shows the error (residuals). Regression aims to draw the best line among all possible lines, which passes through most observations and aiming at minimizing the total residuals. This minimization problem is displayed on Eq. 11.10. Once this line is drawn in the data, the prediction becomes much easier as a formula will be generated. Afterward simply putting the values of all predictors will end up in a calculated response value.

In order to make valuable analysis and highly accurate predictions, there are some important problems which need to be fixed, if existed. Those problems are:

- Multicollinearity Problem (multiple linearly related variables)
- Heteroscedasticity Problem (changing variation)
- Selection Bias (linear or non-linear)

The collected data has 30 parameters (dimensions) and 26,000 observations. However, in most cases, some of the parameters are probably correlated with one another. In such a situation, this causes serious multicollinearity problems. For instance, if one would like to collect gender information and there are only two options: female and male. For that case, if one defines two separate parameters like female which is set to true if the observant is female, and false if it's a male and exactly the opposite for another variable called male then these two variables are highly negatively correlated to each other. Since one can only be either male or female. In such a problem of multicollinearity, two parameters should be reduced to one. For example, an arbitrary value of true to define female and false to define male can be used. To assess, the correlation between pairwise parameters, the Eq. 11.11 is used in our analysis. If high negative or positive correlation values are calculated (anything above 0.5 as absolute value) between the two variables, the model that has higher R-Squared (explains how close the data are to fitted line) was chosen using Eq. 11.15. Moreover, the highly inflated variables that have a value of more than five were identified using the Eq. 11.16, and removed from the model. After fixing the multicollinearity problem, the total number of independent parameters was reduced to 15 uncorrelated dimensions (also known as perpendicular parameters) and one response (dependent) variable. However, the total portion of the variance explained, which can be calculated by Eq. 11.15, shows that there is no drop even after filtering total parameters to almost half. Usage of fewer parameters in the model without sacrificing the R-squared value is beneficial, as those models are simpler to calculate and understand. In the studied dataset, the heating and cooling actions were collectively exhaustive, meaning the machine for such a section is either in cooling or in heating mode. There is no other available condition such as the idle state. If the heating parameter was true, the cooling parameter was false and vice-versa. As these two parameters were negatively and highly correlated, removing either one will fix the problem. The same applies to other parameters that were used in the energy consumption calculations. For example, the torque of the engine, the steam flow rate, etc. is multiplied with a different set of weighting to calculate the total energy consumption as a consumption factor. Since those factors are represented linearly in the calculated cost field, the existence of the fields in the data was not needed, therefore removed from the calculations.

To avoid the second problem of heteroscedasticity, the variability of variance requires to remain constant throughout the range of the values of other parameters. If the heteroscedasticity does not occur, it is called homoscedastic. In order to measure if the data is homoscedastic, the Eq. 11.14 is used and there was no need to make any changes to the data as the observations were satisfying homoscedasticity.

Lastly, to choose linear or non-linear fit of the regression models, different options (linear vs cubic and quadratic) were tested and comparisons between different regressions were calculated using Eq. 11.15. After the above-mentioned operations, the remaining parameters which were selected to generate the model were identified. Those parameters can be categorized in below groups,

- Engine parameters (e.g., temperature, torque, speed, etc.)
- Product parameters (e.g., enter-exit temperature, mixture, flow, etc.)
- Operational parameters (e.g., cooling, heating, mixing, etc.)

The output model with the above variables, which has the lowest mean square prediction error rate, is calculated using Eq. 11.13 is shown in Fig. 11.3. Those parameters have a direct effect on the energy spending calculations, as their change causes the highest impact on the energy spending.

All of the significant parameters are shown in Table 11.1, followed by residual summary on Table 11.2.

The residual standard error rate of 1.55 on 21079 degrees of freedom is calculated. The model has a 41% R-Squared value of total explained, which is calculated using Eq. 11.15.

The response variable that will be calculated is the cost per unit of production. The cost parameter consists of various sub-parameters with certain weights based on their effect which is calculated and shared by the investigated company. The parameters linearly affecting the total consumption are averaged to total consumption is displayed below;

- Electric consumption
- Chiller consumption
- Water flow rates
- Water vapor consumption

As a summary, the model consists of the abovementioned input and output parameters. The normalized weights are taken for each output factor to calculate the average cost per unit of production. For example, the water flow, which not only

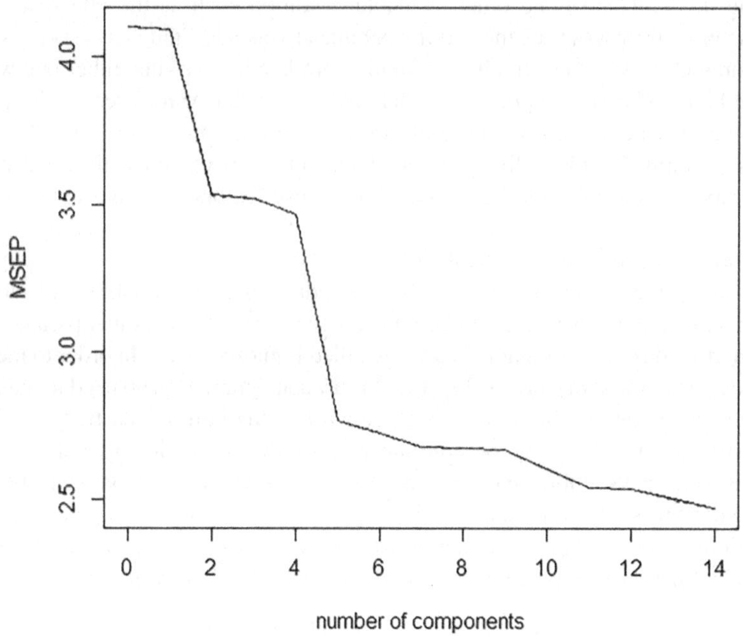

Fig. 11.3 MSPE of components

Table 11.1 Linear regression summary table

Coefficient	Estimate	Std. error	*T*-value	*P*-value
(Intercept)	2.87e+01	1.17e+00	24.6	< 2e−16 *
Parameter 1	1.44e+01	1.41e+00	10.2	< 2e−16 *
Parameter 2	2.98e−03	5.21e−05	57.1	< 2e−16 *
Parameter 3	−9.88e−02	3.09e−03	−31.9	< 2e−16 *
Parameter 4	1.05e−02	6.86e−04	15.3	< 2e−16 *
Parameter 5	1.44e−02	3.52e−03	4.1	4.49e−05 *
Parameter 6	−3.96e−02	2.25e−03	−17.6	< 2e−16 *
Parameter 7	1.20e−01	5.76e−03	20.9	< 2e−16 *
Parameter 8	6.23e−02	4.89e−03	12.8	< 2e−16 *
Parameter 9	−7.80e−01	9.96e−02	−7.8	4.87e−15 *
Parameter 10	−8.53e−01	4.59e−02	−18.6	< 2e−16 *
Parameter 11	4.95e−01	7.82e−02	6.3	2.42e−10 *
Parameter 12	1.37e+00	3.41e−02	40.2	< 2e−16 *
Parameter 13	1.05e+00	2.74e−02	38.2	< 2e−16 *

"*" indicates the significant coefficient, with value less than 0.001

Table 11.2 Summary of residuals

Min	1Q	Median	3Q	Max
−8.17	−0.97	−0.05	0.95	9.27

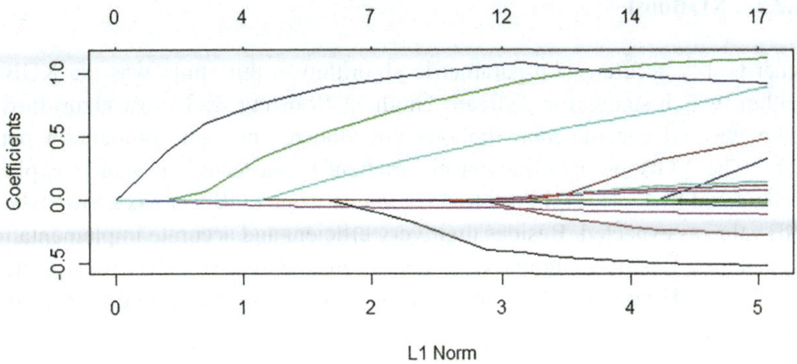

Fig. 11.4 Lasso coefficient

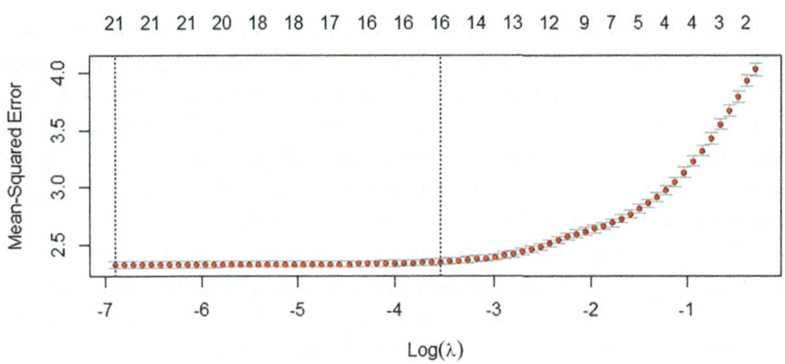

Fig. 11.5 Lasso mean squared error

holds the amount of water used but also the energy consumed to pump the water to the machine, is calculated within that parameter. Electric consumption consists of multiple engine consumption rates, and so forth. Therefore, a normalized weight and unit consumed merged into cost variable and used in the model as a response parameter.

To make the parameter selection, other approaches like ridge regression and lasso were used. Both algorithms aim to reduce parameters and offer the most significant ones to be used. The lasso generates better results in terms of RMSE and is calculated using Eq. 11.3. The non-significant parameters shrink to zero and the Fig. 11.4 shows how each parameter reacts to the algorithm.

In order to avoid much smaller explained values, and since the error doesn't get lower after around 1416 parameters, the cutting happened on that level. The change in the error rate while increasing the parameters is shown in Fig. 11.5. On top of lasso regression which comes in third place, the second most effective algorithm was XGBoost.

11.5.2.2 XGBoost

Another highly accurate non-parametric algorithm in this study was the XGBoost algorithm, which stands for eXtream Gradient Boosting. XGBoost algorithms are well known and became more famous after the recent high competition performances. The XGBoost is an advanced gradient boost technique that is explained and calculated by Eq. 11.4. In our RMSE comparison, XGBoost has a lower (better) RMSE value as expected. Besides their very efficient and accurate implementations, the calculation time is generally very low. Therefore, applying XGBoost is critical and beneficial. However, in this study, as a soloist algorithm, the best performance was obtained with Random Forest.

11.5.2.3 Random Forest

The lowest RMSE value on test data, consequently, the most accurate result was obtained by the Random Forest (RF) algorithm in the studied dataset. RF algorithm, as shown in Eq. 11.5 randomly selects parameters with replacement and grows separate decision trees. As each tree has a significant and non-significant combination of parameters, the prediction accuracy differs. Regular decision tree algorithms normally select the most significant parameters which end up getting highly correlated among each other and typically have low prediction accuracy on the test dataset. Since this approach helps avoid the over-fitting problem, the result on test data is generally higher than the regular decision tree approaches. The prediction vs actual target values are shown in Fig. 11.6. As both RF and XGBoost are non-parametric algorithms, much less assumption and parameter engineering is required by them. As opposed to regression, the parameters are not selected before the application and algorithms choose or eliminate the one based on certain rules. Even though individual algorithms can provide very low RMSE values, it is a good practice to always check ensemble options that take multiple algorithms as input and return a new outcome.

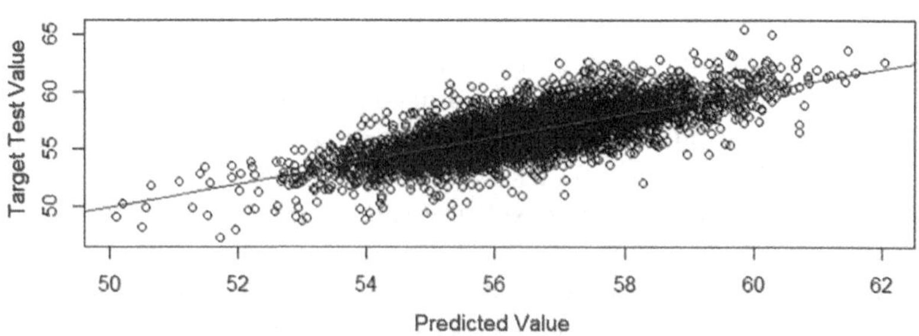

Fig. 11.6 Random forest prediction accuracy

11.5.3 Ensemble Model

As shown in Table 11.3, the study showed close RMSE values for most algorithms and some performed better than others individually. But besides their performances, an ML study cannot be completed without an ensemble model application. In the analysis, both simple average and weighted average techniques were used. The weighted average ensemble technique gave the best RMSE value compared to all soloist algorithms. The RMSE for weighted average was calculated as "1.37". Since random forest has "1.38", each option can be used to predict the energy consumption almost with the same accuracy. However, the ensemble weighted average technique was selected as the preferred method, as it gave the most accurate output and calculation time didn't differ much from the configuration in hand.

To calculate the weighted average ensemble, firstly all algorithms were tested one by one. Later on, the most accurate top three algorithms (with the lowest RMSE values) were selected. A new linear regression model was applied with target test value as response and top three algorithms' predictions as the independent parameters. As the model was generated, the coefficients for top algorithms, in our case Random Forest, XGBoost, and Lasso models' predictions were weighted and standardized to their totals as shown in Eq. 11.17. The new prediction value was calculated using Eq. 11.18. Afterward, the RMSE was calculated based on the weighted average prediction value and the actual test target value.

11.5.4 Results

The cost of energy spending, for each unit produced in the collected data, has an average (mean) of 56 units (not actual dollars spent but a calculated value). The minimum spending for the same product can be as less as 47 units and the highest production cost in terms of energy has 66 units of cost. To compare the outputs of different models, XGBoost, SVM, Decision Trees, Random Forest, Neural-Net, Multiple Regression, Lasso, and ensemble models are listed in Table 11.3. The corresponding RMSE values calculated by Eq. 11.12 in ascending order show the

Table 11.3 Models with RMSE values

Model	RMSE
Weighted average ensemble	1.3668
Random forest	1.3845
Simple average ensemble	1.3901
XGBoost	1.4752
Lasso regression	1.5112
Linear regression	1.5467
Support vector machines	1.5635
Neural network	1.5993
Decision tree	1.6807

test prediction accuracy. The smaller test RMSE value explains the algorithm trained well on the data and test results are of higher inaccuracy. As shown in Table 11.3, the difference in RMSE values range from approximately “1.37” to “1.68”. The weighted average ensemble technique, which combines the top three algorithms, ends up with a top RMSE value of “1.37” which is selected as the main algorithm.

As a result, when the production parameters are pre-configured for the selected product, instead of the average 56 units, the outcome can be reduced to 53.5 units, which is a direct energy saving of 5.4% compared to the average consumption from a single extruder under 95% accuracy. If the highest consumption rate is taken, the reduction can be maximized to 18%. As the extrude machine in this factory consumes 35–45% of the total cost reduction, the savings in actual dollars spent will be critical.

11.6 Conclusion

The study took place at a multinational food producer to reduce the energy consumption of a single extruder machine within the production line. The machine is normally configured and parameters are manually updated by operator inputs. Based on the machine learning technique, multiple algorithms were tested and the weighted ensemble of Random Forest, XGBoost, and Lasso jointly fit the problem most accurately. Based on the given future production parameters, predictions were made with acceptable product quality ranges. For the provided configuration, operators can view the impact of their newly set parameters on energy consumption. When the consumption is above the reasonable threshold, the operator can update the parameters to achieve better results. As the system runs in real-time, the impact on energy consumption can be predicted as soon as the parameters are set or changed. The analyzed extruder machine consumes around 35–45% of all energy used in the factory. IOT based ML study shows a 5–6% reduction in energy consumption when compared to the average energy spending of the same machine. Since there is 24/7 full capacity production, the potential saving is ultimate. This study contributes to merge theoretical analytical tools to real life production environments to maximize efficiency and reduce energy spending. Moreover, the study tests IIOT’s direct impact on production for firms to identify the potential ROI. As the cost of IOT systems with ML capabilities includes multiple aspects, actual figures and exact ROI calculations were out of the scope of this paper.

The future works can be as follows;

- Improve machine learning accuracy
- Implement IOT solution to configure machine automatically
- Extend single analysis to broader product mix
- Extend energy saving throughout the full production line
- Implement ROI calculation for IOT systems

References

1. Aqlan, F., Ramakrishnan, S., & Shamsan, A. (2017). Integrating data analytics and simulation for defect management in manufacturing environments. In *2017 winter simulation conference (WSC)*. IEEE. https://doi.org/10.1109/WSC.2017.8248104
2. Banerjee, S., & Chandra, M. G. (2019). A software framework for procedural knowledge based collaborative data analytics for IoT. In *2019 IEEE/ACM 1st international workshop on Software Engineering Research & Practices for the Internet of Things (SERP4IoT)*. IEEE. https://doi.org/10.1109/SERP4IoT.2019.00014
3. Dai, H.-N., Wang, H., Xu, G., Wan, J., & Imran, M. (2019). Big data analytics for manufacturing internet of things: Opportunities, challenges and enabling technologies [arXiv: 1909.00413]. *Enterprise Information Systems*, 1–25. https://doi.org/10.1080/17517575.2019.1633689
4. Evans, R., & Boreland, M. (2018). Multivariate data analytics in PV manufacturing four case studies using manufacturing datasets. *IEEE Journal of Photovoltaics, 8*(1), 38–47. https://doi.org/10.1109/JPHOTOV.2017.2778571.
5. Feldkamp, N., Bergmann, S., & Strassburger, S. (2015). Visual analytics of manufacturing simulation data. In *2015 winter simulation conference (WSC)*. IEEE. https://doi.org/10.1109/WSC.2015.7408215
6. Ganga, D., & Ramachandran, V. (2018). IoT-based vibration analytics of electrical machines. *IEEE Internet of Things Journal, 5*(6), 4538–4549. https://doi.org/10.1109/JIOT.2018.2835724
7. Gauchi, J.-P., & Chagnon, P. (2001). Comparison of selection methods of explanatory variables in PLS regression with application to manufacturing process data. *Chemometrics and Intelligent Laboratory Systems, 58*(2), 171–193. https://doi.org/10.1016/S0169-7439(01)00158-7
8. Ghosh, A. M., & Grolinger, K. (2019). Deep learning: Edge-cloud data analytics for IoT. In *2019 IEEE Canadian conference of electrical and computer engineering (CCECE)*. IEEE. https://doi.org/10.1109/CCECE.2019.8861806
9. Glen, S. (n.d.). *Statistics how to website*. https://www.statisticshowto.com
10. Groger, C., Niedermann, F., Schwarz, H., & Mitschang, B. (2012). Supporting manufacturing design by analytics, continuous collaborative process improvement enabled by the advanced manufacturing analytics platform. In *Proceedings of the 2012 IEEE 16th international conference on computer supported cooperative work in design (CSCWD)*. IEEE. https://doi.org/10.1109/CSCWD.2012.6221911
11. Jain, S., Lechevalier, D., & Narayanan, A. (2017a). Towards smart manufacturing with virtual factory and data analytics. In *2017 winter simulation conference (WSC)*. IEEE. https://doi.org/10.1109/WSC.2017.8248023
12. Jain, S., Shao, G., & Shin, S.-J. (2017b). Manufacturing data analytics using a virtual factory representation. *International Journal of Production Research, 55*(18), 5450–5464. https://doi.org/10.1080/00207543.2017.1321799
13. James, G., Witten, D., Hastie, T., & Tibshirani, R. (2013). *An introduction to statistical learning* (Vol. 112). Springer.
14. Jun, C., Lee, J. Y., & Kim, B. H. (2019). Cloudbased big data analytics platform using algorithm templates for the manufacturing industry. *International Journal of Computer Integrated Manufacturing, 32*(8), 723–738. https://doi.org/10.1080/0951192X.2019.1610578
15. Kibira, D., Hatim, Q., Kumara, S., & Shao, G. (2015). Integrating data analytics and simulation methods to support manufacturing decision making. In *2015 winter simulation conference (WSC)*. IEEE. https://doi.org/10.1109/WSC.2015.7408324
16. Kozjek, D., Vrabič, R., Rihtaršič, B., Lavrač, N., & Butala, P. (2020). Advancing manufacturing systems with big-data analytics: A conceptual framework. *International Journal of Computer Integrated Manufacturing, 33*(2), 169–188. https://doi.org/10.1080/0951192X.2020.1718765

17. Krumeich, J., Jacobi, S., Werth, D., & Loos, P. (2014). Big data analytics for predictive manufacturing control - a case study from process industry. In *2014 IEEE international congress on big data*. IEEE. https://doi.org/10.1109/BigData.Congress.2014.83
18. Kumaraguru, S., Kulvatunyou, B., & Morris, K. (2014). Integrating real-time analytics and continuous performance management in smart manufacturing systems. In *IFIP advances in information and communication technology*. https://doi.org/10.1007/978-3-662-44733-822
19. Lantz, B. (2013). *Machine learning with r*. Packt Publishing Ltd.
20. Moyne, J., & Iskandar, J. (2017). Big data analytics for smart manufacturing: Case studies in semiconductor manufacturing. *PRO, 5*(4), 39. https://doi.org/10.3390/pr5030039
21. Qin, J., Liu, Y., & Grosvenor, R. (2017). Data analytics for energy consumption of digital manufacturing systems using Internet of Things method. In *2017 13th IEEE conference on automation science and engineering (CASE)*.
22. Ramsey, F., & Schafer, D. (2012). *The statistical sleuth: A course in methods of data analysis*. Cengage Learning.
23. Shao, G., Shin, S.-J., & Jain, S. (2014). Data analytics using simulation for smart manufacturing. In *Proceedings of the winter simulation conference 2014*. IEEE. https://doi.org/10.1109/WSC.2014.7020063
24. Shin, S.-J., Woo, J., & Rachuri, S. (2014). Predictive analytics model for power consumption in manufacturing. *Procedia CIRP, 15*, 153–158. https://doi.org/10.1016/j.procir.2014.06.036
25. Shrouf, F., & Miragliotta, G. (2015). Energy management based on Internet of Things: Practices and framework for adoption in production management. *Journal of Cleaner Production, 100*, 235–246. https://doi.org/10.1016/j.jclepro.2015.03.055
26. Sundaram, R. B. (n.d.). Analytics vidhya. In *2019 8th international conference on*. IEEE-guide-to-understand-the-math. https://doi.org/10.1109/ICITM-behind-xgboost.2019.8710673
27. Tao, F., Bi, L., Zuo, Y., & Nee, A. Y. C. (2017). A Cooperative co-evolutionary algorithm for large-scale process planning with energy consideration. *Journal of Manufacturing Science and Engineering, 139*(6), 061016. https://doi.org/10.1115/1.4035960
28. Vater, J., Harscheidt, L., & Knoll, A. (2019). *Smart manufacturing with prescriptive analytics*.
29. Zuo, Y., Tao, F., & Nee, A. Y. C. (2018). An Inter net of things and cloud-based approach for energy consumption evaluation and analy sis for a product. *International Journal of Computer Integrated Manufacturing, 31*(4–5), 337–348. https://doi.org/10.1080/0951192X.2017.1285429

Part V
Smart Cities

Chapter 12
Hype: A Data-Driven Tool for Smart City Profile (SCP) Discrimination

Yolène Berrou, Eddie Soulier, Philippe Calvez, Babiga Birregah, and Francis Rousseaux

Abstract Over the last decades, Information and Communication Technology (ICT) has become more and more present in city systems as an answer to tackle urbanization issues, leading to the concept of a smart city. However, local authorities may have difficulties handling the complexity of developing pertinent smart city use cases efficiently and matching their strategy to reach urban sustainability. We propose a smart city profile computation as a first step to identify the city's profile, to better know its situation, and whether it is coherent with the followed strategy. The smart city profile is analyzed based on simplicial complexes, topological objects. We present a tool named Hype developed to compute simplicial complexes and Q-analysis, a method that analyzes the structural connectivity of the simplicial complexes. Finally, an application is presented to compute the profile of the city of Paris based on data publicly available.

Keywords Smart city · Data-driven · Simplicial complex · Urban sustainability

---

Y. Berrou (✉)
Informatic and digital society laboratory (LIST3N), University of Technology of Troyes, Troyes, France

Computer Science and Artificial Intelligence Laboratory, ENGIE Lab CRIGEN, Stains, France
e-mail: Yolene.BERROU@engie.com

E. Soulier · B. Birregah
Informatic and digital society laboratory (LIST3N), University of Technology of Troyes, Troyes, France

P. Calvez
Computer Science and Artificial Intelligence Laboratory, ENGIE Lab CRIGEN, Stains, France

F. Rousseaux
CReSTIC EA 3804, team MODECO, University of Reims Champagne Ardenne, Reims, France

G. Kayakutlu, M. Ö. Kayalica (eds.), *Decision Making Using AI in Energy and Sustainability*, Applied Innovation and Technology Management,
https://doi.org/10.1007/978-3-031-38387-8_12

12.1 Introduction

Cities are complex systems that face multiple issues. Since 2018, more than half of the world's population has been living in urban areas, and the United Nations predicts that this number will reach almost 70% by 2050 [1]. The increase of the urban population leads to a rise in the urbanization process along with air and water pollution, traffic congestion, security [2], and/or the increasing energy demand [3]. Over the last decades, Information and Communication Technology (ICT) has become more and more present in city systems as an answer to tackle urbanization issues [2] or to enhance the city and the quality of life [4, 5]. This approach leads to the smart city concept, which does not have a unique definition [6–8]. We define a smart city as a city developing solutions using (more or less) ICT technologies to reach urban sustainability [9].

Local authorities may have difficulties handling the complexity of the city and succeeding in creating smart city projects in an efficient way [10]. Identifying pertinent projects to develop that match their strategy appears difficult for local authorities. Three workshops have highlighted this[1] about smart cities projects strategies from the local authorities' (Paris, Nice Metropolis, La Rochelle, Lyon Metropolis, and Prats-de-Mollo-la-Preste) and the providers' (IBM, Enogrid, GFI group, GRDF) points of view. These actors pointed out the identification of pertinent use cases as an issue. Traditionally, the use cases are part of the software engineering domain and allow to model the functional and technical aspects. According to Bittner and Spence, "A use case describes how an actor uses a system to achieve a goal and what the system does for the actor to achieve that goal." [12]. Those use cases are a functional description of a system to be conceived (level 0), considering the users (level 1), and that is nowadays more and more completed by a strategic and economic analysis (level 2). However, this last level still needs to be more integrated into the use cases.

We hypothesize that use cases must be aligned with the city strategy to be pertinent. In this context, the first step would consist of precisely identifying the city's profile to better know its situation and coherently with the strategy. Multiple heterogeneous features characterize a smart city profile (SCP): projects developed to involve many actors can be related to multiple domains or even multiple levels.

This chapter aims to provide a method to characterize the profile of a smart city, based on the projects it implemented, to help identify pertinent use cases, and to measure the city's coherence strategy concerning the projects already developed. We propose a computation method to determine the SCP based on simplicial complexes.

---

[1]These workshops were conducted by the French National Agency of Territories' Cohesion (ANCT) and the University of Technology of Troyes (UTT), in collaboration with ENGIE Lab CRIGEN CSAI, the University of Le Havre, and the Chronos Group. More information on the workshops can be found in [11].

The first section presents the methodology of the SCP characterization. Section 12.2 offers the computation method and the tool developed to characterize an SCP. Finally, an application is presented in Sect. 12.3.

12.2 Methodology

As aforementioned, we hypothesize that use cases must be aligned with the city strategy to be pertinent and that the smart city first has to be described. However, a great diversity of smart cities develops various projects involving multiple actors that can be related to multiple domains or even to multiple levels. Therefore, there is no unique model that emerges to describe smart cities. A data-driven approach seems more relevant than a model-driven approach to give an accurate description of a smart city. Indeed, the idea is not to provide a generic model at a high level of abstraction but to identify the characteristics of the smart city concerning the projects developed. Given the complexity of what needs to be characterized, this approach should be completed to increase its robustness. Thus, we propose four ideal-type profiles. The computed profile could then be positioned regarding these ideal-types profiles.

This profile computation is the first step of a use case methodology presented in [11] that was inspired by the use case modeling methodology proposed by Gottschalk et al. [13], who applied the standard of the International Electrical Commission on use case methodology for complex systems [14] to smart grids, using the Smart Grid Architecture Model (SGAM) [15]. We proposed a Smart City Model to include strategic layers to the use cases.

12.2.1 Modeling Smart City Profiles (SCP)

Mora and Deakin made a deep analysis of the state of the art of smart cities literature, assuming that "Smart cities are those cities in which a platform of ICT solutions is deployed to meet urban sustainability priorities." [9, p. 19]. The bibliometric analysis they conducted led to identifying two main groups: techno-centric and advocates a more holistic approach to smart city development – both being ICT-driven. They also identified "four main dichotomies" in the strategic principles that smart cities can follow. Each dichotomy can be split into two "hypotheses". Then they analyzed and classified the activities into categories for four "leading examples" of smart European cities: Amsterdam, Barcelona, Helsinki, and Vienna, to identify the "strategic principles" followed.

The two main ICT-driven groups identified by Mora and Deakin in the literature [9] (techno-centric and holistic) constitute our first two SCP.

Since the focus on ICT technologies seems too restrictive, we added two other types. The first is characterized by its social aspect. Green [16] advocated that a city can be smart even without systematically using technology to solve its problems. The social profile can be described by its propensity to promote democratic values, considering the heterogeneity of the citizens to address social needs in line with the specific population needs. Problems are regarded as a hole within this profile, and technology may be only one possible solution [11].

The second profile is between the social and holistic profiles. Cities associated with this profile consider social and human issues as essential and view technology as a tool to address them.

In conclusion, we propose four ideal-type profiles from techno-centric to social-centric: techno-centric (P1), techno-social-centric (holistic) (P2), socio-technical-centric (P3), and social-centric profiles (P4).

To characterize the profiles, we decomposed the categories identified by [9] and enriched them with features identified from the workshops (see introduction) and from the book of Green [16]. Then, these features were linked to hypotheses and then to each profile. We added a 3-helix model and associated it with the techno-social-centric and socio-technical-centric profiles. Table 12.1 shows how features are related to each profile. Round markers represent a needed feature for the profile. Triangle markers represent optional links, which means that these features may be associated with the profile but are not required. They are computed in the same manner as the others, the distinction being made during the result analysis. Finally, star markers mean that only a few of the features can be considered for the profile. For example, as the techno-centric profile (P1) is characterized by a mono-dimensional logic, only one or two application domains may be implemented.

12.2.2 Computing Smart City Profiles (SCP)

An SCP is described by heterogeneous features, highlighted by the activities the city develops. Hereafter, we use "initiative" instead of "activity" because of its comprehensive sense. An initiative is anything a city does to achieve something (such as developing an application, involving citizens, making a roadmap, etc.).

Knowing that profiles can share hypotheses, these features can be linked to one or several hypotheses. Thus, we need a computation method to consider the features heterogeneity and highlight the structure of multiple belonging of initiatives to features to characterize the SCP.

We propose a topological approach based on simplicial complexes, able to consider the multidimensionality of relations between features and initiatives. It is completed with Hype, a tool developed for the computation.

Table 12.1 Features with their links to each profile: techno-centric (P1), techno-social-centric (P2), socio-technical-centric (P3), and social-centric (P4)

P1	P2	P3	P4	Features
●	●	●	●	Actor: government
●	●	●	▲	Actor: industrial
▲	●	●	▲	Actor: research
		▲	▲	Actor: civil society
●	●	●	▲	Development of an application or a service
★	●	●	▲	Application domain: transport and mobility
★	●	●	▲	Application domain: energy
★	●	●	▲	Application domain: waste
★	●	●	▲	Application domain: water
★	●	●	▲	Application domain: housing
★	●	●	▲	Application domain: buildings
★	●	●	▲	Application domain: government
★	●	●	▲	Application domain: environment
★	●	●	▲	Application domain: social
★	●	●	▲	Application domain: quality of life
			▲	No development of service or application
●	●	●	▲	Development of a data platform or infrastructure
	★	●	●	Citizen participation at the beginning
	★	●	●	Citizen participation in the middle
	★	●	●	Citizen participation at the end
●				No citizen participation
	▲	●	●	Promote citizen engagement
	▲	●	●	Promote public awareness and digital literacy
●	▲	●	●	Stimulate user-driven innovation and community-led urban development
●	●	▲		Improve understanding of smart city development and the benefits of an ict-driven approach
	▲	●	●	Other community building
	●	●		No community building
	▲	●	●	Roadmap, program, action plan
	▲	●	●	Workgroups that manage the general course of the smart city development
	▲	●	●	Evaluation measures of results and progress
	▲	●	●	Other strategic framework
●				No strategic framework
			●	Address complex problems
		●	●	Address social needs and advance policy
		●	●	Promote democratic values
		●	●	Use of public data
			●	Make public data actionable
			●	Taking into account the ones in need

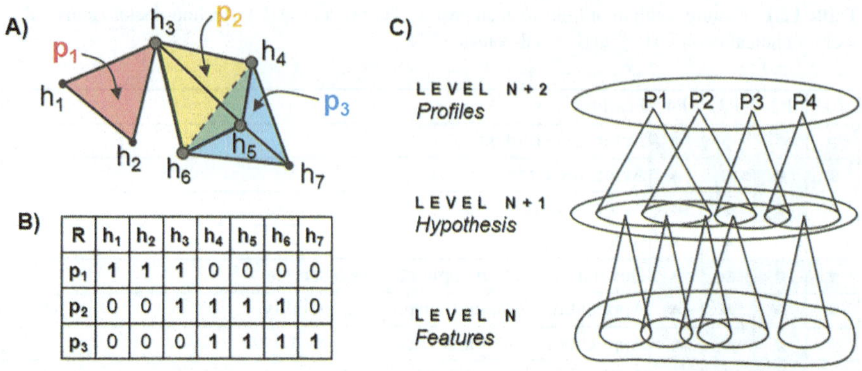

R	h_1	h_2	h_3	h_4	h_5	h_6	h_7
p_1	1	1	1	0	0	0	0
p_2	0	0	1	1	1	1	0
p_3	0	0	0	1	1	1	1

Fig. 12.1 (**a**) A simplicial complex composed by three simplices; (**b**) the incidence matrix of the simplicial complex with rows as simplices and columns as vertices; (**c**) schematic representation of the three hierarchical levels in SCP

12.2.2.1 Simplicial Complexes to Study Connectivity

Simplicial complexes allow studying n-ary relations among sets. They enable the multidimensionality of relations among entities to be calculated and are used in many areas, such as social systems [17], chess [18], or road traffic systems [19].

Simplices can be understood as generalizing the notion of triangles in n-dimension. Considering a set $X = \{x_1, x_2, \ldots, x_n\}$ of vertices, any subset $Y = \{y_1, y_2, \ldots, y_m\}$ of X is a simplex. Its dimension p is given by its number of vertices minus one and is written p-simplex [20]. For example, a point is a 0-simplex, a line containing two vertices is a 1-simplex, a triangle is a 2-simplex, etc.

Simplicial complexes can be viewed as simplices glued to each other, i.e., sharing vertices. The simplicial complex K is written $K_Y(X, R)$ with R the relation between the sets Y and X. The opposite relation can also be analyzed, namely, $K_X(Y, R^{-1})$, called the conjugate complex.

Figure 12.1a shows a simplicial complex composed of three simplices: one 2-simplex (p_1) and two 3-simplex (p_2 and p_3), and Fig. 12.1b represents its associated incidence matrix with rows as simplices and columns as vertices.

Atkin developed a method to study the structural characteristics of simplicial complexes named Q-analysis [17, 21]. He also introduced the concepts of hierarchy and cover set. Hierarchy does not imply any meaning of superiority instead of levels of classification. Considering only two levels in the hierarchy, the $(N + 1)$-level set acts as a cover of the set at the N-level. Elements of the $(N + 1)$-level set are subsets of the N-level set [18, 20].

In our case, we want to analyze the structural connectivity between a set F of features and a set I of initiatives. The relation between the features and the profiles is made with the 3-level hierarchy where features (F) are at the N-level, hypotheses are at the $(N + 1)$-level, and profiles are at the $(N + 2)$-level (Fig. 12.1c).

Table 12.2 Results of the Q-analysis

Dimension	Components	Number of components
$q = 3$	$\{p_2\}$; $\{p_3\}$	2
$q = 2$	$\{p_2, p_3\}$; $\{p_1\}$	2
$q = 1$	$\{p_2, p_3\}$; $\{p_1\}$	2
$q = 0$	$\{p_1, p_2, p_3\}$	1

The Q-analysis consists in analyzing the structural connectivity of simplicial complexes for each dimension of the simplicial complex (from zero to the highest dimension in the complex) [17]. This structural connectivity introduces the notion of q-nearness: two simplices are q-near if they share p vertices, with $q = p - 1$.

For example, let $H = \{h_1, h_2, \ldots, h_7\}$ be a set of hobbies and $P = \{p_1, p_2, p_3\}$ a set of people. There is a relation R between p_i and h_j if the person p_i likes the hobby h_j. It is represented with a 1 in the incidence matrix (Fig. 12.1b). The relation between H and P forms a simplicial complex noted $K_P(H, R)$, composed of three simplices p_1, p_2, p_3.

Table 12.2 gives the results of this Q-analysis. We start with the highest dimension, which is 3 (because p_2 and p_3 have four vertices). At $q = 3$, two simplices exist: p_2 and p_3. However, they are not connected to each other with a dimension of 3, because they are only 2-near (sharing a 2-dimensional face). Thus, they are connected in separated components. At $q = 2$, p_2 and p_3 become connected in the same component $\{p_2, p_3\}$. The simplex p_1 appears in the analysis because it is of dimension 2 but is not in the same component as p_2 and p_3, because p_1 is only 0-near with p_2 (sharing h_3). At $q = 1$, nothing changes, but at $q = 0$ all the simplices are connected in the same component. The simplicial complex is composed of a chain of connected simplices, where each simplex shares at least one vertex with another simplex.

Finally, one important measure that will be used in Sect. 12.3 is the structure vector Q which represents the number of components for each q-dimension of analysis.

12.2.2.2 Hype, a Tool to Compute Simplicial Complexes

To perform calculations on simplices, an application named Hype was co-conceived by the UTT laboratory and ENGIE Lab CRIGEN CSAI and developed by Data Nostra. The main objective of Hype is to provide a model of the simplicial complexes and analyze their structures and dynamics. The simplicial complexes are represented as a directed graph, where vertices and simplices are distinguished according to the direction of the arcs. Three versions were developed. This new version provides more structural metrics and introduces the notion of traffic that will not be developed here.

Hype is composed of three parts: one giving an overview of the data and the possibility to edit it; one for structural analysis ("backcloth"); and one for the dynamic analysis ("traffic").

The first step consists in choosing the direction of the relationship we want to analyze before performing a Q-analysis. The algorithm used consists of three steps: (1) the dimensions are computed for each simplex; (2) the proximity is computed for each simplex by counting the number of vertices in common; (3) for each q dimension and each simplex, the algorithm determines which simplices can be reached. Thus, each simplex has a set with its associated simplices, allowing the components (that is, groups of connected simplices at dimension q) to be found if there are disjoint sets.

The results of the Q-analysis are provided through a visualization of the components, the list of the components for a given q-dimension, and the structure vector – also represented with a chart line.

The visualization is called a q-complex [20]. It shows the simplices existing at a given dimension q. There is a link between simplices if they are at least q-near. Vertices can be accessed by clicking on a simplex. This visualization allows a representation of the q-nearness of the simplicial complex and makes a graph easier to interpret. However, vertices are not visually represented, thus we miss some information – for example, if a simplex is a face of another or if there are holes in the structure.

12.3 Application

This section presents an example of the profile computation for the city of Paris with the tool Hype. We chose Paris because it participated in the workshops (see the introduction) and has publicly available data on their project.

The data collected to build the incidence matrix consisted of project descriptions available on three websites: (1) the DataCity program[2] website [22] presented during the workshops, (2) the city of Paris' website [23, 24], and (3) the "*Budget participatif*" website [25].

The DataCity program is an innovation program created in 2015 by Numa and the city of Paris, which aims to enhance collaboration between various stakeholders to develop and experiment with innovative solutions using urban data to make the city smarter. We enriched these data with the city of Paris' website regarding the strategic roadmap or the development of various services by the city itself. Finally, the "*Budget participatif*" website presents projects selected by the citizens and developed by the city. Due to the vast heterogeneity of these projects, we only considered the projects with the "smart and digital city" tag.

[2] Projects' descriptions were available in 2020. However, only 34 projects of Paris are still available to date.

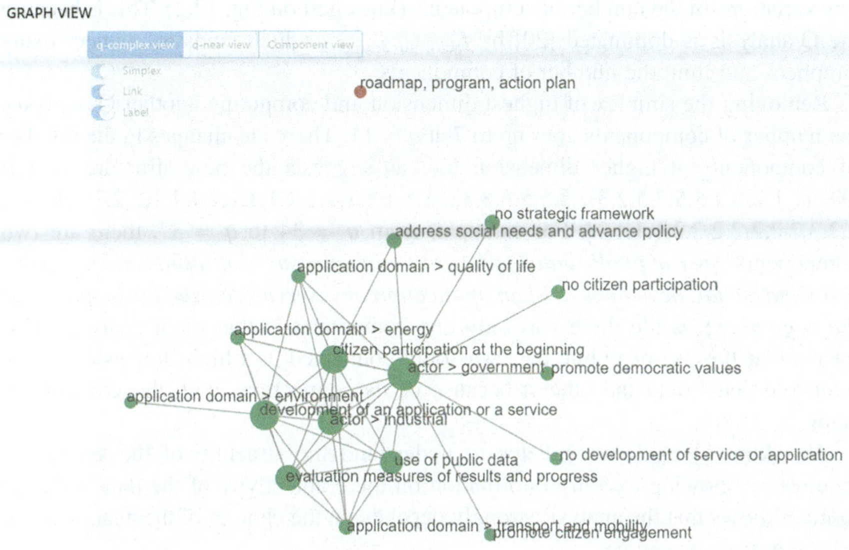

Fig. 12.2 Hype's screenshot of the Q-analysis result at $q = 11$. Simplices are represented as the nodes and are connected if they share at least 12 vertices. There are two disconnected components (green and red)

In total, 67 initiatives related to Paris were analyzed: 40 from the DataCity program, 17 from the website of the city of Paris, and 8 from the website "*Budget participatif*". Two initiatives were added: one for the DataCity program itself and the "*Budget participatif*" program. All the projects were considered as real city initiatives to have enough data, even though not all projects have been implemented in the city, some of them remaining at the proof-of-concept stage.

The data used for the computation comes from an analysis of information publicly available on websites. Thus, it may not completely represent reality, as initiatives or information may be missing, which is particularly the case for the features relating to community building and strategic framework. Moreover, the relation to the features was made from the descriptions of the projects on these websites, and thus may depend on the interpretation and may be incomplete, as the details of the projects are not provided.

In the end, the incidence matrix contains 38 rows or simplices (features) and 67 columns or vertices (initiatives).

Before characterizing the SCP, we can explore the simplicial complex. The simplex with the highest dimension is $f_{\text{actor:government}}$ with a dimension of 61, meaning that there are only five initiatives where the government is not involved over the 67 analyzed. Its dimension is so high that it may tend to connect almost every other simplices, thus hiding some existing gap in the connectivity.

The study of the structure vector Q indicates high connectivity in the simplicial complex. The number of components is limited (up to three), and there are only a few variations of the number of components (line chart on Fig. 12.2). This is because the Q-analysis is dominated [20] by $f_{\text{actor:government}}$, which tends to connect many simplices and limit the number of components.

Removing the simplex of highest dimension and computing another Q-analysis, the number of components goes up to 7 at $q = 11$. There are changes in the number of components at higher dimension too, as suggests the new structure vector: $Q = [1,1,2,1,2,3,5,2,5,2,3,7,5,5,5,3,3,3,2,1,1,1,1,1,1,1,1,1,1,1,1,1,1,1,2,2,2,2,2,1,1,1,1,2,3,2,2,2,2,2,2,2,2,1,1,1,1,1]$. For example, from $q = 34$ to $q = 37$, there are two components: {*use of public data*} and {*evaluation measures of results and progress*; *actor:industrial*; *development of an application or a service*; *citizen participation at the beginning*}, while there was only one component in the initial analysis. This means that the use of public data was only connected at a high dimension to the industrial actor (and the others) because of its connections with the government actor.

This kind of exploration helps to understand the structure of the smart city features by showing exciting information on the connectivity of the data. Furthermore, it shows that the analysis strongly depends on the choices of the data, whether it be simplices or vertices.

Regarding the profile characterization, we first look at the helix model. We see that $f_{\text{actor:industrial}}$ appears at a pretty high dimension of 44. A third actor category seems at a low dimension of $q = 4$: the civil society. Collaboration between government and industrial actors seems quite strong regarding civil society. Thus, the SCP is closer to the two-helix collaboration model than a three-helix collaboration model.

Regarding the integration logic, we look at the applications domains. The initiatives of Paris cover all the application domains (see Table 12.1). Also, the fact that the first simplex of category "application domain," which is $f_{\text{application domain:transport and mobility}}$, appears at a pretty low dimension of $q = 17$ regarding $f_{\text{development of an application or a service}}$ (which is a condition to have a simplex of the application domain type). This suggests that the application domains are quite various. Three application domains stand out from the analysis: transport and mobility ($q = 17$), energy ($q = 16$), and quality of life ($q = 15$). The others appear from $q = 11$ ($f_{\text{application domain:social}}$) to $q = 0$ ($f_{\text{application domain:water}}$). This suggests an integrated intervention logic, even though some domains are rarely addressed.

The features related to community building are associated with the type of approach (top-down or bottom-up) and the kind of strategy (techno-led, holistic, socio-technical, or social). The simplex $f_{\text{citizens participation at the beginning}}$ appears at a high dimension of $q = 50$. However, other community-building features appear at a lower dimension. $f_{\text{no citizen participation}}$ appears at $q = 14$, and $f_{\text{no community building}}$ appears at $q = 8$, which mitigates the appearance of the simplex $f_{\text{citizens participation at the beginning}}$ at $q = 50$. This suggests a will to involve citizens in the development of initiatives. Other community building features appear in the analysis, such as $f_{\text{promote citizens engagement}}$ at $q = 11$, $f_{\text{stimulate user-driven innovation and community-led urban development}}$, or $f_{\text{workgroups that manage the general course of the smart city development}}$ at $q = 8$. Finally, we can say that community building seems to be considered, even though it could be improved and broadened to other features than citizens engagement at the beginning of the initiatives.

In the same way, there is an attempt to construct a global strategic framework, shown with the high dimension of $f_{\text{evaluation measures of results and progress}}$ at $q = 43$ but mitigated by the introduction of the simplex $f_{\text{no strategic framework}}$ at $q = 18$.

These two categories of features, as say community building and strategic framework tend toward a bottom-up approach, mitigated by the prominence of the government and industrial actors.

The social features category does not appear as high as the other categories. Nevertheless, they are significant: more than half of the initiatives analyzed use public data ($f_{\text{use of public data}}$ appears at $q = 37$), which is coherent that the government is strongly involved in the initiatives. It is followed by $f_{\text{address social needs and advance policy}}$ at $q = 17$ and $f_{\text{promote democratic values}}$ at $q = 14$. Moreover, $f_{\text{taking into account the ones in need}}$, which is associated with only one ideal-type profile the social-centric profile, appears at $q = 7$.

Finally, the computation highlights a two-helix collaboration model with quite integrated intervention logic due to various application domains. It seems that Paris will have a global strategic framework, which could be improved. Community building appears to be encouraged. It is precisely the case of the citizens participation, as they participate at the beginning of most of the initiatives. Despite that, other characteristics of community building could be implemented.

Moreover, the social features category is quite significant in the analysis. Those elements suggest that the profile of Paris falls in between the holistic profile and the socio-technical profile. This seems coherent with the Paris presentation at the workshop, even though the strategic framework appeared to be more present during the workshops than in the analysis. Compared to previous work, where Paris fell within the scope of the techno-social-centric profile (holistic) [11] and with what came out from the workshop, the social characteristics seem to be more present in the present analysis. This increase of social characteristics in the SCP can be due to the data collected on the "*Budget participatif*" website, which was not mentioned during the Paris presentation of the workshop. Finally, the fact that the profile has slightly moved towards the socio-technical profile shows that the results are dependent on the data because of the data-driven approach.

Once the SCP is characterized, the local authorities can compare it with its strategy. For example, if the city's strategy is to enhance the consideration of social aspects, it could start defining use cases with features related to the social-technical profile (P3).

This example of the application allowed to test the feasibility of such kind of computation based on simplicial complexes and test the Hype application.

12.4 Conclusion

Identifying use cases is a difficult task for many local authorities, as mentioned during workshops organized about smart cities projects strategies. We propose a computation method to characterize the smart city profile (SCP) to identify relevant use cases aligned with the city strategy.

Since each city is different, the use cases must be adapted to the city specificities. Thus, the first step consists in identifying the SCP. The work of Mora and Deakin and their bibliometric analysis helped us characterize initiatives categories and lay the foundations of the profile's characterization, completed by the book of Green [16], and the workshops, we identified four ideal-type profiles with their associated features.

We proposed an innovative method to compute an SCP based on a topological approach with simplicial complexes. We present the Q-analysis, a method introduced by [17] to analyze how the features are related to the initiatives developed in the smart city. An application named Hype was designed to perform this analysis. Finally, an example of profile computation based on data publicly available is given for heuristic purposes. The results provide information on the city strategy profile and show the feasibility of the computation. Adding more initiatives and features to the computation would improve the precision of the computed profile.

Further analysis could be conducted by increasing the data to have an even more precise profile characterization. Time information could be added to see the evolution of the initiatives' implementation, associated with indicators such as citizens' satisfaction. Further research could be conducted to detect structural holes in the complex – at that time, Hype only detects cycles. This could be done by computing the Betti numbers for the complexes [20, p. 137], [26]. Other research could be conducted to consider spatial information and pursue the integration of time in Hype.

Finally, once the SCP is computed, further research would analyze whether the initiatives align with the city strategy to reach urban sustainability with an indicator. Then, the coherence between the profile and the goal could be measured, leading to modifying the initiatives and the strategies.

Acknowledgments We would like to thank the collaborators the workshops were organized: the French National Agency of Territories' Cohesion (ANCT), the University of Le Havre, and the Chronos Group. We would also like to thank the participants of the workshops. Finally, we would like to thank the Data Nostra team, Guillaume Blot, and Maxime Guery for developing the Hype application used for this research (https://www.datanostra.ai/).

References

1. United Nations. (2019). World urbanization prospects: The 2018 revision [Key facts]. United Nations, New York, (ST/ESA/SER.A/420). Consulté le: févr. 09, 2021. [En ligne]. Disponible sur: https://population.un.org/wup/Publications/Files/WUP2018-Report.pdf

2. Yin, C., Xiong, Z., Chen, H., Wang, J., Cooper, D., & David, B. (2015). A literature survey on smart cities. *SCIENCE CHINA Information Sciences*. https://doi.org/10.1007/s11432-015-5397-4
3. Basu, S., Bale, C. S. E., Wehnert, T., & Topp, K. (2019). A complexity approach to defining urban energy systems. *Cities, 95*, 102358. https://doi.org/10.1016/j.cities.2019.05.027
4. Batty, M., et al. (2012). Smart cities of the future. *The European Physical Journal Special Topics, 214*(1), 1. https://doi.org/10.1140/epjst/e2012-01703-3
5. Silva, B. N., Khan, M., & Han, K. (2018). Towards sustainable smart cities: A review of trends, architectures, components, and open challenges in smart cities. *Sustainable Cities and Society, 38*, 697–713, avr. https://doi.org/10.1016/j.scs.2018.01.053
6. Albino, V., Berardi, U., & Dangelico, R. M. (2015). Smart cities: Definitions, dimensions, performance, and initiatives. *Journal of Urban Technology, 22*(1), 3–21. https://doi.org/10.1080/10630732.2014.942092
7. Camero, A., & Alba, E. (2019). Smart City and information technology: A review. *Cities, 93*, 84–94. https://doi.org/10.1016/j.cities.2019.04.014
8. Caragliu, A., del Bo, C., & Nijkamp, P. (2011). Smart cities in Europe. *Journal of Urban Technology, 18*(2), 65–82. https://doi.org/10.1080/10630732.2011.601117
9. Mora, L., & Deakin, M. (2019). *Untangling smart cities: From utopian dreams to innovation systems for a technology-enabled urban sustainability*. Elsevier.
10. Keh, E., Lawrence, M., Sauz, R., Dadashi, N., & Homayounfar, N. (2021). The Ethical Smart City Framework & Toolkit: An inclusive application of human-centered design and public engagement in Smart City development. *Interaction Design and Architecture(s) Journal - IxD&A, 50*, 63–81.
11. Berrou, Y., et al. (2020). Smart City development strategy profile: Use case modeling based on simplicial complexes. *ISPRS Annals of the Photogrammetry, Remote Sensing and Spatial Information Sciences, VI-4/W2-2020*, 25–32. https://doi.org/10.5194/isprs-annals-VI-4-W2-2020-25-2020
12. Bittner, K., & Spence, I. (2003). *Use case modeling*. Addison-Wesley Professional.
13. Gottschalk, M., Uslar, M., & Delfs, C. (2017). *The use case and smart grid architecture model approach: The IEC 62559-2 use case template and the SGAM applied in various domains*. Springer International Publishing. https://doi.org/10.1007/978-3-319-49229-2
14. IEC 62559-2. (2015). *Use case methodology – Part 2: Definition of the templates for use cases, actor list and requirements list*. International Electrotechnical Commission. [En ligne]. Disponible sur: https://webstore.iec.ch/publication/22349
15. CEN–CENELEC–ETSI. (2012). Smart Grid Coordination Group - Smart grid reference architecture report 2.0. Consulté le: avr. 09, 2020. [En ligne]. Disponible sur: https://www.researchgate.net/publication/263264218_CEN_-CENELEC_-_ETSI_Smart_Grid_Coordination_Group_-_Smart_Grid_Reference_Architecture_Report_20
16. Green, B. (2019). *The smart enough city: Putting technology in its place to reclaim our urban future*. MIT Press.
17. Atkin, R. (1977). *Combinatorial connectivities in social systems: An application of simplicial complex structures to the study of large organizations*. Birkhäuser Verlag.
18. Atkin, R. (1981). *Multidimensional man*. Penguin Books.
19. Johnson, J. (1981). The Q-analysis of road traffic systems. *Environment and Planning. B, Planning & Design, 8*(2), 141–189. https://doi.org/10.1068/b080141
20. Johnson, J. (2014). *Hypernetworks in the science of complex systems*. Imperial College Press.
21. Atkin, R. (1974). *Mathematical structure in human affairs*. Heinemann Educational.
22. Numa et Ville de Paris. (2019). DataCity. Datacity-FR. https://www.datacity-fr.numa.co (consulté le avr. 20, 2020).
23. Ville de Paris. (2018). Le plan stratégique "Paris intelligente et durable. https://www.paris.fr/pages/le-plan-strategique-paris-intelligente-et-durable-2706 (consulté le févr. 19, 2021).
24. Ville de Paris. (2018). Paris, ville connectée à plus d'un titre. https://www.paris.fr/pages/paris-ville-connectee-2341 (consulté le févr. 19, 2021).

25. Mairie de Paris. (2021). Budget Participatif – Paris. Site du Budget Participatif de la Mairie de Paris. https://budgetparticipatif.paris.fr/bp/ (consulté le févr. 19, 2021).
26. Atkin, R. et Casti, J. L.. Polyhedral Dynamics and the Geometry of Systems. IIASA Research Report, Laxenburg, Austria, RR-77-006, mars 1977. Consulté le: août 31, 2020. [En ligne]. Disponible sur: http://pure.iiasa.ac.at/id/eprint/709/

Chapter 13
An Integrated Hesitant Fuzzy Linguistic MCDM Methods to Assess Smart City Solutions

Gülçin Büyüközkan, Merve Güler, and Esin Mukul

Abstract The smart city concept is a new approach to urban planning that aims to connect different elements in a city to each other through technology and data to enhance municipal services and infrastructure, as well as citizens' quality of life by the careful identification and selection of smart solutions. The aim of this study is to propose an analytical approach for the prioritization of suitable smart city solutions. In this context, a smart city model is presented with the help of the literature and industrial reports to define the needs of urban dwellers. The evaluation of the proposed model is considered a Multi Criteria Decision Making (MCDM) problem, which is addressed by the Hesitant Fuzzy Linguistic (HFL) approach. This HFL technique is used to deal with the uncertainty and hesitancy of expert opinions, and a Group Decision-Making (GDM) approach is employed to handle the partiality of the evaluation process. The importance degrees of the identified smart city parameters are calculated with the hierarchical HFL Simple Additive Weighting (SAW) method, whereas the smart city solutions are prioritized with the HFL Additive Ratio Assessment (ARAS) method. The applicability of the presented approach is demonstrated in a case study.

Keywords Hesitant fuzzy linguistic term set · ARAS · Hierarchical SAW · MCDM · Smart city

13.1 Introduction

Nowadays, more than half of the world's population lives in cities. People move to big cities in order to meet their basic needs for security, services, health, education and employment. Overpopulated cities, on the other hand, face the challenge of meeting the ever-changing and growing demands of their people. This exacerbates

G. Büyüközkan (✉) · M. Güler · E. Mukul
Galatasaray University, Industrial Engineering Department, Istanbul, Türkiye
e-mail: gbuyukozkan@gsu.edu.tr

G. Kayakutlu, M. Ö. Kayalica (eds.), *Decision Making Using AI in Energy and Sustainability*, Applied Innovation and Technology Management,
https://doi.org/10.1007/978-3-031-38387-8_13

cities' problems in the environment, health and safety, to name a few. To solve these problems, the smart city concept has emerged.

Digitalization will play a crucial role in addressing these problems and plays a part in this new smart city concept that aims to bring new perspectives and long-term value. Better quality of life, economic and environmental sustainability, security and a just social structure are achievable through digitalization to transform ordinary cities of today into smart cities of tomorrow [1, 2]. The smart city concept offers advantages in the fields of sustainable development, competitiveness and environmental sustainability. For this reason, countries make special efforts to understand this new concept to adapt and direct this dynamic process [3, 4]. There are different smart city solutions in various application areas. Therefore, there is a need for a tool that can compare and prioritize available smart city solutions.

This study aims to present an analytical approach for the evaluation of the proposed smart city models and prioritization of smart city solutions with a Multi-Criteria Decision-Making (MCDM) approach that is based on the hesitant fuzzy linguistic (HFL) technique [5, 6] to deal with the uncertainty and hesitancy of expert opinions. More specifically, the hierarchical HFL Simple Additive Weighting (SAW) [7] and HFL Additive Ratio Assessment (ARAS) methods [8] are integrated with each other to address this decision problem.

The contributions of this study can be listed as:

- A novel smart city model is proposed for prioritizing smart city solutions. It contributes to the literature on smart city solutions by proposing an integrated HFL MCDM methodology for the first time.
- An integrated hierarchical HFL SAW – HFL ARAS methodology is presented for the first time. The HFL SAW method is used to find the important degrees of the smart city components, and the HFL ARAS method is used to prioritize smart city solutions.
- A Group Decision-Making (GDM) approach is applied for handling the partiality of the evaluation process.
- The proposed evaluation methodology is applied on a case study for Istanbul.

The plan of this study is as follows. Section 13.2 presents the smart city concept and the proposed smart city model and solutions. Section 13.3 provides the proposed methodology and the steps of the analytical approach. A case study for Istanbul is given to illustrate the potential of the presented approach in Sect. 13.4. Finally, the conclusion and the future perspectives of the study are provided.

13.2 The Research Subject: Smart City Concept and Smart City Solutions

13.2.1 Smart City Concept

With the fundamental forces shaping today's world, transformation is inevitable. Many central and local governments around the world continue making their cities smarter, and better prepared for the future with a higher quality of life. Considering the rapidly growing adaptation and prevalence of smart city applications, the number of cities benefiting from these applications is increasing day by day.

It is important to thoroughly understand the concept of smart cities. Smart grid systems are one such technology to make urban areas smarter, followed by other technologies such as fixed camera systems, public transport and smart traffic management. Investments in these areas account for half of today's smart city investment opportunities. The first 100 cities that invested the most in smart city initiatives in 2019 accounted for 29% of global spending in this area. According to International Data Corporation (IDC) forecasts, smart city investments are expected to reach $189 Billion by 2023. Seventy percent of the cities in the world offer a great opportunity for the spread of smart city solutions in the future [9]. Smart city investments in the future will primarily focus on connected vehicle solutions, digital twin applications and wearable technologies [3, 4, 9].

In the literature, there are a number of studies on smart city solutions. Bellini et al. [1] proposed a smart city solution for resource description framework stores aggregating pen and private data sets. Chen et al. [10] presented a car park finder project. The smart facility application is a smart city solution that makes use of the advantages of space technology [11]. A smart solution for the development of sustainable construction materials is presented by Ralegaonkar et al. [3]. Basiri et al. [12] presented a smart city solution for sustainable urban development. Internet of Things (IoT) based smart city solution is presented by Ganzha et al. [2]. Smart city solution designs are evolved by leveraging Artificial Intelligence (AI) and sensor fabrics [13]. Al Ridhawi et al. [14] examined fog solutions in wireless networks for energy efficiency. The measurement of the efficiency of smart solutions is explained by Turečková and Nevima [4].

13.2.2 The Proposed Smart City Model and Solutions

Table 13.1 illustrates the proposed model that has been created by reviewing the literature and selected industrial reports, and by gathering additional expert opinions. While creating the smart city model, four main components that are important for Istanbul are taken into consideration.

Technologies on which smart city solutions are based are written in parentheses. A brief explanation of the solutions is provided next.

Table 13.1 Smart city components

Smart city components	References
Smart transportation (C1)	
Accessibility of transportation modes (C11)	[15–17]
Sustainable, innovative and safe transportation services (C12)	[17–19]
Traffic management (C13)	[17, 19, 20]
Public transport priority (C14)	[16, 17, 21]
Smart environment (C2)	
Waste management (C21)	[15, 17, 19, 21]
Energy management (C22)	[16, 17, 21, 22]
Sustainable resource management (C23)	[16, 18, 20, 21]
Ecological footprint (C24)	[18, 19, 23]
Smart living (C3)	
Distance education (C31)	[16–18]
Elderly/Disabled citizen services (C32)	[17, 19, 23]
Cultural facilities (C33)	[18, 22]
Online health services (C34)	[15–17, 22]
Smart governance (C4)	
Public and social services (C41)	[18, 21, 22]
Electronic service delivery (C42)	[15, 18, 20]
Citizen engagement and transparency (C43)	[16, 18, 19]
Public security (C44)	[16, 17, 22]

- *S1. City Net (Virtual Private Networks and IoT)*

City net interconnects public agencies, visitors, residents' IoT devices and businesses with network foundation and cloud-based applications. Its technical requirements are based on its multi-tenancy, scalability, availability, security, simple operations and environmental conditions [15, 16, 20, 24].

- *S2. Municipal Cloud (Big Data and Cloud Computing)*

Municipal Cloud allows smart city application areas by breaking siloes and attaining economies of scale at the same time. Its technical requirements are based on its multi-tenancy, elasticity, business continuity, performance, fluid operations and sustainability [16, 18, 24].

- *S3. Smart Transit (Wi-Fi and Cloud-based Network Management)*

Smart transit solutions provide appropriate solutions for smart shelters and on-board communications. These solutions aim to increase operational efficiency by realizing real-time fleet management. Its technical requirements are based on its hardening, location, Wi-Fi, cloud management and voice support [17, 18, 20, 24].

- *S4. Municipal Wi-Fi (Smart Sensors)*

Municipal Wi-Fi delivers facilities to visitors and residents. Different application areas for Wi-Fi-enabled sensors and devices can be listed as parking, smart lighting and waste management. Its technical requirements are based on its scalability, flexible deployment, quality of experience, performance and analytics [16, 19, 21, 24].

- *S5. Smart Civic Venue (Augmented Reality/Virtual Reality/Data Analytics)*

The smart civic venue aims to make civic venues such as museums and conference centres smarter. Location and way-finding allow application areas such as geo-fenced alerts, visitor guides and more. Its technical requirements are based on its low maintenance, accuracy, simple management and monitoring, people tracking, multiple Bluetooth Low Energy options and analytics [18, 22, 24].

- *S6. Smart Roads and Highways (Wireless Technology/IoT)*

Smart roads and highways utilize information and communications technologies to avoid accidents and decrease congestion by keeping drivers informed in real-time. Networks on the system provide communications for all intelligent transport systems and their sub-systems. Its technical requirements are based on its availability, virtualization, scalability, performance, security and environmental conditions [16, 18, 22, 24].

13.3 The Proposed Integrated Research Methodology

Decision processes in the real world are not always straightforward and simple. The uncertain environment frequently complicates the available choices and how to select them. Linguistic information can aid in the management of uncertainty. A linguistic method, hesitant fuzzy sets, were proposed for the first time in 2010 [5]. A model presenting linguistic expressions via a set of Hesitant Fuzzy Linguistic Term Sets (HFLTS) was presented by Rodriguez et al. [6] for enhancing the richness of linguistic flexibility in the MCDM process.

In HFLTS, decision makers (DMs) utilize comparative linguistic expressions for evaluating the defined criteria and alternatives. HFLTS is a strong and helpful technique to overcome uncertainty and hesitancy. It provides flexibility with linguistic expressions, as well as realistic results [5, 6]. In this study, HFL SAW is preferred thanks to its shorter computation time, practicality and simplicity. HFL ARAS calculates the complex relative efficiency of suitable alternatives with the help of a utility function. It is preferred thanks to its capability to solve highly complex problems, effectiveness, and ease of use for application [25].

An HFLTS, H_s, is a subset of the linguistic elements of the set $S = \{s_0, \ldots, s_g\}$ [6]. The upper bound H_{s+} and lower bound H_{s-} of the HFLTS are described as

$$H_{s+} = \max(s_i) = s_j, s_i \in H_S \text{ and } s_i \leq s_j \ \forall_i \tag{13.1}$$

$$H_{S-} = \min(s_i) = s_j, s_i \in H_S \text{ and } s_i \leq s_j \forall_i \tag{13.2}$$

The E_{GH} function is used to convert comparative linguistic term sets into HFLTS [5, 6].

The envelope of the HFLTS, env(H_S), is a linguistic interval with the upper bound (max) and the lower bound (min) as shown below [26]:

$$\text{env}(H_S) = [H_s^-, H_s^+], H_s^- \leq H_s^+ \tag{13.3}$$

The steps of HFL SAW and HFL ARAS are explained in Fig. 13.1 [7, 8]:

13.4 Application

City governments, industry leaders, subject experts, banks, research groups and other smart city stakeholders are brought together around smart city projects. Smart cities are gaining popularity in the literature, both in the world, as well as in Turkey. In this study, the smart city components and solutions for Istanbul are evaluated. As of 2021, 3 of the 81 projects implemented within the scope of the European Innovation Partnership Smart Cities and Communities Program were carried out in Istanbul, Manisa and Eskişehir [17]. Among these cities, metropolitan Istanbul, which has many components, stands out due to its cosmopolitan structure.

In this study, there are three DMs who evaluate the proposed smart city components and solutions. All three DMs are sufficiently experienced and knowledgeable in the area of smart cities. The weights of the DMs are accepted as equal.

In the first phase, the main smart city components given in Table 13.1 are evaluated by DMs with respect to their experience and knowledge by using the linguistic scale shown in Table 13.2. Table 13.3 shows the DMs' decision matrix for main components.

The fuzzy envelope is used to transform the linguistic terms in Table 13.3 into trapezoidal fuzzy numbers. These trapezoidal fuzzy numbers from three different DMs are aggregated. Then, the sub-components given in Table 13.1 are evaluated by DMs by using the linguistic scale shown in Table 13.2. The decision matrix for sub-components is shown in Table 13.4.

The assessments in Table 13.4 are converted into the trapezoidal fuzzy numbers and these numbers are aggregated. These values are multiplied by the weights of the main components to obtain the hierarchical structure. Then, the obtained trapezoidal fuzzy number is defuzzified with the Eq. (6). These defuzzified numbers are then normalized. Table 13.5 illustrates the weights of the final components.

According to the obtained results, the most appropriate smart city component is "Sustainable, innovative and safe transportation services (C12)". This component is followed by "Accessibility of transportation modes (C11)", "Public security (C44)"

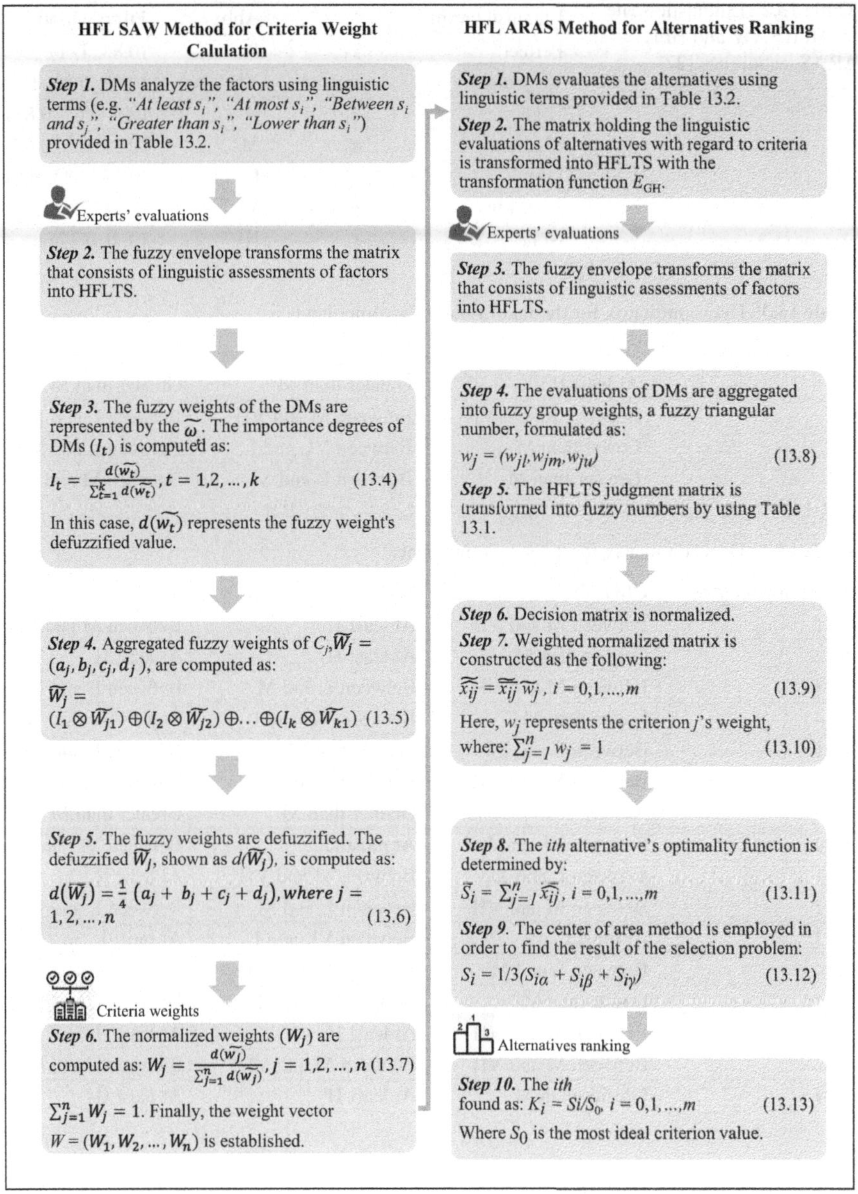

Fig. 13.1 The flowchart of the proposed research methodology

and "Sustainable resource management (C23)". These four components stand out with their weights compared to other criteria.

In the second phase, DMs first evaluated the proposed smart city solutions by using the linguistic scale given in Table 13.2. The evaluation of DMs for the

Table 13.2 Linguistic scale for HFL SAW and HFL ARAS techniques [27]

Linguistic term	S_i	Abb.	Fuzzy numbers
Perfect	s_3	P	(0.83, 1, 1)
Very high	s_2	VH	(0.67, 0.83, 1)
High	s_1	H	(0.5, 0.67, 0.83)
Medium	s_0	M	(0.33, 0.5, 0.67)
Low	s_{-1}	L	(0.17, 0.33, 0.5)
Very low	s_{-2}	VL	(0, 0.17, 0.33)
None	s_{-3}	N	(0, 0, 0.17)

Table 13.3 Decision matrix for the main smart city components

Main components	DM1	DM2	DM3
C1	At least H	Greater than M	Greater than M
C2	Between M and VH	Between M and VH	Between L and M
C3	Lower than M	Between VL and L	Between L and M
C4	Greater than M	Between L and M	Between M and VH

Table 13.4 Decision matrix for the sub-components

Sub-components	DM1	DM2	DM3
C11	Between M and VH	At least H	Between M and VH
C12	Greater than M	At least H	At least H
C13	Between M and VH	Between L and M	Between L and M
C14	Lower than M	Between VL and L	Lower than M
C21	Between L and M	Lower than M	Between L and M
C22	Between M and VH	Between L and M	Lower than M
C23	At least H	Greater than M	Greater than M
C24	Between VL and L	At most L	Between VL and L
C31	Between L and M	Between M and VH	At least H
C32	Between M and VH	Between L and M	Between L and M
C33	At most L	Between VL and L	At most L
C34	Between L and M	Between M and VH	At least H
C41	Lower than M	Between L and M	Between L and M
C42	Between M and VH	At least H	At least H
C43	Between M and VH	Between M and VH	Between L and M
C44	Greater than M	At least H	At least H

accessibility of transportation modes (C11) criterion is provided in Table 13.6 as an example.

The steps of the HFL ARAS method are applied by using the Eqs. (8)–(13). The ranking of the smart city solutions is illustrated in Table 13.7.

According to the results of this case study, the most suitable smart city solution for Istanbul is "S3. Smart Transit", followed by "S1. City Net", and "S6. Smart Roads and Highways", respectively. Table 13.7 indicates that these three smart city

Table 13.5 The weights of the smart city components

Main Comp.	Weights	Sub-Comp.	Aggregated fuzzy weights	Defuzzified weights	Norm. weights	Ranking
C1	(0.500,0.670, 0.830,1.000)	**C11**	(0.193,0.373, 0.600,0.887)	0.513	0.102	2
		C12	(0.278,0.485, 0.736,1.000)	0.625	0.124	1
		C13	(0.112,0.259, 0.462,0.723)	0.389	0.078	6
		C14	(0.000,0.114, 0.274,0.500)	0.222	0.044	10
C2	(0.277,0.443, 0.613,0.777)	**C21**	(0.031,0.123, 0.272,0.476)	0.226	0.045	9
		C22	(0.046,0.148, 0.307,0.371)	0.218	0.043	11
		C23	(0.170,0.344, 0.579,0.833)	0.481	0.096	4
		C24	(0.000,0.050, 0.170,0.344)	0.141	0.028	15
C3	(0.057,0.223, 0.387,0.557)	**C31**	(0.019,0.112, 0.258,0.464)	0.213	0.042	12
		C32	(0.013,0.086, 0.215,0.403)	0.179	0.036	14
		C33	(0.000,0.013, 0.086,0.215)	0.079	0.016	16
		C34	(0.019,0.112, 0.258,0.464)	0.213	0.042	12
C4	(0.333,0.500, 0.667,0.833)	**C41**	(0.038,0.138, 0.296,0.511)	0.246	0.049	8
		C42	(0.148,0.307, 0.518,0.786)	0.440	0.088	5
		C43	(0.092,0.222, 0.409,0.647)	0.343	0.068	7
		C44	(0.186,0.362, 0.591,0.833)	0.493	0.098	3

solutions have very close results to each other. Therefore, all these three solutions can be regarded as a priority for Istanbul's smart city initiatives.

Table 13.6 Alternatives' evaluation matrix for the C11 criterion

Accessibility of transportation modes (C11)			
Smart city solutions	DM1	DM2	DM3
S1	At least VH	Between H and VH	At least VH
S2	Between L and M	Between L and M	Between VL and L
S3	At least P	At least P	At least P
S4	Between M and H	Between VH and P	Between M and H
S5	Between VH and P	Between H and VH	Between VH and P
S6	At least P	At least P	At least VH

Table 13.7 The ranking of the smart city solutions

Smart city solutions	$S_{i\alpha}$	$S_{i\gamma}$	$S_{i\beta}$	S_i	K_i	Ranking
S0	0.254	1.114	3.465	1.611	1	–
S1	0.133	0.826	3.164	1.374	0.853	2
S2	0.062	0.618	2.760	1.147	0.712	5
S3	0.146	0.866	3.155	1.389	0.862	1
S4	0.142	0.885	3.303	1.302	0.808	4
S5	0.089	0.608	2.666	1.121	0.696	6
S6	0.133	0.834	3.060	1.342	0.833	3

13.5 Conclusion

The global agenda is occupied with the economic, social, environmental and cultural needs of future generations more than ever. This agenda is closely tied to creating smarter cities that produce value-added solutions for growing urban populations. Smart cities collect data and make use of digital technologies for the greater good of their citizens in terms of safety, quality of life, public property and infrastructure, and rapid access to health and transportation systems, among others.

In this chapter, an integrated HFL MCDM approach is presented to evaluate the proposed smart city model and prioritize the smart city solutions. The proposed model is constructed based on literature review, industry reports and expert views. The weights of the smart city components are determined with the hierarchical HFL SAW method. The smart city solutions are then prioritized with HFL ARAS. An application for Istanbul illustrates the potential and usefulness of the proposed integrated approach.

The case study identified sustainable, innovative and safe transportation services; accessibility of transportation modes; public security and sustainable resource management to be the most important criteria for smart cities. For Istanbul, the most appropriate smart city solutions are then found as "Smart Transit; City Net; and Smart Roads and Highways".

For future studies, it can be interesting to extend this analysis to other cities and compare the results. Another future direction for this study can be the integration of new evaluation criteria and alternatives. Moreover, other HFL MCDM methods can be applied to compare their results in the future.

Acknowledgements The authors would like to express their gratitude to the experts who gave assistance, advice and feedback. This work has been supported by the Scientific Research Projects Commission of Galatasaray University (Project No: FOA-2021-1059 and FOA-2023-1181).

References

1. Bellini, P., Bertocci, L., Betti, F., & Nesi, P. (2016). Rights enforcement and licensing understanding for RDF stores aggregating open and private data sets. In *2016 IEEE international smart cities conference (ISC2)* (pp. 1–6). IEEE.
2. Ganzha, M., Paprzycki, M., Pawlowski, W., Szmeja, P., & Wasielewska, K. (2018). Identifier management in semantic interoperability solutions for IoT. In *2018 IEEE international conference on communications workshops (ICC workshops)* (pp. 1–6). IEEE.
3. Ralegaonkar, R. V., Madurwar, M. V., & Sakhare, V. V. (2019). Sustainable construction materials. In *Architecture and design: Breakthroughs in research and practice* (pp. 658–687). IGI Global.
4. Turečková, K., & Nevima, J. (2020). The cost benefit analysis for the concept of a smart city: How to measure the efficiency of smart solutions? *Sustainability, 12*(7), 2663.
5. Torra, V. (2010). Hesitant fuzzy sets. *International Journal of Intelligent Systems, 25*(6), 529–539.
6. Rodriguez, R. M., Martinez, L., & Herrera, F. (2012). Hesitant fuzzy linguistic term sets for decision making. *IEEE Transactions on Fuzzy Systems, 20*(1), 109–119.
7. Chou, S. Y., Chang, Y. H., & Shen, C. Y. (2008). A fuzzy simple additive weighting system under group decision-making for facility location selection with objective/subjective attributes. *European Journal of Operational Research, 189*(1), 132–145.
8. Zavadskas, E. K., Turskis, Z., & Bagočius, V. (2015). Multi-criteria selection of a deep-water port in the Eastern Baltic Sea. *Applied Soft Computing, 26*, 180–192.
9. IDC Worldwide Semiannual Smart Cities Spending Guide, 2018H2 (May 2019).
10. Chen, D., Teoh, S. H. A., & Yong, S. L. C. (2016). Car park finder–presumptive design brings the best out of it! In *International conference on human-computer interaction* (pp. 347–353). Springer.
11. Termizi, A. A. A., Ahmad, N., Omar, M. F., Wahap, N. A., Zainal, D., & Ismail, N. M. (2016). Smart facility application: Exploiting space technology for smart city solution. In *IOP conference series: Earth and environmental science* (Vol. 37, No. 1, p. 012049). IOP Publishing.
12. Basiri, M., Azim, A. Z., & Farrokhi, M. (2017). Smart city solution for sustainable urban development. *European Journal of Sustainable Development, 6*(1), 71–71.
13. Money, W. H., & Cohen, S. (2019, May). Leveraging AI and sensor fabrics to evolve Smart City solution designs. In *Companion proceedings of the 2019 world wide web conference* (pp. 117–122).
14. Al Ridhawi, I., Aloqaily, M., & Boukerche, A. (2019). Comparing fog solutions for energy efficiency in wireless networks: Challenges and opportunities. *IEEE Wireless Communications, 26*(6), 80–86.
15. Sharifi, A., Khavarian-Garmsir, A. R., & Kummitha, R. K. R. (2021). Contributions of Smart City solutions and technologies to resilience against the COVID-19 pandemic: A literature review. *Sustainability, 13*(14), 8018.
16. Kumar, H., Singh, M. K., Gupta, M. P., & Madaan, J. (2020). Moving towards smart cities: Solutions that lead to the Smart City Transformation Framework. *Technological Forecasting and Social Change, 153*, 119281.

17. EY. (2021). Değer yaratmak için Akıllı Şehirler: Akıllı şehirler nasıl değer üretiyor? Türkiye ve dünyadan uygulama örnekleri.
18. Giffinger, R., & Pichler-Milanović, N. (2007). *Smart cities: Ranking of European medium-sized cities*. Centre of Regional Science, Vienna University of Technology.
19. Alabdulatif, A., Khalil, I., Kumarage, H., Zomaya, A. Y., & Yi, X. (2019). Privacy-preserving anomaly detection in the cloud for quality assured decision-making in smart cities. *Journal of Parallel and Distributed Computing, 127*, 209–223.
20. Lee, J. H., Hancock, M. G., & Hu, M. C. (2014). Towards an effective framework for building smart cities: Lessons from Seoul and San Francisco. *Technological Forecasting and Social Change, 89*, 80–99.
21. Ministry of Urban Development Government of India. (2015). Smart City Mission statement& Guidelines.
22. Cohen, B. (2013). Smart city wheel. Retrieved from SMART & SAFE CITY: http://www.smartcircle.org/smartcity/blog/boyd-cohen-the-smart-city-wheel
23. Deloitte Report. (2015). *Smart Cities-How rapid advances in technology are reshaping our economy and society*.
24. Alcatel-Lucent. (2019). *Smart City solution guide*.
25. Büyüközkan, G., & Güler, M. (2020). Smart watch evaluation with integrated hesitant fuzzy linguistic SAW-ARAS technique. *Measurement, 153*, 107353.
26. Liu, H., & Rodríguez, R. M. (2014). A fuzzy envelope for hesitant fuzzy linguistic term sets and its application to multicriteria decision making. *Information Sciences, 258*, 220–238.
27. Beg, I., & Rashid, T. (2013). TOPSIS for hesitant fuzzy linguistic term sets. *International Journal of Intelligent Systems, 28*(12), 1162–1171.

Chapter 14
Presence of Renewable Resources in a Smart City for Supplying Clean and Sustainable Energy

Oveis Abedinia, Ali Ghasemi-Marzbali, Sina Gouran-Orimi, and Mehdi Bagheri

Abstract Nowadays, the world heavily depends on the use of fossil fuels such as oil and coal, which can be a serious threat to the future of the planet and humanity. Fossil fuel consumption can cause irreversible effects such as air pollution, environmental degradation, and climate change. According to studies, carbon dioxide (CO_2) content in the air is currently at its highest level and global warming has recently increased unprecedentedly. Moreover, glacier melting and sea level rise are among other problems of climate change. Accordingly, using renewable energy sources is of great importance and has received considerable attention. Renewable energy sources are clean and do minimum damage to the environment. Furthermore, these sources are inexhaustible and can be used for many years to come. Presently, the world is moving towards increasing the use of renewable energy. Undoubtedly, renewable energies will play an important role in supplying human energy in the near future. Applying renewable energy can be among the most effective ways for having a sustainable city and providing opportunities for energy security, socioeconomic development, and reduction of climate change and environmental impact. Deep and rapid penetration of renewable energy technologies into a smart city is convenient at different scales and can provide a secure foundation for a modern society with low-carbon economy (LCE). The smart urban network is an effective way for integrating renewable energy sources with the existing urban network system, which could bring forth environmental benefits as well as sustainable development and sustainability as well as diversity of energy resources. The present study aims to integrate renewable energies and examines how their presence can provide a sustainable solution in a smart city.

O. Abedinia · M. Bagheri
Department of Electrical and Computer Engineering, Nazarbayev University, Nursultan, Kazakhstan

A. Ghasemi-Marzbali (✉) · S. Gouran-Orimi
Department of Electrical and Biomedical Engineering, Mazandaran University of Science and Technology, Babol, Iran
e-mail: ali.ghasemi@ustmb.ac.ir

G. Kayakutlu, M. Ö. Kayalica (eds.), *Decision Making Using AI in Energy and Sustainability*, Applied Innovation and Technology Management,
https://doi.org/10.1007/978-3-031-38387-8_14

Keywords Smart city · Renewable resources · Security and sustainability · Social welfare

14.1 Introduction

Nowadays, energy is considered as one of the most important solutions for global challenges such as climate change, environmental impact, and sustainable development. Warnings issued by international organizations about climate change and high greenhouse gas emissions worldwide have considered energy as a challenging issue for governments and researchers. Developing a plan should be the main priority of governments in different societies in order to ensure energy security and achieve long-term energy sustainability. Therefore, efforts should be made to grow the use of renewable energy resources in line with power supply as well as socioeconomic development [1]. The member states of United Nations Environment Programmer (UNEP) as well as European Energy Union have identified the fight against environmental pollution and climate change as a common interest. According to a common vision, a low-carbon economy model was developed by the EU until 2050. Some goals such as reducing CO_2 emissions by 40%, enhancing energy efficiency by 27%, and increasing renewable energy consumption by 27% have been set by 2030 in order to achieve a comprehensive and cost-effective plan [2]. Considering the finite nature of fossil fuels and the resulting environmental pollution caused by using this type of fuel, the necessity of consuming renewable energy has become more apparent. Thus, an integrated and modern system should be developed for the optimal use of renewable resources in order to achieve high technical and economic efficiency. Nowadays, cities are facing many challenges due to the uncontrolled migration as well as rising population. The world's population has significantly grown in urban areas. Nowadays, approximately 55% of the world's population lives in urban areas. According to the UN, the urban population is expected to account for 66% of the world's population by 2050. In the past, most of the people around the world lived in small communities. Over the past few centuries, especially in recent decades, this issue has dramatically changed with the migration of massive population from rural to urban areas. Figure 14.1 illustrates the population living in rural and urban areas worldwide in 2000–2020. The number of people migrating to cities has significantly increased over time, especially in recent years. The UN considers this event as a turning point that has intensified since 2007 [3].

Population growth and urban development can have catastrophic effects including excessive energy consumption and environmental degradation, as this development in urban areas requires optimization of the existing resources such as electricity, water, housing, transportation, etc., in order to meet the citizens' needs. Thus, conventional management methods no longer can meet the communities' needs and novel strategies should be adopted to deal with the negative effects of population growth. Accordingly, the idea of smart city was proposed to solve the existing problems, properly manage resources, and prevent energy waste and

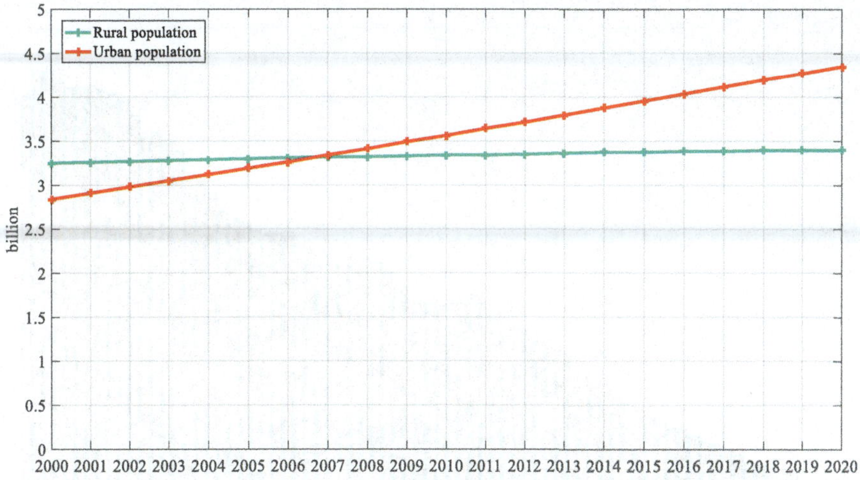

Fig. 14.1 Number of people living in urban and rural areas worldwide in 2000–2020

environmental challenges in communities. Smart city is based on a wide range of technologies such as information and communication, electronics, intelligent transportation, intelligent building systems, and advanced control structures. Smart and digital cities are technology-driven. A city can be defined as a "smart" one when it is organized, interconnected, self-healing, decision-maker, and healthy. Integrating different renewable energy systems in an urban energy system and optimizing this system can create a secure integrated energy system for the region or city [4]. Studies have revealed local organizations such as municipalities are designing a roadmap to convert their region's energy into a sustainable energy system, e.g., Copenhagen aims to be carbon neutral by 2025. Also, they plan to use 100% renewable energy in urban energy management by 2050 as an alternative program. Furthermore, New York City in the United States aims to reduce carbon emissions by 80% by 2050 [5].

The present study aims to develop and integrate renewable resources in the smart energy system related to smart cities. Then, characteristics of these resources and their role in sustainable development are discussed, and the smart energy system that will play an important role in smart cities is introduced. Finally, applying renewable energies such as solar, wind, and geothermal energies as the most reliable sources of sustainable energy is addressed.

14.2 Renewable Resources and Sustainable Development

Renewable energy refers to the type of energy, the source of which, unlike non-renewable (fossil) energies, is capable of being regenerated by nature in a short period of time. In other words, renewable energy resources can be easily

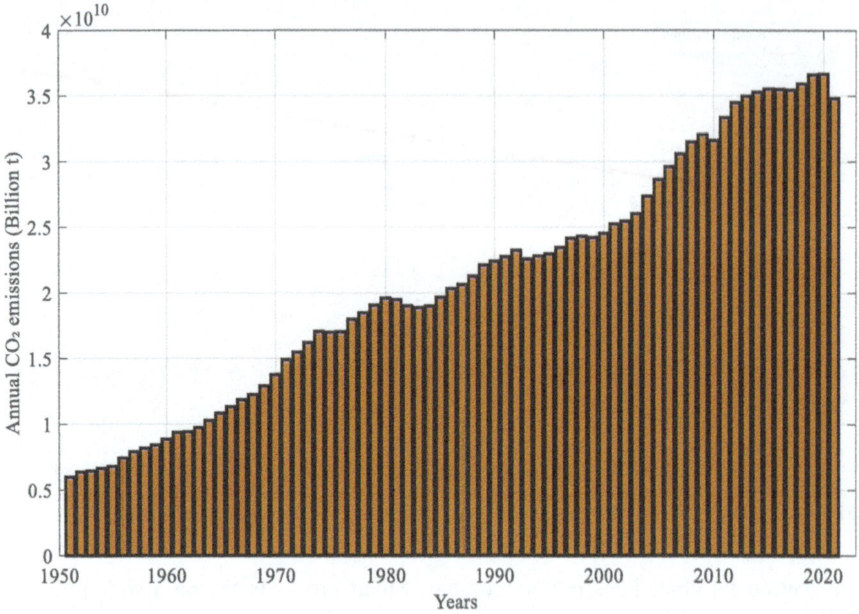

Fig. 14.2 Annual greenhouse gas emissions

replaced after consumption. Solar, geothermal, and wind energies; geothermal electricity; hydroelectric power; hydropower; biomass; etc. are examples of renewable energy. Sustainable development refers to a trade-off between development and environment. Sustainable development was first introduced in a report by International Union for Conservation of Nature (IUCN) in 1980. In this report, entitled "Natural Resources Conservation Strategy," the organization used this term to describe that development is not harmful, but useful, to the nature. Sustainable development is the intersection of society, economy, and environment. The world is rapidly becoming a global village due to the increasing daily demand for energy. The need for energy and related services to satisfy socioeconomic development, welfare, and health is increasing. All communities demand energy services to meet basic human needs. Supplying energy and preventing its negative role in climate change are among the challenges of the energy sector in the path of sustainable development. Fossil fuel consumption has significantly increased as the result of improving the life quality, industrializing the developing countries, and global population growth. Excessive consumption of fossil fuels can increase depletion rate of fossil fuel resources, have harmful impacts on the environment and cause dramatic changes in the earth's climate [6]. Figure 14.2 shows the amount of CO_2 emissions released from fossil fuel combustion from 1950 to 2020. Before the Industrial Revolution, greenhouse gas emissions were significantly lower than the present time. In 1950, 6 billion ton of CO_2 was emitted in the world. This rate approximately quadrupled by 1990, reaching more than 22 billion ton. Greenhouse gas emissions resulting from burning fossil fuels have continued to rapid growth. More than 34 billion ton of CO_2 is currently produced per year [7].

Nowadays, climate change has become among the greatest challenges of the twenty-first century. Serious consequences of climate change can be avoided if efforts are made to transform the current system into a modern clean energy system. Renewable energy resources with their special capacity are considered useful for resolving problems of using fossil fuels and reducing climate change.

Excessive consumption of fossil fuels has led to global warming due to CO_2 emissions. Fossil fuels and global warming are inextricably linked to sustainability and energy. Therefore, Kyoto protocol was adopted to prevent global pollution and monitor greenhouse gas emissions. Today, renewable energy and sustainable development are widely discussed. The majority of the experts believe that renewable energy is a necessity for sustainable development. Renewable energy is directly associated with sustainable development through its impact on human development and economic efficiency. Renewable energy resources provide opportunities for energy security, socioeconomic development, energy access, health and mitigation of climate change, and environmental impacts [8].

14.2.1 Energy Security

Given that the correlation between the energy consumption and economic growth is estimated to be greater than the global average in most of the countries with fossil fuel resources, efforts should be made to maintain energy security in the long run considering sustainability and development. Applying renewable energy-related technologies can diversify resources and reduce the economic damage associated with energy. Furthermore, energy can be supplied in hard-to-reach areas which are far from the grid using these resources.

14.2.2 Socioeconomic Development

Investigations have demonstrated a direct correlation between the economic growth and energy consumption. Globally, per capita income has a positive correlation with per capita energy consumption. Economic growth can be considered an important factor for increasing energy consumption in recent decades, resulting in the increased employment. Applying renewable energy resources can have various positive social effects such as local employment, better health, job opportunities, and consumer choice [9]. Following the operation of renewable energy projects, local labor can be employed in different stages of the project, which can ultimately lead to the use of local capacities, shareholders as well as regional and local insurance, and banking services and expand business. Moreover, it is possible to sell the generated energy such as electricity by investing in distributed generation

(DG) projects. Therefore, applying renewable energy resources can strengthen the economy and social prosperity of the people in the region [10].

14.2.3 Energy Access

Another goal of sustainable development is to ensure the cleanliness, cost-effectiveness, usability, and availability of energy for all the people; this can only be achieved through renewable energy resources that are distributed around the world.

14.2.4 Climate Change

Climate is considered an important category in the human life. For this reason, various countries attempt to maintain the ecological balance using knowledge and technology, and prevent severe climate change by efficient planning, thus accelerating the process of achieving sustainable development. Rapid population growth in recent decades, increasing use of energy, and emission of polluted particles and gases in the atmosphere have reduced air quality and increased various diseases; air pollution has caused damage to ecosystems that can affect food, water, sustainable development, and global economy.

Nowadays, meteorologists, climatologists, and environmental scientists try to reduce the harmful effects of human activities on the climate by making the necessary predictions, analyzing the distribution, and transfer of suspended particles and gaseous air pollutants as well as providing appropriate solutions. Climate change is among the most fundamental human concerns in the twenty-first century. These changes may endanger health in some ways; for example, cold air decreases due to the increase in frequency and intensity of heat waves and leads to death. In the last century, the average surface temperature of the earth has sharply increased. Rising global temperature has led to an annual increase in sea levels in the last century and polar sea ice has become 40% thinner. Industrial companies are both directly and indirectly involved in the emission of 37% of the greenhouse gases resulting from 80% of the energy consumption. If the efficient use of renewable resources is promoted in energy meetings, there is a broad prospect for minimizing greenhouse gas emissions and achieving sustainable development [11].

14.3 Smart Energy System

Energy costs account for a large share of the public capital in developing societies; so, energy problems are among the main obstacles to economic growth in these countries. Given the limitations of energy capacity, the issue of energy security and sustainability is among the most important challenges in this area. To this end, the future energy system will rely on renewable energy resources such as wind and solar energy. Using renewable resources in future energy systems will have its related challenges; however, applying renewable energy in future energy systems is on the agenda due to the irreparable harms of fossil fuels as well as numerous benefits of using renewable resources. The main issue that should be considered for using these resources is that cost-effective and sustainable energy should be flexibly provided simultaneously [2]. This innovative and efficient method of using renewable resources is called smart energy system. The term "smart energy systems" with a specific perspective beyond smart grids puts the focus of energy policy-making and modeling activities on areas other than electricity, including heat, refrigeration, industry, buildings, transportation, etc. and provides more coherent and flexible solutions to move towards using renewable energy. Due to recent advance in the field of energy production, storage, and conversion systems, smart energy systems have received great attention in order to achieve sustainable energy security as well as reduce environmental pollution [12]. Examples of smart energy are as follows:

A. Electric vehicles (EVs)

EVs are considered an essential part of combating climate change by reducing carbon emissions. However, if all the people who own a vehicle prefer to have an electric one and, then, plug in their vehicles to be charged after they get home at 6 pm, controlling the over pressure becomes crucial in the current urban infrastructure. Thus, IoT technology can solve such problems and give consumers more insight and control over energy consumption.

B. Smart meters

Smart meters are another example of smart energy in smart cities. These devices are designed to make a direct connection between the home power or gas meter and power supply. This connection in real time suggests that consumers can accurately monitor how much energy they consume and how much it costs. Therefore, they can make informed decisions about what to use and how to behave more cost-effectively. Applications can provide accurate bills based on the real-time energy consumption as well as control and balance supply and demand.

Figure 14.3 shows an example of the smart energy system consisting of different technologies. An integrated energy system be able to separate into five key sectors: energy resources, generation, distribution, exchange, and energy storage systems along with end users. In the generation sector, the system generates clean and sustainable energy and tries to meet the demand of end users. Energy generation systems convert energy into other forms, e.g., electricity and heating and cooling

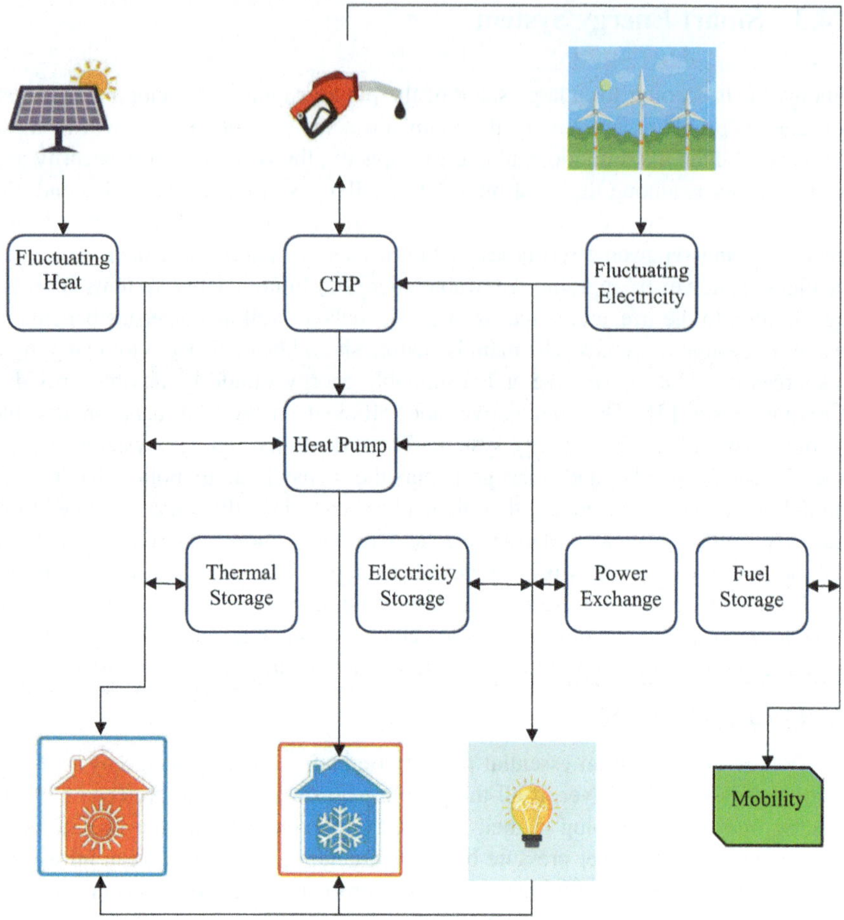

Fig. 14.3 Sample smart energy system

energies. Generally, generation units include a photovoltaic (PV) system, wind turbines (WTs), and a combined cooling and heating power (CCHP) structure. In smart energy systems, renewable resources have a special place over fossil fuels due to their numerous benefits and applications. Uncertainties in sustainability is among the disadvantages of renewable energy resources. Thus, fossil fuels are employed only as a supplement to ensure system performance. In the distribution sector, the output energy is transferred to consumers and energy storage units seek to stock extra energy. The general system and its various parts aim to achieve sustainable performance for reducing fossil fuel consumption as well as harmful environmental pollutants.

The general structure is controlled by developed data as well as information technology of smart management and planning of energy system, which perceives the effective and sustainable performance of the system by optimizing operations and reduces fossil fuel consumption and pollutant emissions. Smart energy grids are classified into three groups: smart power grid, smart thermal grid, and smart gas grid [13].

14.3.1 Smart Power Grid

Smart power grid includes electrical equipment that covers all the generation and consumption sectors in order to supply sustainable and highly reliable energy. Smart grid is sometimes recognized by the public as smart networks operating in the field of power generation. Energy systems can be optimized due to the possibility of consumer participation in load management programs in a smart power grid. Nowadays, creating a smart power platform is required to balance storage, supply, and demand. There, buildings or smart homes can play an active role as smart energy systems [14].

14.3.2 Smart Thermal Grid

In several studies, the perception of smart grid is employed for thermal power units, so that high energy efficiency could be achieved by the optimal management of heating and cooling systems. The required heat can be supplied through renewable energy resources in the long run by proper coordination and structure between different equipment of this grid. To supply cooling demands, cooling consumption of the system can be monitored by growing the thermal ability of the flow and enhancing the temperature change among the supply and return water. In general, the smart thermal grid consists of a network of pipes connecting buildings, cities, and their suburbs together in order to supply consumers' cooling and heating demands in these areas by the surrounding power plants, distributed generation resources, or distribution centers.

14.3.3 Smart Gas Grid

This grid includes equipment that covers all the generation and consumption sectors in a smart manner in order to supply sustainable and highly reliable gas energy. The smart gas grid will play a very important role in future systems. Therefore, it is expected that the current natural gas grid could interact with different types of renewable gases. The smart grid has the ability to store energy in the long run and can be very efficient in supplying energy in the smart system by converting power into gas and liquid. The smart gas system can be adapted to supply the fuel of a part of the transportation system that does not use power as the primary fuel.

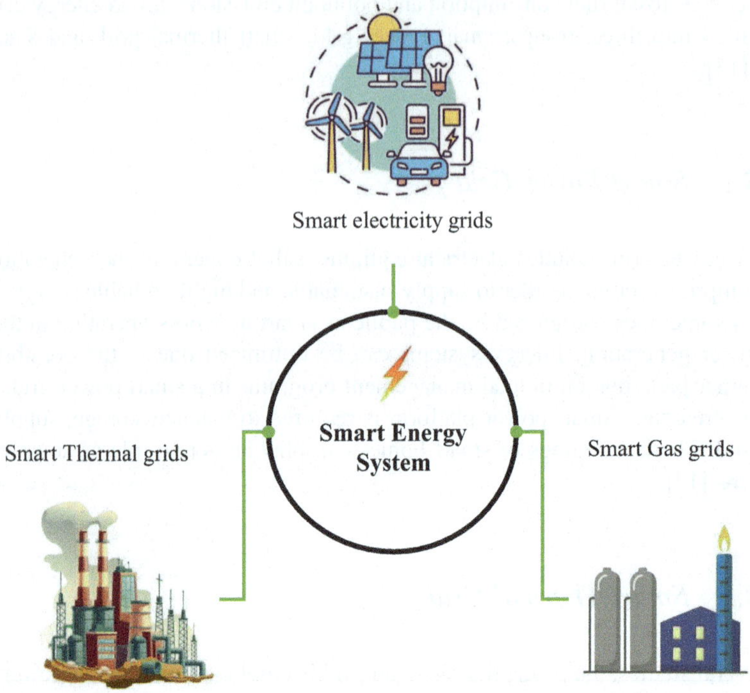

Fig. 14.4 The main parts of the smart grid in the smart energy system

If any of the three domains of power, heat, and gas is considered separately, this issue can cause excessive costs, complexity, and failure to gain proper energy efficiency in the target areas and make it difficult to achieve the great goal of sustainable development. Smart energy systems make the most optimal solution for achieving the appropriate efficiency by combining these three smart domains on the demand side and participation of all the three domains. Figure 14.4 illustrates the key components of the smart grid structure in the smart energy system [15].

Smart grids present a fundamental role in achieving sustainable energy. An optimal integrated approach can be developed for using renewable energy by modernizing systems and applying automated progress monitoring equipment that can analyze user consumption as well as energy generation and consumption costs. In smart energy systems, renewable energy resources can supply 100% of the power demand and have more cost-effective performance.

14.4 Smart Energy Network for Smart City

In this section, renewable resources are examined in the smart power system in the smart city. Given the importance of smart energy, efforts have been made to increase energy efficiency in the smart city and create urban green spaces for improving the quality of life. Renewable energy resources play a special role in this regard. As shown in Fig. 14.5, solar, wind, and geothermal energies; biofuel; biomass; hydrogen; and waste are among the effective renewable energy resources in the smart energy system. In this study, solar, wind, and geothermal energies are introduced. For further studies on other energy resources, refer to [16].

14.4.1 Solar Energy

Solar energy as one of the reliable sources has a special place due to its special nature and ease of installation. Solar or photovoltaic (PV) cell is a semiconductor crystalline and non-crystalline silicon that absorbs sunlight and converts photon energy into electrical energy. Solar energy can be used to supply power, heating and cooling

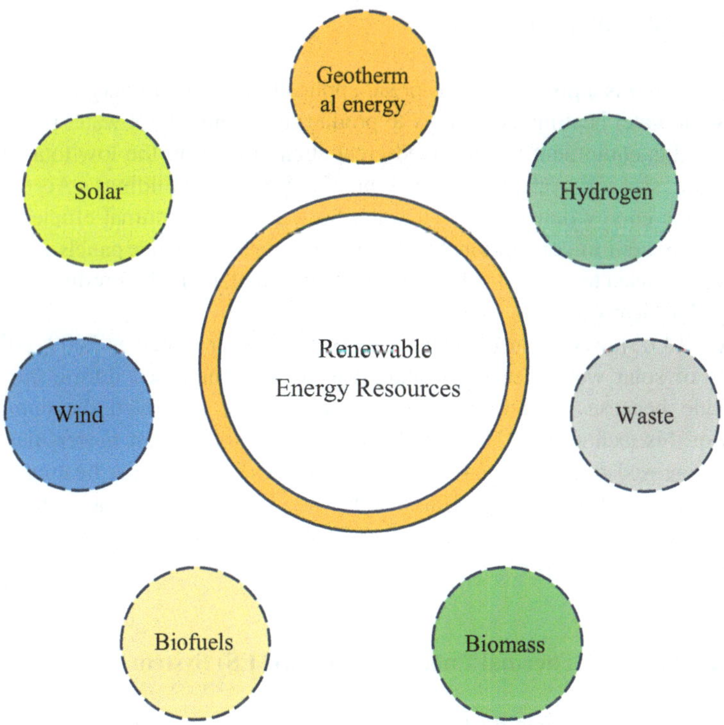

Fig. 14.5 Renewable energy resources

demands, as well as the required energy of EVs. The capacity of solar sources significantly depends on geographical locations and direct solar radiation reaching their surface [17]; for example, India is one of the countries through fast growing in the metropolitan area, a large area of which receives more than 200 W/m^2 of direct solar radiation per year. Operation of solar power plants in different parts of the world is increasing due to the valuable potential of solar energy, e.g., China as one of the most economically powerful countries works on the availability and possibility of using PV systems in different cities of China. China has the largest PV market in the world, where the installed capacity of PV reached more than 200 GW in 2019. However, this country aims to increase the installed wind and solar capacity to 1200 GW by 2030 [18].

In the urban environment, several factors affect the optimal use of PV systems. Proper sunlight, surrounding landscape and location of the building are very effective in the utilization of solar energy resources, e.g., tall buildings have greater potential for integrated solar energy applications. In the building design, one should think about directing solar radiation to the building, and it is better that the volume and elongation of the building be along the east-west axis. This is a general factor and has the best result in each climate; however, it can have different results in some other climates.

14.4.1.1 Solar Water Heating

Heat generation is among the widespread applications of solar energy. It is possible to use solar water heating technology to produce the required hot water in buildings. Applying this equipment has received great attention due to the low total cost and appropriate size of systems installed on the roofs of buildings. According to research, the energy density in solar water heaters has the optimal efficiency of up to 80% compared to conventional water heating methods. Solar panels installed on buildings are used to convert solar energy into heat and, then, the produced energy is employed to heat water.

Figure 14.6 shows the installed capacity of solar water heaters in 2009–2020. The capacity of solar water heating systems has steadily increased during this period worldwide because the adopted novel policies have improved the competitive market of this technology. The installed capacity of solar water heaters has significantly increased from 203 to 482 GW by 2018. The global solar heating capacity increased by 501 GW in 2020. China is the largest market for solar water heating. The following figure shows the capacity of solar water heating systems worldwide from 2009 to 2020 [19].

14.4.1.2 Seasonal Thermal Energy Storage (STES) System

Nowadays, there is a great deal of interest in solar energy systems for storing solar energy when access to sunlight is not possible. Thermal energy storage (TES) is a

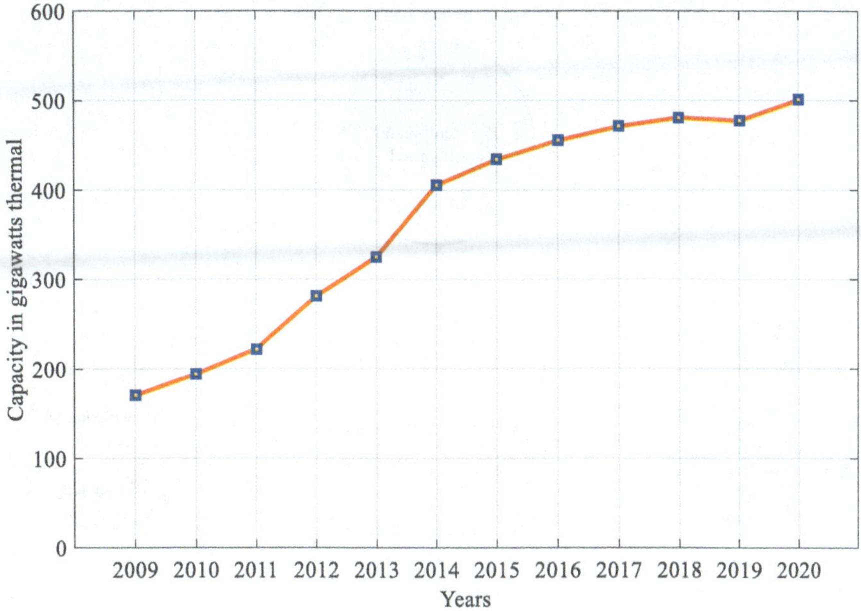

Fig. 14.6 Installed capacity of solar water heater

technology used to balance the mismatch between supply and demand in heating and cooling systems. It recognized as STES, permits the group and TES for solar heating during the summer in order to release the stored heat at the required time or place including the cold winter months.

A large amount of heat is usually generated in the industry that is wasted. However, the generated heat can be stored by applying this technology and reused in appropriate conditions. The same situation is possible in the cold months of the year, i.e., the natural cold can be stored and used for air conditioning in summer [20]. Given that this energy storage system can balance supply and demand and improve temperature fluctuations, it can be optimally used in buildings, industrial processes, and centralization of solar power plants. There is a specific type of STES with good inter-seasonal thermal energy storage capacity in Alberta, Canada, which receives 97% of its annual heat from the district heating system through solar heat generated by solar heating panels installed on the roof of car parks. Furthermore, 91% of total heating demand is supplied by this particular technology in Richmond, Virginia [21]. The increased efficiency, better reliability, more optimal economy, and less environmental pollution (less carbon dioxide) are among the benefits of using TES system. Thermal energy storage is becoming a growing business. According to reports, the global energy storage market is expected to reach $12.50 billion by 2025. Selecting the desired thermal energy storage system related on several factors, e.g., storage duration, economic efficiency, and operating conditions. Thermal energy storage can be classified according to the type of application, technology, and stored material. Figure 14.7 shows TES system classification.

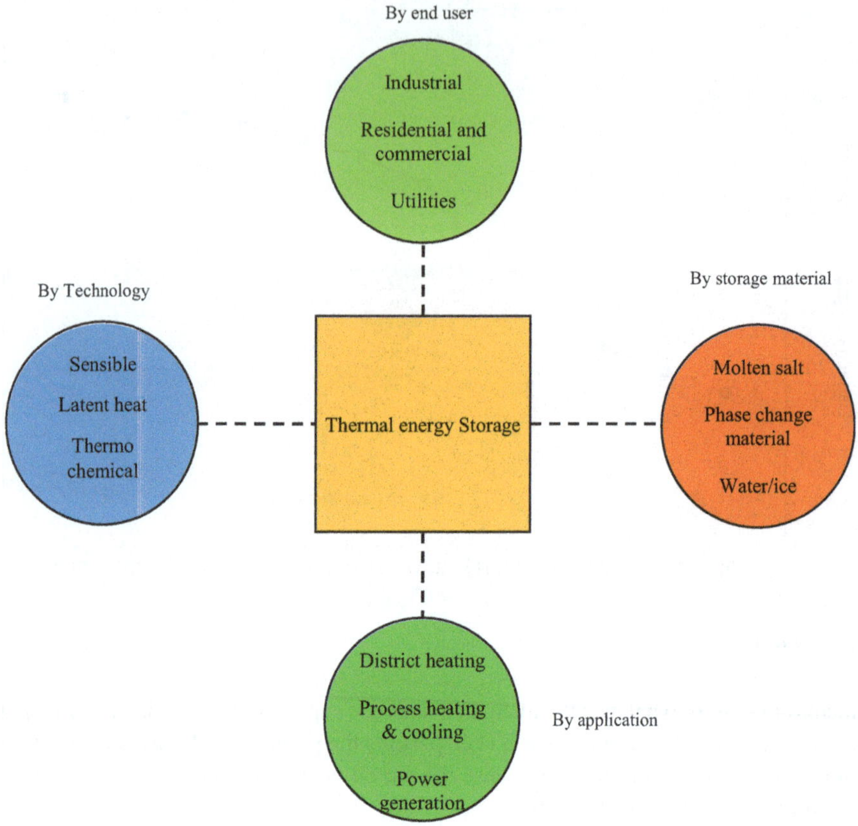

Fig. 14.7 Classification of thermal energy storage systems

14.4.2 Wind

With recent advance in the design and production of wind turbines, power is generated by urban wind turbines installed in high-rise buildings and offshore that are affected by offshore winds. Wind turbines do not require fossil fuels and are pollution-free. It is estimated that there is about 10 million MW of energy in the earth's winds. Wind turbines are divided into horizontal and vertical axes based on the rotation axis orientation. Horizontal-axis wind turbines (HAWTs) should be positioned exactly in the direction of the wind to achieve maximum efficiency; thus, horizontal axis turbines should include a system to be adjusted against the wind.

Access to strong winds is among the advantages of these turbines because the height of the tower allows access to strong winds. In some places, wind speed and efficiency increase by 20% and 34%, respectively, for every 10 m increase in height. These turbines can be used on the shores due to their long pedestals and advanced mechanism. Furthermore, they have high efficiency due to their direct exposure to the wind.

In vertical-axis wind turbines (VAWTs), the main rotor is mounted vertically. The most important advantage of VAWTs is that they do not need to be adjusted to the wind direction and can also be used at low altitudes.

Studies have revealed some modern wind turbine technologies can provide energy to the customer at the cost of EUR 0.05 per kWh, which saves EUR 1102.50 per year compared to the current European average cost of EUR 0.21 per kWh for the power grid and, finally, prevent the annual emission of 1.5 ton of CO_2 into the environment [22]. Studies on the urban wind flow characteristics have indicated VAWTs have better performance for urban power generation. Research has shown VAWTs could supply 33% of the power demand of residential homes in major urban areas in New Zealand. Moreover, the installed vertical turbines at the top of the Bahrain World Trade Center in Manama could supply $11–15 of the building's power demand. Residential buildings in Sicily could generate 40% of their power demand through wind energy [23].

14.4.3 Geothermal Energy

Nowadays, the optimal use of future energy is moving toward smart cities and smart rural communities. Smart communities can reduce the impact of environmental problems using an integrated hybrid system and renewable energy resources, and improve the quality of life. Geothermal energy is a type of renewable energy produced and stored in the earth's crust. The thermal energy available in the earth's crust is continuously generated due to the nature of this planet and radioactive decay of materials. Geothermal energy plays an important role in future smart grids because it has the potential to supply heating, cooling, and power demands.

Ground temperature at shallow depths is affected by seasonal variations of the land surface air temperature; however, the temperature is more stable at greater depths. Groundwater temperature usually does not significantly change in deeper zones between summer and winter. The ground temperature is lesser than the surface ambient temperature during the summer. Therefore, the ground can be used as a suitable source for efficient heat dissipation in cooling systems. Similarly, the ground temperature is greater than the surface air temperature during the winter. So, the ground can be employed as a good source for absorbing heat for heating. In geothermal pumps, a low-temperature ground energy source is usually used for heating and cooling.

Geothermal systems can be applied in small and large systems to provide heating, cooling, and domestic hot water. They are considered the ideal solution for nearly zero-energy buildings (NZEBs). Geothermal pumps if designed, installed, and used properly can produce the required energy on a large scale and reduce greenhouse gas (CO_2) emissions [23]. Figure 14.8 shows the energy transfer between the system and ground in winter and summer.

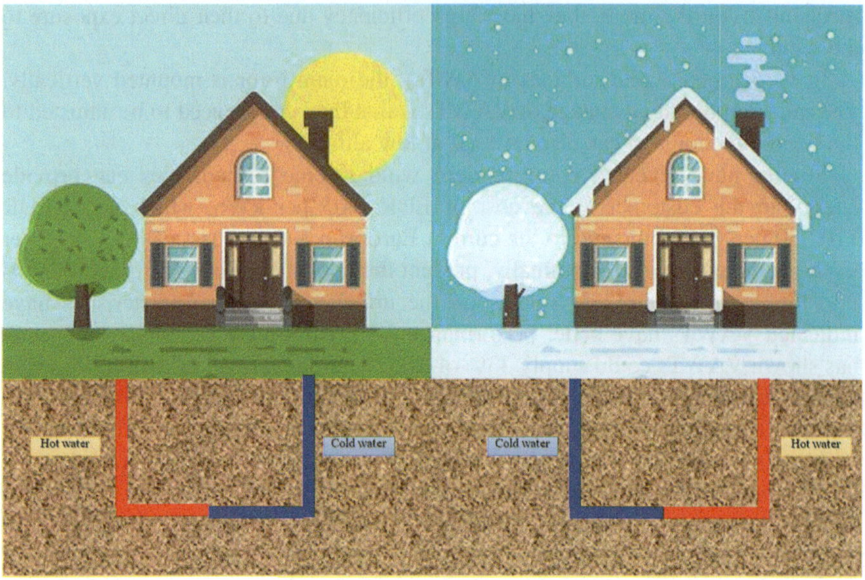

Fig. 14.8 Energy transfer between the system and ground in winter and summer

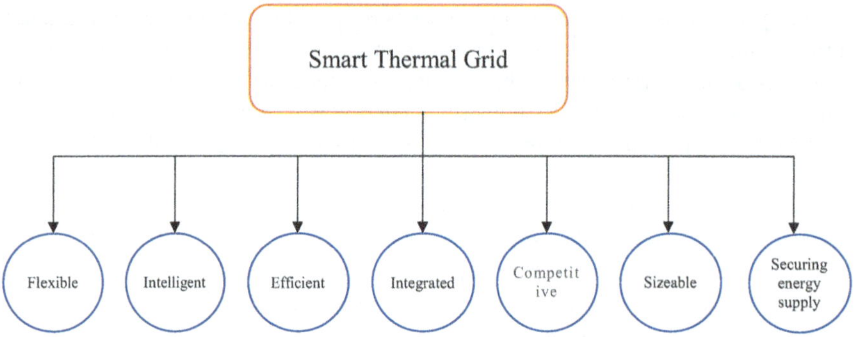

Fig. 14.9 Impacts of using geothermal energy in the smart thermal grid

Geothermal energy has positive effects on the smart thermal grid, such that it provides affordable and reliable heating and cooling systems for customers. Geothermal energy in the smart grid can lead to flexibility, intelligence, efficiency, integration, competitiveness, energy security, and ultimately larger system. Figure 14.9 shows the effects of using geothermal energy in the smart heating grid.

Investigations have revealed 3.4% of the Europe's population lives in urban areas, consuming 70% of the EU's energy and emitting greenhouse gases at the same rate. In the last decade, the installed capacity of geothermal energy has increased worldwide and reached 14 GW in 2020. The United States, Indonesia, and the Philippines are at the top in terms of the installed capacity of geothermal power plants. Many geothermal projects are under development in these countries and some of the largest geothermal power plants in the world exist in these countries. After completing several new geothermal projects, Turkey increased its installed geothermal capacity (232 MW) in 2019 [24]. According to studies on buildings in China, researchers have estimated ground source heat pumps can supply 10–15% and 25–30% of the heating and cooling demands of residential and public buildings, respectively. Currently, this energy source is used as the heating source for residential, commercial and industrial buildings in more than 60 countries [17, 25].

Smart city is a new drift in urban scheduling that strives for complete urban managing and great quality of life. The main components of smart cities are developed technologies, e.g., communication and information technology (ICT), Internet of Things (IoT), big data analytics, social networking, cloud computing, and artificial intelligence. Smart cities provide benefits to citizens using these technologies. Technologies embedded in smart cities enable the collection, processing, and sharing of big data to make informed decisions and increase the effectiveness of urban services [26]. These essential knowledges are currently being applied in numerous urban sectors other than smart city expansion; for instance, IoT is used to collect energy utilization data to optimize energy distribution and consumption. Given that smart city is a holistic approach for turning cities into better places, it is important to consider smart city development for smart energy.

14.5 Conclusion

In recent years, crises caused by excessive consumption of fossil fuels, pollution, energy consumption problems, population growth, social activities, difficulties related to accessibility, urban communication and transportation, etc. have highlighted the importance of using renewable resources. Certainly, these resources will play an important role in energy supply in smart cities in the near future. In other words, applying renewable energy can be among the most effective solutions for having a sustainable and clean city, which provides opportunities for energy security, socioeconomic development, energy access, health and reduction of climate change, and environmental impacts. The urban smart grid can improve economic status of the society and is beneficial for the environment. Furthermore, this system can achieve sustainable development and diversification of energy resources. In order to develop smart city-related technologies in an effective way, proper and up-dated technologies should always be selected in line with the goals of industry and society, and their development plan should be effectively formulated. There are different power and energy systems in smart cities, some of which interact with each other and use different technologies. Technologies should be selected in a smart manner and evaluated properly; an acquisition program should be developed considering their properties; for example, automatic vehicle location (AVL) scheme which is activated by the global positioning system (GPS) can be used for transportation in order to reduce fuel consumption and travel time. Also, transportation data sharing can reduce congestion. Using essential knowledges in a smart city be able to increase power efficiency, reduce energy consumption and CO_2 emissions, and, ultimately, support energy transfer. Given that smart city and energy transfer have some aspects in common, smart city development can improve energy transfer.

References

1. Dadashi-Rad, M. H., Ghasemi-Marzbali, A., & Ahangar, R. A. (2020). Modeling and planning of smart buildings energy in power system considering demand response. *Energy, 213*, 118770.
2. Ceglia, F., et al. (2020). From smart energy community to smart energy municipalities: Literature review, agendas and pathways. *Journal of Cleaner Production, 254*, 120118.
3. https://ourworldindata.org/urbanization
4. Tomar, P., & Kaur, G. (Eds.). (2019). *Green and smart technologies for smart cities*. CRC Press.
5. Thellufsen, J. Z., et al. (2020). Smart energy cities in a 100% renewable energy context. *Renewable and Sustainable Energy Reviews, 129*, 109922.
6. Huang, S., & Abedinia, O. (2021). Investigation in economic analysis of microgrids based on renewable energy uncertainty and demand response in the electricity market. *Energy, 225*, 120247.
7. https://ourworldindata.org/co2-emissions#global-co2-emissions-from-fossil-fuels-global-co2-emissions-from-fossil-fuels
8. Wang, C., Zhang, Z., Abedinia, O., & Farkoush, S. G. (2021). Modeling and analysis of a microgrid considering the uncertainty in renewable energy resources, energy storage systems and demand management in electrical retail market. *Journal of Energy Storage, 33*, 102111.
9. Lund, H. (2007). Renewable energy strategies for sustainable development. *Energy, 32*(6), 912–919.
10. Candra, O., Chammam, A., Alvarez, J. R. N., Muda, I., & Aybar, H. Ş. (2023). The impact of renewable energy sources on the sustainable development of the economy and greenhouse gas emissions. *Sustainability, 15*(3), 2104.
11. Owusu, P. A., & Asumadu-Sarkodie, S. (2016). A review of renewable energy sources, sustainability issues and climate change mitigation. *Cogent Engineering, 3*(1), 1167990.
12. Zhao, Y., et al. (2021). *The development of regional smart energy systems in the World and China: The concepts, practices, and a new perspective* (p. e1409). Wiley Interdisciplinary Reviews: Data Mining and Knowledge Discovery.
13. Lund, H. (2014). *Renewable energy systems: a smart energy systems approach to the choice and modeling of 100% renewable solutions*. Academic.
14. Xu, Y., et al. (2020). Smart energy systems: A critical review on design and operation optimization. *Sustainable Cities and Society*, 102369.
15. Lund, H., et al. (2017). Smart energy and smart energy systems. *Energy, 137*, 556–565.
16. Hatti, M. (Ed.). (2018). *Renewable energy for smart and sustainable cities: Artificial intelligence in renewable energetic systems* (Vol. 62). Springer.
17. Hoang, A. T., & Nguyen, X. P. (2021). Integrating renewable sources into energy system for smart city as a sagacious strategy towards clean and sustainable process. *Journal of Cleaner Production*, 127161.
18. Wang, Y., He, J., & Chen, W. (2021). Distributed solar photovoltaic development potential and a roadmap at the city level in China. *Renewable and Sustainable Energy Reviews, 141*, 110772.
19. https://www.statista.com/statistics/260688/forecast-capacity-of-csp-plants-worldwide-until-2050/
20. Hailu, G. (2018). Seasonal solar thermal energy storage. In *Thermal energy battery with nano-enhanced PCM*. IntechOpen.

21. Terziotti, L. T., Sweet, M. L., & McLeskey, J. T., Jr. (2012). Modeling seasonal solar thermal energy storage in a large urban residential building using TRNSYS 16. *Energy and Buildings, 45*, 28–31.
22. https://cordis.europa.eu/article/id/244881-wind-turbine-for-city-rooftops
23. Gagliano, A., et al. (2013). Assessment of micro-wind turbines performance in the urban environments: an aided methodology through geographical information systems. *International Journal of Energy and Environmental Engineering, 4*(1), 1–14.
24. Jardeby, Å. (2013). Developing geothermal heat pumps in smart cities and communities.
25. Huttrer, G. W. (2020). Geothermal power generation in the world 2015-2020 update report. In *Proceedings world geothermal congress* (Vol. 2020).
26. Kholerdi, S. S., & Ghasemi-Marzbali, A. (2021). Interactive time-of-use demand response for industrial electricity customers: A case study. *Utilities Policy, 70*, 101192.

Chapter 15
Syrian Household Energy Consumption Behavior Analysis in Turkey: Bayesian Belief Network

Ebru Acuner and M. Özgür Kayalica

Abstract Consumers' adoption of efficient use of energy varies depending on the demographic, situational, and behavioral dynamics in their households. Migrants influence the energy consumption patterns of their host countries because of having different energy usage behavior. The number of migrants, especially from Syria, has been increasing in Turkey. Meanwhile, energy consumption has been increasing, rapidly. Providing equity of supply and demand can be a major issue, especially in the electricity market, causing more stability and preventing scarcity thus leading to reliable forecasts. In this context, this study aims to identify the main driving forces that form the migrated household's behavior towards electricity consumption in Turkey. For this purpose, a detailed survey was prepared and conducted between April and June 2019 in the districts of Zeytinburnu and Fatih. There are 503 participants, 251 from Fatih and 252 from ZEYTINBURNU. Then, answers to the survey are investigated by using NETICA software considering Bayesian Belief Networks. According to obtained results, BBN shows that in the current situation migrated households consume electricity inefficiently with 73%. To change habits, they should be aware of how to use electric appliances more efficiently and financial support by distribution and/or retail companies should be considered for tariff/bill arrangements and providing more efficient electrical appliances for these households.

Keywords Migrated household · Electricity consumption behavior · Bayesian Belief Network

---

E. Acuner (✉)
Istanbul Technical University, Energy Institute, Istanbul, Türkiye
e-mail: acuner@itu.edu.tr

M. Ö. Kayalica
Istanbul Technical University, Energy Institute, Istanbul, Türkiye

Istanbul Technical University, Faculty of Management, Istanbul, Türkiye
e-mail: kayalica@itu.edu.tr

G. Kayakutlu, M. Ö. Kayalica (eds.), *Decision Making Using AI in Energy and Sustainability*, Applied Innovation and Technology Management,
https://doi.org/10.1007/978-3-031-38387-8_15

15.1 Introduction

International Organization for Migration (IOM) defines "migrant" as "*a global term, describing a person who moves away from his/her usual residence, within a country or through an international border, either temporarily or permanently, and for a variety of reasons*" [32]. In addition, United Nations (UN) states "migrant" as "*any person who moves, within a country or through an international border, away from his/her habitual residence, regardless of the legal status, being voluntary or involuntary, reasons for the movement or the length of the stay*" [65]. Therefore, "migration" can be defined "*the movement of a person or a group of people from his/her/their homes to settle in another*" [49]. Migration is a normal human activity. People can move for better work or economic opportunities, for family reasons, or for studying. Nonetheless, others are forced to move due to terrorism, conflicts, persecution, or human rights violations. In addition, as a new driving force, some people escape from the adverse effects of natural disasters, climate change, and/or other environmental issues [65]. The 2020 global data show that in the world, approximately 281 million international migrants are constituting 3.6% of the global population. Generally, it can be claimed that over the past five decades, the number of international migrants has increased. For instance, it was around 128 million in 1990, meaning that 119.5% more in the current situation [32].

For the first time, among 17 Sustainable Development Goals (SDGs), 11 of them cover migration or mobility either his/her preference or forced to move from their homes. The fundamental principle can be defined as "leave no one behind," including migrants and/or displaced people. SDG 10 entitled "Reduced Inequalities," contains "*to facilitate, safe, responsible and ordered migration or mobility of people, incorporating through the implementation of well-planned/managed migration policies*" as a seventh target. Some cross-cutting targets refer to remittances, trafficking, international student mobility, etc. [65].

Likewise, migration/mobility and the energy supply/demand relationship can be considered as another key issue for the world considering demographic, social, economic, and technological changes. Among SDGs, SDG 7, called "Clean and Affordable Energy" is directly related to energy policies and strategies. Target 7.3 specifies that "*it is expected to double the energy efficiency improvement, globally by 2030*" and target 7.b comprises "*developing infrastructure and improving technology to provide sustainable energy services for all*". Similarly, under economic drivers of migration, reaching better utility (electricity, gas, water) services can be identified. It can be emphasized that for well-being, a high amount of energy consumption per capita is used as one of the main indicators. World final energy consumption rises to 9983 million tonnes of oil equivalent (Mtoe) in 2019, of which 19.7% and 16.4% are comprised of electricity and natural gas, respectively, as declared by the International Energy Agency (IEA). Furthermore, about 30% of the electricity and natural gas is consumed in residential buildings [31]. It is predicted that residential electricity consumption will increase by 2% annually whereas natural gas consumption by 20% because of the increase in urbanization, owing to both domestic and international migration [30]. From these figures, it can be concluded that for sustainable energy systems, the impact of migration on energy consumption should be explored considering acceptability, availability, accessibility, and affordability issues [29].

When the migration phenomenon is considered in Turkey, the DG of Migration Management (DGMM) under the Ministry of Interior Relations announce that there are approximately four million migrants, especially those looking for international protection. This constitutes 4.88% of the total population and it is worth mentioning that it was 3.54% in 2017. Most of the migrants are from Syria (3,600,000) and 98% live in urban areas. For example, 3.68% of the population in Istanbul is Syrian migrants [11]. Hence, economic, demographic, and social strategies are very important considering these people. As a specific example, accessibility to utility services together with their consumption behavior, education, and other socio-economic facilities are accepted as prior actions by the Ministry of Development (MoD) [62]. Additionally, the "Regional Refugee & Resilience Plan for Turkey 2021–2022" summarizes the principal actions for the migrants as listed below:

- Making better conditions in camps and their households
- Accessing both formal and informal education programs
- Providing public utility (electricity, gas, and water) services, continuously
- Ensuring social unity/consistency with Turkish counterparts
- Reducing the effect of population increase due to many reasons on natural resources [64]

When energy is of concern, in Turkey as an import-dependent country, the total energy supply (TES) is about 147 Mtoe in 2020 and residential buildings consume around 27% of TES. Total energy supply increases 1.38% to meet the increase in consumption as compared to 2017 [14, 31]. National Energy Efficiency Action Plan (NEEAP) for Turkey aims to decrease primary energy consumption by 14% in 2023 as compared to 2017. In other words, 23.9 Mtoe energy saving is targeted which requires about 11 billion $ investment [44]. Moreover, a new study comprising energy demand projection for Turkey claimed that one of the main reasons for the consumption increase will be due to the growing service/commercial sector (hotels, shopping malls, etc.) demand and the growing migrated population within the country [45].

Concerning a sustainable energy system, for reliable future predictions to meet the targets, migrants' energy consumption should be investigated. In this respect, the objective of this study is to define the key drivers, affecting the electricity consumption behavior of the migrants in Turkey (mainly Istanbul). For this purpose, after Sect. 15.1, the Sect. 15.2 contains explanations of the main drivers shaping energy consumption behavior, migrated households' consumption behavior analyses, and Bayesian Belief Network (BBN) usage for the subject. In the Sect. 15.3, migrated households' survey, statistical, and BBN analyses are described. Then, the Sect. 15.4 comprises the results obtained by statistical analyses and BBN of the migrated households' survey. Lastly, in the Sect. 15.5, derived conclusions, and recommendations for more efficient energy consumption by the migrated households are listed.

15.2 Literature Review

In this section, literature is reviewed considering the main drivers shaping energy consumption behavior, migrated households' behavior analysis, and the usage of BBN for consumer behavior analysis.

15.2.1 Main Drivers Shaping Energy Consumption Behavior

A consumer can be defined as a reasonably acting individual that is willing and talented to take decisions to buy goods and to use services [42]. Concerning the role of the consumer, where all the potential actions are surrounded by equitable rights and responsibilities through the society should be improved based on the energy consumption practices together with their effects [26]. Consumers' energy consumption behaviors can be affected by both current and new policies, strategies, and technologies as well as the socio-economic situation. To support sustainable behavioral change through energy efficiency, besides conventional instruments, like laws, subsidies, and taxes [35], adding more variety using commitments/goal-setting and showing the consequences through feedback, rewards, and punishments can increase consumer engagement [48]. For instance, an analysis considering consumers' attitudes to buying green electricity in China confirms that willingness is associated with energy security and energy efficiency [24]. Similarly, it is revealed that in the United States, consumers are concerned about the effect of energy consumption on the environment and feedback is beneficial to change behaviors about the decrease in their energy consumption [10].

Numerous studies and projects, using quantitative research tools (i.e., surveying, semi-structured or open interviews, focus group meetings), were carried out to explore different characteristics of European consumers' energy consumption behaviors. As an example, the "USmartConsumer" project target to affect European households in the partner member states by informing them on energy efficiency. It is declared that project partner member states were selected to provide diversified data related to the topic of interest from the European Union (EU) for covering different geographical areas and cultural conditions [2, 66]. Moreover, 65 projects were identified comprising consumer engagement-related studies from the EU Joint Research Centre (JRC). One of the objectives is to gather information on the consumption trends based on consumers' needs and experiences. Another important objective is to give information to consumers on efficient technologies and to examine strategies for sustainable behavioral change [41].

In Turkey, through governmental policies, it is indicated that determining the knowledge level and current behaviors of consumers towards energy efficiency and revealing their needs for efficient consumption practices are very essential. For example, research conducted by the Ministry of Energy and Natural Resources, DG of Energy Efficiency and Environment (MENR – EVCED) in 2020 aims to define "Key Performance Criteria" (KPI) (Energy Efficiency Information Index and Energy Efficiency Behavior Index) necessary to decide the level of knowledge, interest, and behavior of the households about energy efficiency to establish an "Energy Efficiency Awareness Index". In this study, a total of 3000 people from different households in 26 cities of Turkey were questioned. The obtained data were examined by using the Statistical Package for the Social Sciences (SPSS) program [1]. According to the obtained results, 87.4% of households know the importance of using energy efficiently. On the other hand, 3.4% of the households do not believe that energy efficiency provides real savings, 3.4% state that they do not need it, and 2.2% think that it reduces comfort conditions [18]. Based on these results, for Turkish consumers, the specified indicators were calculated as follows:

- "Energy Efficiency Information Index" is determined as 177.9. This value indicates that the public has a high – low level of knowledge on energy efficiency.
- "Energy Efficiency Behavior Index" has a score of 137.5. This value implies that the behavioral dimension in energy efficiency is at the middle-low level. The difference between information and behavior indexes shows the area of development that should be worked on with all relevant stakeholders.
- "Energy Efficiency Awareness Index" is determined as 157.7. This value indicates that the public has a moderate level of awareness about energy efficiency [18].

15.2.2 Studies Concerning Migrants

In the literature, most of the consumption behavior analysis concerns mainly domestic consumers. However, there are vulnerable consumers which are defined as a "*consumer group, having difficulties in accessing products and services for their needs because of their specific conditions*" [28]. One of the vulnerable consumers can be considered as migrants due to various factors starting from their homes to other settlements.

From the studies in the literature (Table 15.1), it can be derived that migrated consumers usually face energy poverty, caused by poor living conditions, low household income, and high utility bills [59]. To deal with the energy vulnerability/poverty issues regarding migrants, policies and related actions should cover legislative, financial, social, and technical strategies.

15.2.3 Bayesian Belief Network Applications on the Energy Consumption

BBN which is also known as the "Bayes network or Bayesian model" can be defined as a strategic decision model to handle probabilistic events and to explain

Table 15.1 Migrants' energy consumption behavior studies [3]

Study	Objective	Method	Results
Lehne et al. [38]	Forecasting overall scale of energy poverty and three scenarios for how to improve energy access for lighting and cooking	Demand estimates by end-use accounting approach on energy consumption profiles of households	Basic solar lanterns and efficient cooking appliances will save 303 million $ in a year requiring totally 334 million $ investment
Muye et al. [47]	Determine household energy use of low-skilled internal migrants in Beijing	Total of 1300 questionnaires were used to analyze	Energy types used by internal migrants (from rural to urban areas) differ from urban residents, but the total consumption is similar
Komatsu et al. [36]	Investigating migration effects on household energy use and analyzing related emissions in Hanoi, Vietnam	Empirical research – application of Propensity Score Matching [70]	Policies and strategies should be different whether urbanization is driven by population growth or by migration
Shrestha et al. [57]	Analyzing energy consumption profiles of slum-dwellers In Bangkok and Khon-Kaen, Thailand, considering energy expenditures, and the main drivers for accessing electricity	Survey consisted of 100 households in both Bangkok (January 2007 to February 2007) and Khon-Kaen (September 2007 to October 2007)	16% and 26% of income are for monthly energy expenditures in Bangkok and in Khon-Kaen, respectively. To increase the affordability of the energy services, as a policy alternative, price subsidies, the application of an electrification program and a decrease in the service charge for the low electricity-consuming households can be proposed
Fuguitt et al. [20]	Including energy consumption as an impact of community population change based on sociological perspective	Regression analysis	Metropolitan migrants increase heterogeneity as compared to the rural community, together with high energy demand

uncertainties for a problem that is under discussion. Hence, BBN can be used for prediction, anomaly detection, diagnostics, reasoning, and decision-making under uncertainty (javatpoint.com). For this reason, in the energy field, BBN can be applied for demand optimization, energy management, energy consumption, energy efficiency, and energy-saving programs [3].

Table 15.2 represents the summary literature on BBN applications in the energy field. It can be stated that as a new evolution, consumer behavior analysis is getting more and more important. Studies, summarized in Table 15.2, explore the energy

Table 15.2 Literature on BBN applications in the energy field [3]

Study	Objective
Hawarah et al. [25]	Present a methodology considering the peak consumption times to analyze the consumers' energy needs in a household automation system
Cinar and Kayakutlu [9]	Propose a decision model for presenting different factors and their interactions affecting investments in renewable and nuclear energy in Turkey
Morris et al. [46]	Implement electricity-saving program in Australia Determine the variables that are important for decreasing the peak electricity household demand in Australia
Carbonari et al. [7]	Design an energy management system in order to define household strategies for providing the desired comfort conditions with minimum cost Develop an energy demand modelling approach basically comprising temperature, humidity, radiation, and carbon dioxide (CO_2) concentration
Hammer et al. [22]	Propose the "User Trust Model", to save energy in smart energy systems by means of automatic decisions

consumption pattern to show the similarities and differences, and also to determine the principal issues having effects on the energy consumption behavior. In this case, BBN can produce results that support model the of energy consumption, concerning consumers' behavior effect.

15.3 Methods

For the determination of relations among household characteristics and energy consumption profiles, household surveys conducted by the national statistical institutions of the country, questionnaires, and interviews together with monitoring electricity consumption through energy bills provided by the energy suppliers can be used [2].

In this study, using a literature survey concerning the energy consumption behavior analysis of the households, especially in Turkey, migrated households survey was conducted. Subsequently, descriptive statistical analyses for migrated households' data were presented. Then, electricity consumption models were constructed and a t-test comparison was made to see how migrated households differ. Finally, BBN analysis for migrated household survey was made to understand the perceptions of the migrated people on energy consumption.

15.3.1 Survey on Migrated Households

Several survey analyses have been realized to understand the effects of drivers on the behavior analysis of Turkish consumers (Table 15.3). By examining these surveys, questions for the migrated consumers were prepared.

Table 15.3 Household energy surveys were conducted in Turkey. (Adapted from Ediger et al. [13])

Study	Subject	Respondent
Erten [17]	Energy saving behavior of students	970 students from seventh, eighth and ninth grades
MENR [43]	Energy efficiency awareness situation	2531 households from seven geographical regions
Ediger and Kentmen [12]	Social dimension of energy choices	Investigation based on 1005 citizens between 2005 and 2006
Oz [50]	Energy consumption behavior in relation to environmental concerns	Households in Bursa
Energy Efficiency Association [16]	Awareness of women on energy efficiency	2929 women in 20 cities
Istanbul Metropolitan Municipality [27]	Energy efficiency in daily life	1124 participants
Diyarbakir Chamber of Electrical Engineers [60]	Relationship between electricity usage and consumer behaviors	Households in Diyarbakir
Özcan et al. [52]	Demographic and economic factors affecting household energy consumption	Household Budget Survey data for the period 2002 and 2006 comprising a totally 181,548 households
Sahin and Köksal [56]	Households' electricity consumption and saving potential	260 households in Ankara
Emeç et al. [15]	Energy poverty and energy preferences	Data from the Turkish Statistical Institute
Ulukavak-Harputlugil and Harputlugil [63]	Households consumer behavior on environmental issues and energy conservation	2578 households
Tumbaz and Moğulkoç [61]	Attitudes and behaviors of Turkish households concerning energy efficiency	526 households
Ediger et al. [13]	Households' energy consumption, related policy preferences with environmental concerns	Turkish Public Preferences Energy Survey, 1204 respondents

In the survey, there are two parts. In the first part, the main aim is to collect the energy consumption and related data and in the second part to determine the profile of consumption behavior of the migrated consumers. The main topics covered in these two parts can be listed, as follows:

- First part:
  - General information: district, duration of the residence
  - Information about the building: type, age, insulation situation
  - Information on the household characteristics: owner or tenant, area, number of rooms, number of the occupants

  - Energy consumption information: energy bills, heating, and water-heating systems, electric appliances, lighting system
- Second part:
  - Typical energy consumption behaviors
  - Knowledge already gained and desired to be learned about efficient energy consumption

15.3.2 Bayesian Belief Network

BBN [33, 54] is defined as a probabilistic model for integrating both quantitative and qualitative data, particularly for data-limited circumstances, i.e., to represent expert opinions, to identify significant uncertainties, and to determine relations among variables [8, 23, 37].

BBN has both a qualitative component, a directed acyclic graph (DAG) representing dependencies among variables, and a quantitative component, Conditional Probability Tables (CPTs) providing the strengths of this dependency [5]. The directed acyclic graph covers nodes, indicating the system variables, and arrows, demonstrate the interactions among these nodes. A node can be either continuous or discrete, and the beliefs regarding these nodes are clarified using probability distributions. Moreover, conditional probability tables enumerate the probability distribution of variables, based on Bayes' theorem designated as:

$$P(A\backslash B) = \frac{P(B\backslash A).P(A)}{P(B)} \tag{15.1}$$

$P(A)$, $P(B)$: the probabilities of observing A and B (which are independent of each other)
$P(A\backslash B)$: conditional probability of A, given B
$P(B\backslash A)$: conditional probability of B, given A
$P(B\backslash A)/P(B)$: "Bayes factor" [55]

It should be noted that $P(A\backslash B)$ can be found if $P(A)$, $P(B)$, and $P(B\backslash A)$ are known.

Apart from presenting which variables are relevant, without any numerical calculations, BBN can form a statistical base, in case of data limitation. Accordingly, BBN can be stated as a beneficial tool for modelling current knowledge and helping the decision-makers, because it describes uncertainties and complexities using the initial state of the system [19, 53]. Additionally, more benefits of BBN can be summarized as follows:

- Due to being transparent in nature, relationships among variables can be obvious [4].
- After constructing a model, the probability distribution of a specified node can be obtained by using its parents. It is also possible that the parent nodes' probability

distribution is found by using their child nodes. In other words, BBN demonstrates feedback loops [69].

- BBN can be used in the environmental and social sciences, to reflect experts' opinions as a valuable contribution [21, 40, 68].
- BBN can succeed in missing inputs and estimate them accurately [6, 51, 67, 71].

Consequently, BBN can be regarded as a beneficial tool under uncertainty, i.e., limited and/or missing data, to integrate numerous system components and to give opportunity for reflecting experts' and/or stakeholders' opinions where the relations among variables are non-linear or complex [4].

To construct the BBN model in the study, Netica (version 5.12) software is used. Netica contains a set of nodes and links showing variables and their cause-effect relationships, respectively, and a set of probabilities indicating the beliefs in a given state of connecting nodes [58]. In addition, the direction of arrows represents that of causality [39].

15.4 Results and Discussions

BBN can be classified as artificial intelligence (AI) method that supports decision-makers to decide and take action. In this study, it is used to see the efficiency of the energy consumption behavior of the migrated households. Based on the conducted survey, the main categories are determined as knowledge, habit, knowledge demand, from whom they desire to gain knowledge, method of the knowledge demand, and types of financial support that they want (Fig. 15.1).

From Fig. 15.1, it can be claimed that in the current situation the energy consumption profile of the migrated consumers is inefficient at 73.1%. They have basic knowledge of efficient lighting and standby power. They unconsciously use single-term tariffs. They desire to learn effective usage of not only electricity but also natural gas and water utilities, with a ratio of 63%. Mostly, they want to contact the related non-governmental organizations (NGOs), municipality representatives, and also teachers/academicians and they desire brochures in their languages. Fifty-five percent of the migrated households want a reduction in their utility bills as a financing mechanism.

In a study conducted by Acuner and Kayalica [2], experts' opinions were taken by the prepared questionaries to find the key variables, for modeling migrated household electricity consumption behaviors. In this survey, it is desired to decide on the efficient usage of electricity by migrated consumers concerning socio-economic factors, actors in the market, strategies that can be used, tools for improvement, and financing methods to support the energy efficiency applications (Fig. 15.2).

From the experts' opinions, to be 73% efficient while consuming electricity by the migrated households the following results can be derived:

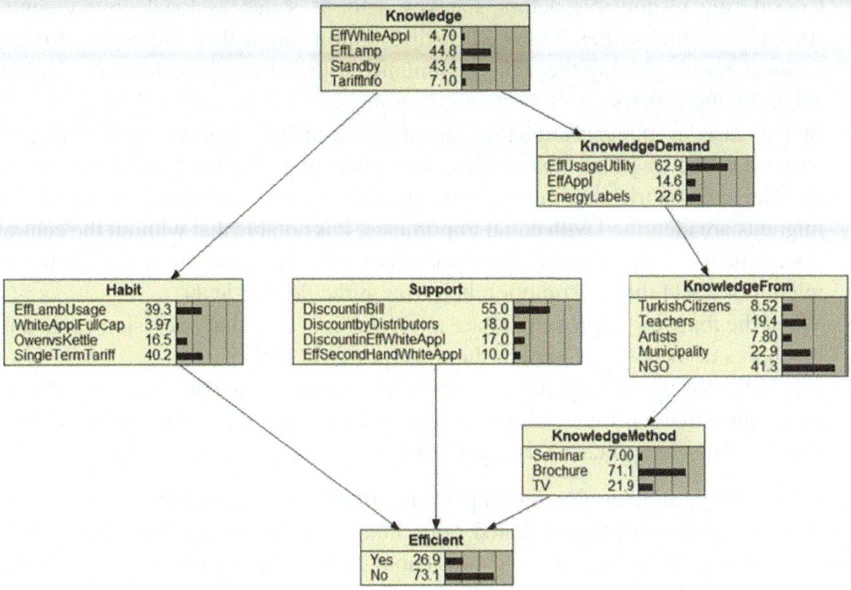

Fig. 15.1 BBN analysis of the migrated households' energy consumption behavior

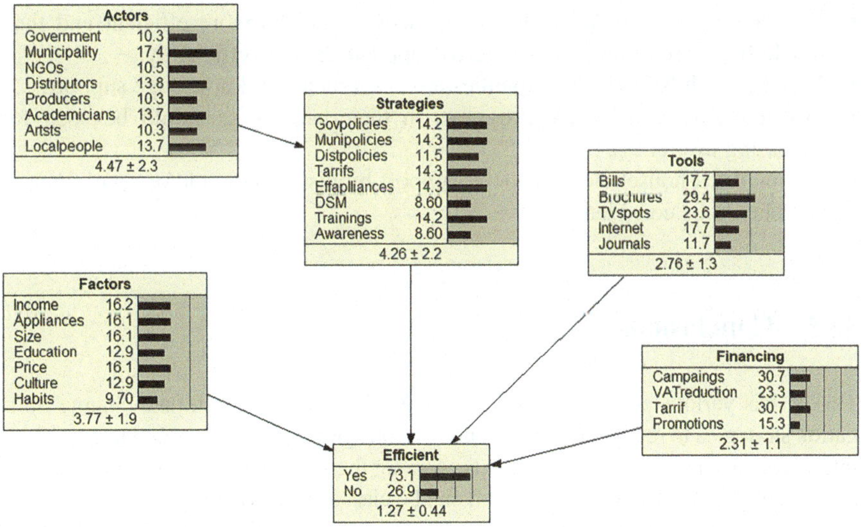

Fig. 15.2 BBN based on experts' opinions

- When socio-economic factors are analyzed, income level, number of residents, market electricity prices, and the status of electric appliances are observed as equivalently significant.

- Considering responsible actors, the key actor is stated as local governments, basically, municipalities, followed by distribution companies. Moreover, domestic/local people sharing the same community with the migrants are stated as key actors by the experts.
- In the case of policies/strategies, local government's policies and strategies specific to the migrated households, arrangements of energy tariffs, promotion of efficient electric appliances, and specific training activities designed for migrants are identified with equal importance. It is notable that without the central government's support using appropriate policies/strategies, the listed strategies might not affect the consumption behavior at the desired level.
- When the tools for improvement are analyzed, it is seen that television spots and brochures in their languages are the most powerful tools.
- It is indicated by the experts that energy efficiency campaigns and tariff adjustments are effective financial tools. Also, value-added tax reduction while purchasing efficient appliances is significant.

Additionally, to determine the important variables to increase the efficient electricity consumption among migrated households, the expert opinion survey BBN results are applied to the migrated household survey BBN analysis in which the probability distribution of parent nodes can also be determined by using their child nodes.

According to Fig. 15.3, the following results can be obtained:

- To change habits, their electric appliances should be more efficient and they should be aware of how to use electric appliances more efficiently.
- Together with NGOs, municipalities are also crucial for knowledge support.
- Brochures, as well as TV programs in their languages, should be used for increasing awareness.
- For migrated households, financial support by distribution and/or retail companies should be considered.

15.5 Conclusions

Turkey has various energy efficiency, renewable energy, and climate change mitigation strategies to reduce its dependence on foreign energy sources and to cope with international standards and agreements. Among these strategies, households' energy consumption behavior analysis, which can be classified as a no or low-cost measure, can be regarded as an effective tool. If consumers are informed about different measures/mitigation strategies to save energy, consumers' awareness will increase leading to behavioral changes and causing a reduction in energy consumption by between 15% and 25% [34].

BBN was used for taking opinions of migrated households on energy consumption and analyzing the behaviors of migrants. There is also an expert opinion survey conducted by Acuner and Kayalica [2]. According to the energy experts' opinions,

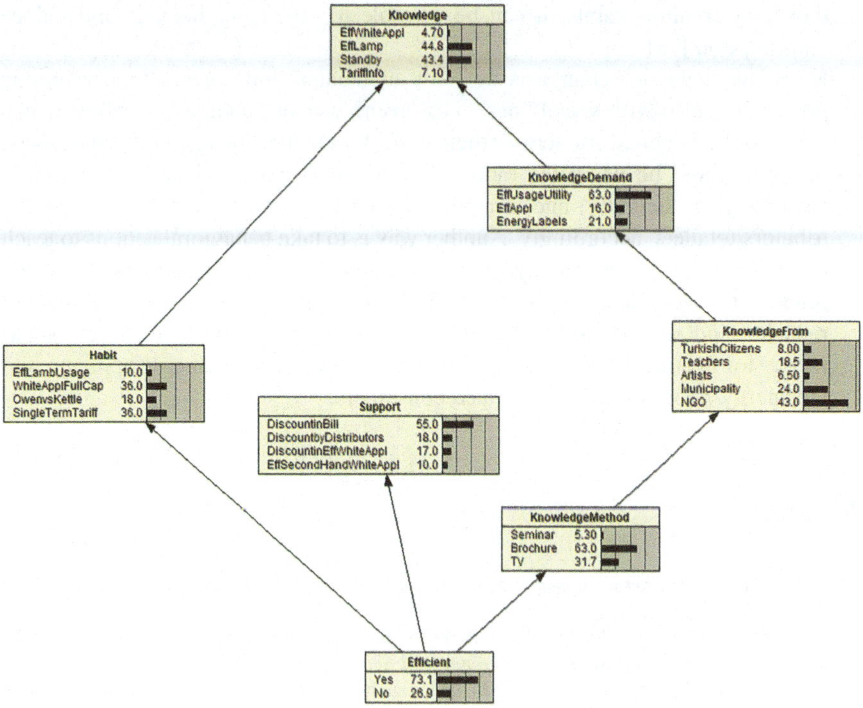

Fig. 15.3 BBN results according to an expert opinion survey for migrated households

by means of effective central and local government policies and strategies, arrangement of tariffs, training for efficient use of white appliances through brochures, and TV spots in their languages are crucial depending on migrants' income and education level, culture from their countries, habits gained in Turkey through campaigns concerning efficient white appliances and utility bill reductions. While migrated households know efficient lighting and standby power together with a single-term tariff (they may only hear about it), they desire to learn efficient usage of all utilities from basically NGOs (NGOs may give some promotions and can speak their languages) than the municipalities by means of brochures and TV spot announcements in their languages. In the end, they want to see a reduction in their utility bills.

As a conclusion, the following recommendations can be stated:

- For the behavior change in favor of efficiency, one of the key subjects is the non-energy benefit (NEB), which is any benefit other than energy-saving received from an energy efficiency activity. It can be specified that 2.5 times more non-energy benefits can be obtained as compared to energy benefits. Although NEB is not a new thing and has been studied for more than 25 years, the concept has been very slow to be recognized [72]. For example, behavior change can be regarded as an NEB. By providing consumers with accurate information about their consumption habits, it can be possible to efficiently manage and reduce consumption [73].

- Moreover, behavior change is not only associated with energy and economic policies but also with social ones, "i.e., ensuring poor/vulnerable consumers can reach to the benefits of energy efficiency". To identify energy-poor consumers, one way could be using the ratio between energy cost and income (monthly, seasonally, or annually) and arranging the tariffs together with the energy distributors/retailers, accordingly. Another way is to take behavioral actions to teach efficient usage to the energy-poor/vulnerable/migrated consumers by social funds, arranged in cooperation with the local governments, energy distributors/retailers, and related NGOs. In this way, they can be used to distribute smart meters for the true arrangement of tariffs and more efficient electric appliances to the energy-poor/vulnerable/migrated consumers.

References

1. Acuner, E. (2020). *Syrian household energy consumption in Istanbul* [Ph.D. thesis]. Istanbul Technical University.
2. Acuner, E., & Kayalica, M. O. (2018). A review on household energy consumption behavior: How about migrated consumers? *Environmental Economics, 9*(4), 8–21.
3. Acuner, E., & Kayalica, M. O. (2020). Behavior analysis of refugees' electricity consumption in developing countries. *IEEE Transactions on Engineering Management, 69*, 1206–1215.
4. Aguilera, P. A., Fernandez, A., Fernandez, R., Rumi, R., & Salmeron, A. (2011). Bayesian networks in environmental modelling. *Environmental Modelling & Software, 26*(12), 1376–1388.
5. Borsuk, M. E., Stow, C. A., & Reckhow, K. H. (2004). A Bayesian network of eutrophication models for synthesis, prediction, and uncertainty analysis. *Ecological Modelling, 173*, 219–239.
6. Bressan, G. M., Oliveira, V. A., Hruschka, E. R., & Nicoletti, M. C. (2009). Using Bayesian networks with rule extraction to infer risk of weed infestation in a corn-crop. *Engineering Applications of Artificial Intelligence, 22*, 579–592.
7. Carbonari, A., Vaccarini, M., & Giretti, A. (2014). Bayesian networks for supporting model-based predictive control of smart buildings. In *Dynamic programming and Bayesian inference, concepts and applications* (pp. 3–39). InTech.
8. Cheon, S. P., Kim, S., Kim, J., & Kim, C. (2008). Learning Bayesian networks-based diagnosis system for wastewater treatment process with sensor data. *Water Science and Technology, 58*(12), 2381–2393.
9. Cinar, D., & Kayakutlu, G. (2010). Scenario analysis using Bayesian networks: A case study in the energy sector. *Knowledge-Based Systems, 23*, 267–276.
10. DeCicco, J., Yan, T., Keusch, F., Munoz, D., & Neidert, L. (2015). U.S. consumer attitudes and expectations about energy. *Energy Policy, 86*, 749–758.
11. DG of Migration Management (DGMM). (2019). *Statistics on migration*. Ministry of Interior Relations.
12. Ediger, V. Ş., & Kentmen, Ç. (2010). Social dimension of energy and energy choices of Turkish public. *Mülkiye Dergisi, 34*(268), 281–300.
13. Ediger, V. Ş., Kirkil, G., Çelebi, M., Ucal, Ç., & Kentmen, Ç. (2018). Turkish public preferences for energy. *Energy Policy, 120*, 492–502.
14. EIGM, Ministry of Energy and Natural Resources DG of Energy Affairs. (2020). *General energy balances 2020*. https://enerji.gov.tr/enerji-isleri-genel-mudurlugu-denge-tablolari

15. Emeç, H., Altay, A., Aslanpay, Y., & Özdemir, M. O. (2015). Energy poverty and energy choice profile in Turkey. *Finans-Politik & Ekonomik Yorumlar, 52*(608), 9–22.
16. ENVER. (2012). *Evaluation of women's awareness on energy efficiency*. Enerji Verimliliği Derneği.
17. Erten, S. (2002). Investigation of energy-saving behaviors of the students. *Hacettepe University Journal of Education, 22*, 67–73.
18. EVCED, DG of Energy Efficiency and Environment. (2020). *Energy efficiency awareness index report* (Yayın No: 18/V1). Enerji Verimliliği ve Çevre Dairesi Başkanlığı.
19. Forio, M. A., Landuyt, D., Bennetsen, E., Lock, K., Nguyen, H. T., Ambarita, M. N. D., et al. (2015). Bayesian belief network models to analyse and predict ecological water quality in rivers. *Ecological Modelling, 312*, 222–238.
20. Fuguitt, G., Heberlein, T. A., & Rathbun, P. R. (1991). Migration consequences for household energy consumption in a nonmetropolitan recreation-retirement area. *Rural Sociology, 56*(1), 56–69.
21. Haapasaari, P., & Karjalainen, T. P. (2010). Formalizing expert knowledge to compare alternative management plans: The sociological perspective to the future management of Baltic salmon stocks. *Marine Policy, 34*(2010), 477–486.
22. Hammer, S., Wissner, M., & Andre, E. (2015). Trust-based decision-making for smart and adaptive environments. *User Modeling and User-Adapted Interaction, 25*(3), 267–293.
23. Hanea, M. D., Jagtman, H. M., Van Alphen, L. M., & Ale, B. J. (2010). Quantitative and qualitative analysis of the expert and non-expert opinion in fire risk in buildings. *Reliability Engineering & System Safety, 95*(7), 729–741.
24. Hast, A., Alimohammad, B., & Syri, S. (2015). Consumer attitudes towards renewable energy in China—The case of Shanghai. *Sustainable Cities and Society, 17*, 69–79.
25. Hawarah, L., Ploix, S., & Jacomino, M. (2010). User behavior prediction in energy consumption in housing using Bayesian networks. In *10th international conference artificial intelligence and soft computing* (pp. 372–379). Springer.
26. Huijts, N. M. A., Molin, E. J. E., & Steg, L. (2012). Psychological factors influencing sustainable energy technology acceptance: A review-based comprehensive framework. *Renewable and Sustainable Energy Reviews, 16*(1), 525–531.
27. IBB. (2012). *Istanbul Metropolitan Municipality: "Energy efficiency in daily life"*. http://www.ibb.gov.tr/_layouts/download.aspx?SourceUrl=%2FtrTR%2FDocuments%2Fhaber%2Focak2011%2Fanketsonuclar.doc
28. INSIGHE. (2015). *"Energy poverty and vulnerable consumers in the energy sector across the EU": Analysis of policies and measures*. Retrieved: September, 29, 2018 from http://www.insightenergy.org
29. International Energy Agency (IEA). (2011). *IEA model of short-term energy security (MOSES)*. https://www.iea.org/media/freepublications/oneoff/moses_paper.pdf
30. International Energy Agency (IEA). (2019). *Key world energy statistics*. https://www.iea.org/publications/freepublications/publication/KeyWorld2017.pdfIEA
31. International Energy Agency (IEA). (2021). *Key world energy statistics*. https://www.iea.org/reports/key-world-energy-statistics-2021/final-consumption
32. International Organization for Migration (IOM). (2020). *World migration report*. https://publications.iom.int/system/files/pdf/wmr_2020.pdf
33. Jensen, F. V., & Nielsen, T. D. (2007). *Bayesian networks and decision graphs*. Springer.
34. Kama, Ö., & Kaplan, Z. (2012). A recommendation for increasing energy efficiency in Turkey: Incentives for electric appliances. *Hukuk ve İktisat Araştırmaları Dergisi, 4*(1), 201–211.
35. Kolk, A. (2012). The role of consumers in EU energy policy. *Carbon Management, 3*(2), 175–183.
36. Komatsu, S., Ha, H. D., & Kaneko, S. (2013). The effects of internal migration on residential energy consumption and CO2 emissions: A case study in Hanoi. *Energy for Sustainable Development, 17*, 572–580.

37. Landuyt, D., Broekx, S., D'Hondt, R., Engelen, G., Aertsens, J., & Goethals, P. L. M. (2013). A review of Bayesian belief networks in ecosystem service modelling. *Environmental Modelling & Software, 46*, 1–11.
38. Lehne, J., Blyth, W., Bazilian, M., & Grafham, O. (2016). Energy services for refugees and displaced people. *Energy Strategy Reviews, 14*, 134–146.
39. Li, J., Song, G., Semakula, H. M., & Zhang, S. (2019). Climatic burden of eating at home against away-from-home: A novel Bayesian Belief Network model for the mechanism of eating-out in urban China. *Science of the Total Environment, 650*, 224–232.
40. McDowell, R. W., Nash, D., George, A., Wang, Q. J., & Duncan, R. (2009). Approaches for quantifying and managing diffuse phosphorus exports at the farm/small catchment scale. *Journal of Environmental Quality, 38*, 1968–1980.
41. Mengolini, A. M. (2017). *Prosumer behavior in emerging electricity systems* [Doctoral dissertation]. Retrieved from https://doi.org/10.6092/polito/porto/2675327
42. Micklitz, H. W., & Reisch, L. (2011). An introduction to the special issue on behavioral economics, consumer policy and consumer law. *Journal of Consumer Policy, 34*, 271–276.
43. Ministry of Energy and Natural Resources (MENR). (2010). *Report on energy efficiency in households*. Enerji ve Tabii Kaynaklar Bakanlığı.
44. Ministry of Energy and Natural Resources (MENR). (2019). *Progresses about Turkish National Energy Efficiency Action Plan*. Ministry of Energy and Natural Resources.
45. MMO, Chamber of Mechanical Engineers. (2019). *Energy outlook of Turkey*. Turkish Chamber of Mechanical Engineers.
46. Morris, P., Vine, D., & Buys, L. (2013). Application of a Bayesian network complex system model to a successful community electricity demand reduction program. *Energy, 84*, 63–74.
47. Muye, R., Shu, T., Kirk, S., Guofeng, S., Huizhong, S., Ye, H., Han, C., Yilin, C., Xi, C., Junfeng, L., Bengang, L., Xilong, W., & Canfei, H. (2015). Direct energy consumption associated emissions by rural-to-urban migrants in Beijing. *Environmental Science and Technology, 49*(22), 13708–13715.
48. Ofgem. (2016). *Retail energy markets in 2016*. https://www.ofgem.gov.uk/system/files/docs/2016/08/retail_energy_markets_in_2016.pdf
49. Oxford Dictionary. https://www.ourmigrationstory.org.uk/about/what-is-migration.html
50. Oz, M. E. U. (2011, March 13–16). Relationship between residential energy consumption and environmental factors: Bursa case. In *Proceedings of 10th National plumbing engineering congress* (pp. 26–33).
51. Ozbay, K., & Noyan, N. (2006). Estimation of incident clearance times using Bayesian networks approach. *Accident Analysis & Prevention, 38*, 542–555.
52. Özcan, K., Gülay, E., & Üçdoğruk, Ş. (2013). Economic and demographic determinants of household energy use in Turkey. *Energy Policy, 60*, 550–557.
53. Pavan, A. M., Mellit, A., De Pieri, D., & Kalogirou, S. A. (2013). A comparison between BNN and regression polynomial methods for the evaluation of the effect of soiling in large scale photovoltaic plants. *Applied Energy, 108*, 392–401.
54. Pearl, J. (1986). Fusion, propagation and structuring in belief networks. *Artificial Intelligence, 29*, 241–288.
55. Phan, D. T., Smarta, J. C. R., Capon, S. C., Hadwen, W. L., & Oz, S. (2016). Applications of Bayesian belief networks in water resource management: A systematic review. *Environmental Modelling & Software, 85*, 98–111.
56. Sahin, M. C., & Köksal, M. A. (2014). Standby electricity consumption and saving potentials of Turkish households. *Applied Energy, 114*, 531–538.
57. Shrestha, S., Ram, M., Kumar, S. M., & Dhakal, A. (2008). Modern energy use by the urban poor in Thailand: A study of slum households in two cities. *Energy for Sustainable Development, 12*(4), 5–13.
58. Song, G., Semakula, H. M., & Palmer, P. F. (2018). Chinese household food waste and its' climatic burden driven by urbanization: A Bayesian Belief Network modeling for reduction possibilities in the context of global efforts. *Journal of Cleaner Production, 202*, 916–924.

59. Thomson, H., & Snell, C. (2013). Quantifying the prevalence of fuel poverty across the European Union. *Energy Policy, 52*, 563–572.
60. TMMOB EMO. (2012). *Evaluation of electricity consumption profiles of Diyarbakir households and determination of problems concerning electricity usage*. Elektrik Mühendisleri Odası.
61. Tumbaz, M. N. M., & Moğulkoç, H. T. (2018). Profiling energy efficiency tendency: A case for Turkish households. *Energy Policy, 119*, 441–448.
62. Turkish Informatics Foundation (TIF). (2016). *Evaluation report of smart city concept for Turkey*. Turkish Informatics Foundation.
63. Ulukavak-Harputlugil, G., & Harputlugil, T. (2016). Household behavior analysis on energy saving and environmental comfort. *Journal of the Faculty of Engineering and Architecture, 31*(3), 695–708.
64. UNHCR, UN Refugee Agency. (2021). *Regional refugee & resilience plan: Turkey country chapter 2021–2022*. https://www.unhcr.org/tr/wp-content/uploads/sites/14/2021/03/3RP-Turkey-Country-Chapter-2021-2022_EN-opt.pdf
65. United Nations (UN). (2020). https://www.un.org/en/global-issues/migration
66. USmartConsumer FP7 Project. (2015). https://ec.europa.eu/energy/intelligent/projects/en/projects/usmartconsumer
67. Uusitalo, L. (2007). Advantages and challenges of Bayesian networks in environmental modelling. *Ecological Modelling, 203*, 312–318.
68. Wang, Q. J., Robertson, D. E., & Haines, C. L. (2009). A Bayesian network approach to knowledge integration and representation of farm irrigation: 1. Model development. *Water Resources Research, 45*(2), 1–18.
69. Wooldridge, S., & Done, T. (2004). Learning to predict large-scale coral bleaching from past events: A Bayesian approach using remotely sensed data, in-situ data, and environmental proxies. *Coral Reefs, 23*, 96–108.
70. Rosenbaum, P. R., & Rubin, D. B. (1983). The central role of the propensity score in observational studies for causal effects. *Biometrika, 70*(1):41–55.
71. Nadkarni, S., & Shenoy, P. P. (2004). A causal mapping approach to constructing Bayesian networks. *Decision Support Systems, 38*, 259–281.
72. Gudbjerg, E. (2019). Non-energy benefits and behavior change to EE, Energy Efficiency International Search Conference, Ankara, Turkey, October 4.
73. Belotto, J. M. (2019). EE directives and standardization, Energy Efficiency International Search Conference, Ankara, Turkey, October 4.

Part VI
Modelling the Sustainable Future

Chapter 16
Informativeness in Twitter Textual Contents for Farmer-Centric Pest Monitoring

Shufan Jiang, Rafael Angarita, Stéphane Cormier, Julien Orensanz, and Francis Rousseaux

Abstract Data mining in social media has been widely applied in different domains for monitoring and measuring social phenomena, such as opinion analysis towards popular events, sentiment analysis of a population, detecting early side effects of drugs, and earthquake detection. Social media attracts people to share information in open environments; however, new forms of bias and nuances can emerge. Facing the newly forming technical lock-ins and the loss of local knowledge in agriculture in the era of digital transformation, the urge to re-establish a farmer-centric precision agriculture is urgent. The question we ask is that whether social media like Twitter can help farmers to share their observations at the field-level towards the constitution of agricultural knowledge bases. To answer that question, in this work we chose several scenarios to collect tweets, then we applied different natural language processing techniques to measure the informativeness of tweets in French as a complementary source for pest monitoring.

Keywords Crowdsensing · Social media · Smart Agriculture · NLP

S. Jiang (✉)
Institut Supérieur d'Electronique de Paris, LISITE, Paris, France

Université de Reims Champagne Ardenne, CReSTIC, Reims, France
e-mail: jiang.chou.fan@gmail.com

R. Angarita
Institut Supérieur d'Electronique de Paris, LISITE, Paris, France
e-mail: rangarit@parisnanterre.fr

S. Cormier · F. Rousseaux
Université de Reims Champagne Ardenne, CReSTIC, Reims, France
e-mail: francis.rousseaux@univ-reims.fr

J. Orensanz
Cap2020, Gradignan, France

G. Kayakutlu, M. Ö. Kayalica (eds.), *Decision Making Using AI in Energy and Sustainability*, Applied Innovation and Technology Management,
https://doi.org/10.1007/978-3-031-38387-8_16

16.1 Introduction

Facing the challenge of growing population and changing alimentary habits, precision agriculture emerges to increase food production sustainability. An important subject in precision agriculture is to measure natural hazards within global and local aspects, and then improve the risk prevention tasks while limiting the usage and the impact of pesticides. We can classify mainstream real-time monitoring technologies of natural hazards into two categories [15]: (i), indirect monitoring scheme by analysing environment parameters produced by sensor networks and Internet of Things (IoT) devices to infer the probability of phytosanitary[1] risks [38]; and (ii), direct monitoring scheme by processing images [40]. However, current precision agriculture technologies favour large-scale monoculture practices that are unsustainable and economically risky for farmers [17]. Moreover, according to the Food and Agriculture Organization of the United Nations, farms of less than 2 hectares accounted for 84% of all farms worldwide in 2019, and most of these small farms are family farms [33].

We suggest that observation data from precision agriculture cannot represent all forms of farms, especially small farms. Recently, how to encourage the participation of farmers to share their knowledge and observations is drawing the attention of researchers [23, 25]. The knowledge we can acquire from actual precision practices is solid scientific knowledge; however, it has not gained enough support [20], This results in the loss of legitimacy of local and traditional knowledge and, therefore, in the vanishing of traditional knowledge. As [20] points out that farmer's tacit knowledge is seen to rely on social process for knowledge exchange, the reducing number of farmers and the individualism weaken the local ties of blood and neighbourliness for knowledge acquisition. The diversification of professions in the agricultural domain also destabilizes traditional structures of professional sociability [48]. Moreover, some traditional knowledge, such as solar terms in traditional Chinese lunisolar calendars, fail to work due to the impact of climate change and of intensive industrialized farming on the ecosystem [39]. Researcher studied the role of Twitter in farmer-to-farmer and in farmer-to-rural-profession knowledge exchange and suggests that the use of Twitter among rural professionals and farmers is well evolved with open participation, collaboration (retweeting) and fuller engagement (asking questions, providing answers/replies) dominating one-way messaging (new/original tweets).

Following the *social sensing* paradigm [50], individuals – whether they are farmers or not – have more and more connectivity to information while on the move, at the field-level. Each individual can become a broadcaster of information. In this sense, real-time hazard information is published in social networks such as Twitter. Indeed, Twitter enables farmers to exchange experience among them, to subscribe to topics of interest using hashtags and to share real-time information about natural hazards. Compared to paid applications, information on Twitter,

[1] Relating to plant diseases especially in agricultural crops.

presented in form of text, image, sound, video or a mixture of the above, is more accessible to the public but less formalized or structured. More and more farmers get involved in online Twitter communities by adding hashtags as #AgriChatUK[2] or #FrAgTw,[3] to their posts on Twitter [10]. Thus, Twitter we can see Twitter as an open source of farmer to farmer knowledge exchange.

In social media such as Twitter, farmers publish in natural written language their empirical knowledge of farmers, motivations of their holistic decisions, judgement and prediction for the yield. Given the nature of such publications, it is not straightforward to efficiently and effectively take advantage of the information they contain, let alone doing it automatically and relating these data to data coming from other sources such as sensors or other information systems. To handle, process, and make these data useful, it is necessary to start by classifying its textual content automatically. This paper tackles the following question: which health information about plants can we automatically extract from textual contents on Twitter, and what is the quality of this information?

16.2 Related Works

16.2.1 Crowdsensing for Agriculture

Crowdsensing is a sensing paradigm that empowers ordinary people to contribute with data sensed from or generated by their sensor-enhanced mobile devices [7, 13, 14]. It introduces a new shift in the way we collect data by permitting to acquire local knowledge through smart devices carried by people, such as smartphones, tablets, smartwatches, among others. Benefiting from human intelligence, crowdsensing processes allow to leverage enhanced sensors of smartphones in a fast and economical way, in contrast to more expensive traditional methods like wired sensor networks. Driven by the increasing recognition of the central role of farmers and the diversity of farm types, farming styles, and producer characteristics in the digitization of agriculture [27], we have witnessed the emergence of crowdsensing applications for smart farming [35]. The advantage of such applications is that as they predefine the data they collect and the role of users, they could gather structured and meaningful information for decision-making process. For example, [46] designs a set of combo boxes and a text field for farmers to report pest/disease incidence to agricultural experts. However, as these applications are used by a specific group of registered users, and they are built for collecting specific observations, in the use case of crop health monitoring, these applications lack of the adaptivity of for emerging issues and the interoperability between coexisting applications.

[2] http://www.agrichatuk.org/

[3] https://franceagritwittos.com/

16.2.2 NLP for Twitter-Based Crowdsensing

Comparing to specific applications, social media such as Twitter can provide a massive community and an "open" publish/subscribe infrastructure to integrate sensors and smart devices for a more ubiquitous crowdsensing [11, 51]. Though the openness may come with the loss of information accuracy and completeness, Twitter has been used for monitoring environment, infrastructure, and social issues. Depending on studied objects, Twitter-based crowdsensing applications can be categorized as: (a) feature-based applications, which focus on behaviours and interactions among tweets and users, such as trend analysis [22] and influence detection [8] and (b) content-based applications, which focus on the semantic of textual or imaginary contents, such as topic detection and tracking [3, 18], sentiment analysis [24, 36] and event summarization [30].

In our case, as the number of tweets in agricultural domain is small, all the users are not always engaged in sharing information, the visibility and the influence of the tweets is low, we work on the textual content of tweets to observe crops through the eyes of farmers. Regarding the tools for textual data analysis, statistical language models, such as bag of word, Word2Vec [16] and S. Jiang et al. BERT [12], capture latent semantic features of the tweets in vectorial representations [3]; semantic resources such as ontologies and controlled vocabularies provides domain knowledge for information extraction and the interface for data integration [19, 31]; linguistic tools such as Nooj [44] and Gate [9] allow processing tweet at the orthographical, lexical, morphological, syntactic and semantic levels [37]; machine learning algorithms combined with vectorial text representation or linguistic features, supervised or unsupervised, such as classification, clustering, topic modelling, frequent pattern mining and matrix factorization help to automatically detect and categorize events on Twitter [18, 24, 30].

16.3 Use Cases and Methodology

We focus on detecting anomalies concerning crop health events. Possible anomalies include the time of the event, e.g., too early in the year, the place of the event or the path taken by the pest and the intensity of the attacks. In collaboration with experts in the agricultural domain from Cap2020 (https://www.cap2020.online/) and Arvalis (https://www.english.arvalisinstitutduvegetal.fr/index.jspz), we collected tweets concerning the following issues as observation cases:

- "pyrale du maïs" (*corn borer* in French), representing an insect pest
- "rendement OR moisson OR récolt", for the yield of straw cereals, representing an agricultural activity
- "JNO" (abbreviation for *jaunisse nanisante de l'orge*, barley yellow-dwarf virus (BYDV) in French), representing plant diseases

- "corvidé OR corbeau freux OR choucas de tour OR corneille", for the damage caused by corvids, representing emerging unusual damages

To study these issues, we conceived the following methodology:

1. Collection and filtering of tweets with an initial set of keywords and a prior knowledge of the contexts of events such as cause, results, date and region.
2. Plotting of the historical distribution of the collected tweets to verify whether the topic popularity corresponds to prior knowledge or documented data.
3. Examination of collected tweets with unsupervised algorithms: Latent Dirichlet Allocation [6] and K-Means [45] to extract concepts and to refine the scope of the topic.
4. For the cases with a voluminous collection of tweets such as "pyrale", "JNO" and "corvids", we extract a subset of tweets (between 500 and 3000 distinct text values) to label: whether the text is about general information or a contextualized observation. From the labelled tweets, we build a classifier for event detection.

16.3.1 Data Collection

We use the full-archive search endpoint of the Twitter API[4] to collect tweets. This endpoint allows us to retrieve public tweets from the complete archive dating back to the first tweet in March 2006. It can return up to 500 tweets per request in reverse-chronological order, and pagination tokens are provided for paging through large sets of matching tweets. The API allows us to send 300 requests per 15 min, and for each HTTP request, we can add a query of 1024 characters maximum to the body of the request message. The query is a JSON object that allow us to filter tweets on certain pre-defined fields, like keywords, period, place, language and hashtag, combined with logic operators. When using the API, the matching of keyword is applied to not only the text field of a tweet, but also the username of the author or the text content of the referenced tweet. Moreover, accented and special characters are normalized to standard Latin characters, which can change meanings in foreign languages or return unexpected results. For example, "maïs", which is *corn* in French, will also match "maïs" which means "but" in English. Thus, we isolated accented keywords as special cases: for each accented word, we pulled all the normalized word-filtered tweets from the Twitter API, and then we filtered them again with the accented word in our database. We saved original tweets as well as re-tweets. We have collected in total 16,345 tweets about "pyrale du maïs", 3302 tweets about "JNO", 50,902 tweets about straw cereals and 38,903 tweets about corvids.

[4] https://developer.twitter.com/en/docs/twitterapi/tweets/search/introduction

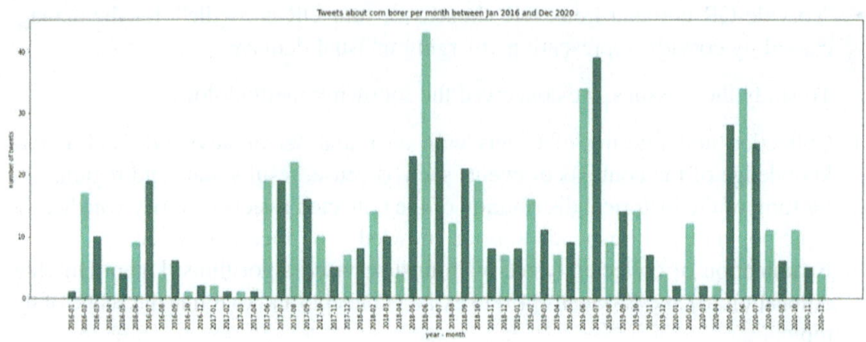

Fig. 16.1 Number of tweets containing "pyrale" and "maïs" by month between 2016 and 2020

16.3.2 Histogram by Mention of Keywords

In this very early stage of information extraction, we do not know what can be found from the mass of tweets. We start with the keywords, as the Twitter Search API asks for a query phase. Indeed, depending on the scope of monitoring, keywords-based search could give useful information, for example, TwitterMonitor [34] used busty keys as entry points for trend detection in real-time.

Use case: corn borer, *pyrale du maïs* in French, are moths that lay their eggs on leaves of maize plant. Their larvae, issued from these eggs, move towards the stem and the corncobs, weaken the plant and eventually causes loss in the yield. It is a challenge to distinguish the larvae of corn borers from the larvae of other moths using language models to learn its context. Another task is to track their propagation timeline. First, we want to confirm that people do talk about corn borers on Twitter. We used "pyrale" as keyword to retrieve tweets from Twitter API, then we kept the tweets that contain "maïs" to construct the dataset. We plot the number of tweets by month and by year in Fig. 16.1, and compare it with records of average corn borer number by trap from Arvalis (see Fig. 16.2). In both figures, we can clearly observe peaks of corn borer between May and August. There is an exception in Fig. 16.1 since there are minor peaks in February, which correspond to the Paris International Agricultural Show (https://en.salon-agriculture.com/) when there people discussed about technologies to fight corn borers. Such exception show that tweets collected by keywords is not precise enough.

Use case: yield of cereals We repeated the same data collection process for the yield of wheat. In this case, we used all the cereals in the French Crop Usage Thesaurus [42] to collect tweets (*céreales `a pailles* in French). As these words are quite frequent, we add conditions to retrieve tweets containing "récolte", "moisson" or "rendement" (harvest or yield in English) and to remove tweets containing "recette" or "farin" (recipe or flour in English) to construct the final dataset of 54,326 tweets between 2015 and 2020. Considering more and more people are engaged in broadcasting information about cereal production, we normalize the

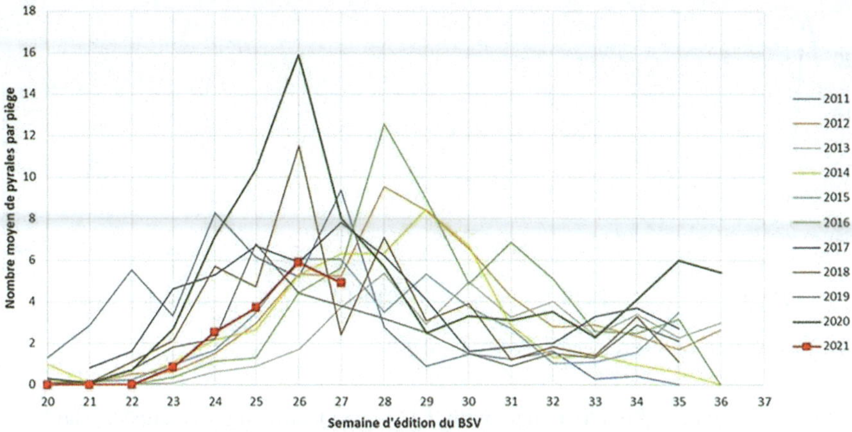

Fig. 16.2 Recorded averaged corn borer number by trap from [1]

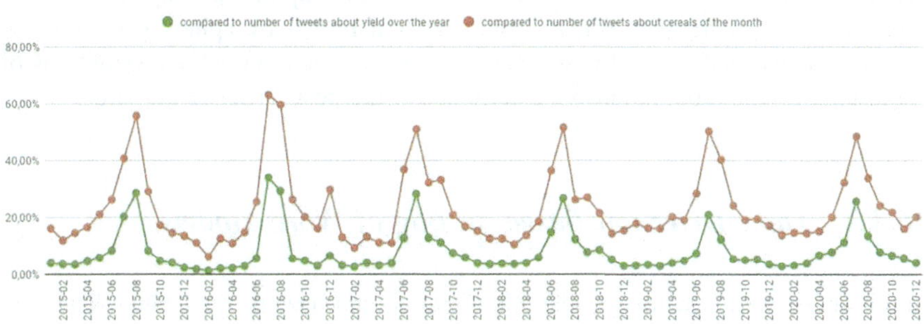

Fig. 16.3 Percentage of tweets concerning cereal yield between 2015 and 2020

counts by using percentage of tweets mentioning cereal yields per month against the total mentions of each year and against the accumulated mentions of each month in 6 years (see Fig. 16.3). Both curves show peaks between June and September each year, which correspond to the harvest season. We can also see that the peak in 2016 is higher than the other years. This abnormal popularity corresponds to the extreme yield loss in France in 2016 due to heavy rainfalls [5]. This case shows that people tend to post more tweets when bad things happen than when everything goes well, which confirms the interest of using Twitter as a source of crop health monitoring. We also plot tweets counts since this catastrophic yield containing the keywords "récolte" and "2016" as keywords in Fig. 16.4. We found that this event is recalled in 2020, when people had a negative prediction for yield. We suggest that the reference of yield loss in 2016 reflects a collective memory on social media.

We executed the same step using LDA topic modelling with the document term matrix. Both exercises succeed in distinguishing tweets in English and tweets about "Johny's Net Online". We can also see that the tweets in English are classified to a

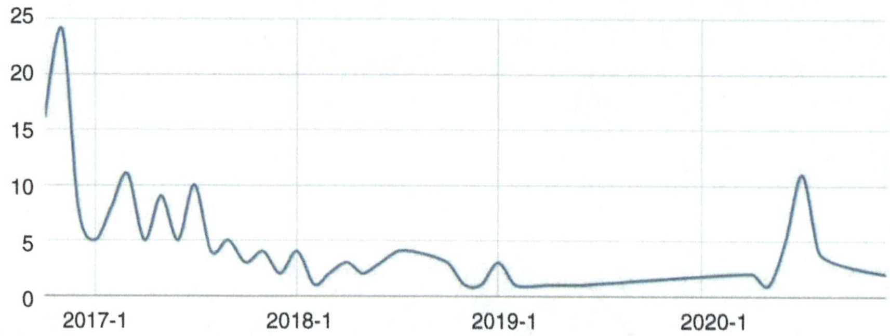

Fig. 16.4 Counts of tweets mentioning yield and 2016

topic or a cluster. An advantage with LDA is visualizing the clusters and the relevance of each term to the topic with an interactive tool LDAVis [43]. We can observe "brassicole" and "hirondella" in a topic or a cluster, these are barley species that resist the BYDV. We can also see "puceron" (aphids in English) in both exercises. This is because BYDV could be transmitted by aphids [2]. This use case represents a category of observation, where we need to track the symptoms of the disease on the crop, but also the activities of the pest carrier of the virus.

16.3.3 Topic Detection Based on Bag of Word Models

In this stage, we know that people talk about natural hazards on Twitter. The goal now is to explore more details about each subject without consuming too much computing resources. We can see this task as an unspecified event detection task [3]. Survey on topic detection [18] discussed different categories of unsupervised learning classification algorithms for topic detection, including clustering techniques like K-Means or DBSCAN, matrix factorization techniques like singular value decomposition (SVD), probabilistic models like Latent Dirichlet Allocation (LDA) [6]. These algorithms have been created to automatically divide a collection of data into groups of similarity for browsing, in a hierarchical or partitional manner [47]. Most of the measures of similarity, such as Euclidean distance or cosine distance, can be only applied on data points in a vectorial space [21]. A simple way to project a collection of documents into vectors is to create a document-term matrix, which describes the frequency of terms that occur in a collection of documents. Term frequency is also called bag-of-words (BoW). In a BoW model, a document is considered as a bag of words. TF*IDF score is a more normalized feature than raw frequency. Short for term frequency–inverse document frequency, TF*IDF is a formal measure of how important a term is to a document (a tweet in our case) in a collection [41].

TF*IDF is defined as follows. Given a collection of N documents, define f_{ij} as the frequency of term (a word) t_i in document d_j. Then define the term frequency TF_{ij} of term t_i in document d_j to be f_{ij} normalized by dividing it by the frequency f_{kj} of the maximum frequency of any term in this document:

$$\mathrm{TF}_{ij} = \frac{f_{ij}}{\max_{k} f_{kj}}$$

The IDF of a term describes how much information the term provides. Suppose term t_i appears in n_i documents, then $\mathrm{IDF}_i = \log_2(N/n_i)$. The TF*IDF score for term t_i in document d_j is defined to be $\mathrm{TF}_{ij} \times \mathrm{IDF}_i$. The terms with highest TF*IDF scores are often the most relevant terms to represent the topic of the document. TF matrix and TF*IDF matrix are widely used for describing the features of the document [4].

Use case: barley yellow dwarf virus (BYDV), "jaunisse nanisante de l'orge" or "mosaïque jaune" in French, commonly known as JNO – there are ten times as many original tweets containing JNO as tweets containing "jaunisse nanisante de l'orge". However, JNO can be abbreviations of other things, such as "Johny's Net Online". Moreover, to obtain as many tweets as possible, we used all the synonyms of "jaunisse nanisante de l'orge" presented in [49]. This list also includes the keyword "BYDV", which brings also tweets in English. Therefore, we need to look into the topics in these tweets. Topics can be identified by finding the feature words that characterize tweets about the topic. At this stage, we do not know what are the topics among the tweets nor how many topics there are, so we cannot use keywords to filter undesired tweets. In this sense, we isolate the irrelevant tweets with the help of a clustering method as follows:

1. Removal of stop words.
2. Calculation of the TF*IDF vector for each tweet. To get a reasonable vocabulary size, we ignore terms that have a document frequency higher than 0.7 or lower than 0.01.
3. Feeding TF*IDF vectors to *K*-Means [47], for *K* between 2 and 20, find the best cluster number *K* using elbow method [28].
4. Calculation of the TD*IDF matrix for each cluster, examination of the 20 terms with the highest TD*IDF scores, and removal undesired clusters.
5. Repeat steps 2–4 till all the clusters talk about BYDV. An example of the final state of this cleaning process is shown in Table 16.1.

We executed the same step using LDA topic modelling with the document term matrix. Both exercises succeed to distinguish tweets in English and tweets about "Johny's Net Online" from tweets about the barley yellow dwarf virus. We find that tweets in English are classified to an isolated topic or cluster. We can observe "brassicole" and "hirondella" in a topic or a cluster; these are barley species that resist the BYDV. We can also see "puceron" (aphids in English) in both experiences. This is because BYDV could be transmitted by aphids [2]. Thus, should this use case represents a category of observation, where we need to track the symptoms of the disease on the crop, but also the activities of the pest carrier of the virus.

Table 16.1 Top TF*IDF scored words in clusters in final state of K-Means based cleaning

Cluster	Top TF*IDF scored words
0	'année', 'blés', 'céréales', 'date', 'date semis', 'faire', 'faut', 'fin', 'jno orge', 'orge', 'précoce', 'pucerons', 'rt', 'variétale"
1	'dégâts', 'jno blé', 'orge', 'pucerons', 'rt', 'symptômes', 'virus'
2	'hiver', 'orge', 'orge hiver', 'pucerons', 'rt'
3	'année', 'automne', 'céréales', 'jno céréales', 'orge', 'pucerons', 'rt', 'traitement', 'virus', 'virus jno'
4	'année', 'brassicole', 'brassicole tolérante', 'brassicole tolérante jno', 'ceuxquifontlesessais', 'comportement', "d'hiver", "d'hiver rangs", 'hirondella', 'jno reconnue', 'jno reconnue brassicole', "l'orge", 'moisson', 'nouvelle', 'orge', 'orge brassicole', 'orge brassicole tolérante', "orge d'hiver", 'orges', 'pucerons'
5	'automne', 'blés', 'hiver', 'jno orges', 'orges', 'orges hiver', 'parcelles', 'printemps', 'pucerons', 'rt'
6	'essais', 'faire', 'orge', 'orges', 'pucerons', 'rt', 'tolérantes', 'tolérantes jno', 'variétés orge', 'variétés tolérantes', 'variétés tolérantes jno'
7	'blé', 'combinaison', 'issue', 'issue combinaison', "jaunisse nanisante l'orge", 'jaunisse nanisante orge', 'jno jaunisse', 'jno jaunisse nanisante', 'jno maladie', "jno maladie l'orge", "l'automne", "l'orge issue", "l'orge issue combinaison", "l'orge jno", "l'orge jno maladie", 'maladie', "maladie l'orge", "maladie l'orge issue", "nanisante l'orge", "nanisante l'orge jno"

16.3.4 Text Classification Based on Pre-trained Language Models

After filtering and cleaning the collected tweets, we can be almost certain that our tweets talk about pests. However, for the purpose of pest monitoring, there is still the need for more precision. A limit of BoW model is that it does not represent the meaning of a word.

A better feature representation technique for text classification is word embedding such as Word2Vec [16], where words from the vocabulary are mapped to N dimension vectors. Such vectors can be pre-trained on a large corpus and re-used for text classification tasks. The comparison between these vectors can be used to measure the similarity between words. Although word embedding may capture syntax and semantics of a word, it cannot keep the full meaning of a sentence [29].

Recent advancements in Bidirectional Encoder Representations from Transformers (BERT) [12] have showed important improvements in NLP, the multihead attention mechanism seems to be promising for contextual representation. The efficiency of fine-tuned BERT models for multi-Label tweets classification has been proved in disaster monitoring [52].

Use case: corvids and other emerging issues in general In the scenario of pest monitoring, the incompleteness of farmers' observations on Twitter, partially resulting from the constraint on the text length, made the observation information unusable. Prior research on understanding farm yield variation [23] proposes to

Table 16.2 Classification based on TF*IDF, with fivefold cross-validation

Dataset	Accuracy	Precision	Recall	*f1*
1	0.767123	0.539823	0.802632	0.645503
2	0.782759	0.566667	0.871795	0.686869
3	0.813793	0.620253	0.680556	0.649007
4	0.844291	0.702381	0.756410	0.728395
5	0.724138	0.536232	0.831461	0.651982

value them by bringing together observations from farmers and precise characterization of environmental conditions. To interconnect observation information on Twitter and other data sources, our first step is to extract tweets about observations. We define an observation as: a description of the presence of a pest in a field in real-time. These tweets may be missing essential information, such as precise location, impacted crop, the current developing status of the pest, the prediction of farmers about upcoming damage, or suggestions of the treatment. The pest might be uncommon, as in the case of corvids, so this kind of damages are getting attention only since 2018. Thus, we can no longer filter tweets using known keywords. This observation detection is a binary classification task.

Given a small set of n labelled tweets $T = \{s_{t1}, s_{t2}, ..., s_{tn}\}$ and a language model LM. Each s_{ti}, $s_{ti} \in T$, is annotated with a label o_i, $o_i \in [0,1]$ indicating whether it is of an observation. s_t can be seen as a sequence of words $s = (w_1 w_2 ... w_l)$, $s \in S$, $T \subset S$, $B \subset S$, where l is the length of the sequence, w is a word in natural language. To capture the features of s, we project S to a vectorial representation X using a *LM*. LM $(S) \rightarrow X$ can be seen as a tokenizer $f(s)$ plus an encoder $g(s^{'})$. The tokenizer contains the token-level semantics:

$f(s) \rightarrow s^{'}$ maps sequences of words $s = (w_1 w_2 ... w_l)$ to a sequence of token $s' = \left(w_1' w_2' \cdots w_{l'}'\right)$, where $w^{'}$ is the index of the token in its built-in dictionary, $l^{'}$ is the length of this sequence of tokens. The encoder $g(s^{'}) \rightarrow x$, $x \in X$ transforms $s^{'}$ to a continuous vectorial representation x [12]. Finally, we trained a softmax classifier with X and labels of T.

To verify if the BERT model can capture enough features of such observations, we invited domain experts to label 1455 tweets concerning core borer, BYDV and corvids. Then we used pre-trained CamemBERT base model [32] to encode tweets and train the classifier. We set the max sequence length to 128, batch size to 16. We use AdamW [26] for optimization with an initial learning rate of $2e^{-5}$. For evaluation, we plotted the precision-recall-threshold curve to find the best threshold to maximize the *f1* score. To compare CamemBERT representations with BoW models, Table 16.2 shows the results of fivefold cross validation of sigmoid classifier based on TDIDF vectors, Table 16.3 shows the results of fivefold cross validation of sigmoid classifier based on CamemBERT vectors. We can see that the accuracy of the latter is quite satisfactory. We use our classifier to predict tweets concerning pests that never appeared in the training set such as "taupin" (wireworms in English), and we observed that it detects when "taupin" is referred to a French name or wireworms. For observation such as *Pris en flagrant délit ... M.Taupin, vous êtes en état*

Table 16.3 Classification based on CamemeBERT, with fivefold cross-validation

Dataset	Accuracy	Precision	Recall	*f*1
1	0.883562	0.759036	0.828947	0.792453
2	0.914384	0.857143	0.835443	0.846154
3	0.893836	0.775000	0.837838	0.805195
4	0.924399	0.913043	0.807692	0.857143
5	0.886598	0.843373	0.786517	0.813953

d'arrestation #maïs #masseeds, even though "M.Taupin" looks like announcing to somebody; the classifier correctly classifies it to be an observation. Which means that the polysemy of "taupin" is properly handled in the contextualized embedding of the tweets, and that the classifier focuses on the sense of the text beyond considering the pest's name only as a keyword. Hence, the classifier enables us to filter all the tweets instead of only tweets containing by keywords.

16.4 Conclusion

In this chapter, we demonstrated the potential of Twitter in the domain of smart agriculture by using NLP techniques to extract information from tweets. The BoW model-based data clustering proves the possibility of semi-automatically browsing topics on Twitter with explainability. The language model-based supervised tweet classification experience demonstrates that, for a given concrete NLP task, language models have the potential to capture their contextual information, which can reduce manual labelling work for specific information extraction. In our scenario of pest monitoring, the extracted tweets containing observations of farmers allow us to monitor pests at the field-level. Thus, we open the possibility of conducting farmer-centric researches such as analysing and addressing the diversity of concerns and decision-making processes of different farmers. Furthermore, we can generalize our approach for the monitoring of other events on Twitter.

References

1. ARVALIS: Figure 2: Evolution du nombre moyen de pyrale par piègeselon l'année, https://www.arvalis-infos.fr/_plugins/WMS_BO_Gallery/page/getElementStream.jspz?id=72073&prop=image
2. ARVALIS: Jaunisse Nanisante de l'Orge (JNO) – Maladie virale sur Blé tendre, blé dur, triticale. (2013). http://www.fiches.arvalis-infos.fr/fiche_accident/fiches_accidents.php?mode=fa&type_cul=1&type_acc=7&id_acc=53
3. Asgari-Chenaghlu, M., et al. (2021). Topic detection and tracking techniques on Twitter: A systematic review. *Complexity, 2021*, 1–15.
4. Bafna, P., Pramod, D., & Vaidya, A. (2016). Document clustering: TF-IDF approach. In *2016 International conference on electrical, electronics, and optimization techniques (ICEEOT)* (pp. 61–66). IEEE.

5. Ben-Ari, T., et al. (2018). Causes and implications of the unforeseen 2016 extreme yield lossin the breadbasket of France. *Nature Communications, 9*(1), 1627.
6. Blei, D. M., Ng, A. Y., & Jordan, M. I. (2003). Latent Dirichlet allocation. *Journal of Machine Learning Research, 3*, 993–1022.
7. Boubiche, D. E., Imran, M., Maqsood, A., & Shoaib, M. (2019). Mobile crowd sensing–taxonomy, applications, challenges, and solutions. *Computers in Human Behavior, 101*, 352–370.
8. Cossu, J. V., Dugué, N., & Labatut, V. (2015). Detecting real-world influence through Twitter. In *2015 Second European network intelligence conference* (pp. 83–90). IEEE.
9. Cunningham, H. (2002). Gate, a general architecture for text engineering. *Computers and the Humanities, 36*(2), 223–254.
10. Defour, T. (2018, February). EIP-AGRI Brochure agricultural knowledge and innovation systems, https://ec.europa.eu/eip/agriculture/en/publications/eip-agri-brochure-agricultural-knowledge-and
11. Demirbas, M., et al. (2010). Crowd-sourced sensing and collaboration using twitter. In *2010 IEEE International symposium on "A world of wireless, mobile and multimedia networks" (WoWMoM)* (pp. 1–9). IEEE.
12. Devlin, J., Chang, M. W., Lee, K., & Toutanova, K. (2019). Bert: Pre-training of deep bidirectional transformers for language understanding. In *Proceedings of the 2019 conference of the North American chapter of the Association for Computational Linguistics: Human language technologies, volume 1 (Long and short papers)*. Association for Computational Linguistics.
13. Du, Y. (2020, July). Collaborative crowdsensing at the edge. Theses, Sorbonne Université, https://hal.archives-ouvertes.fr/tel-02913750
14. Ganti, R. K., Ye, F., & Lei, H. (2011). Mobile crowdsensing: Current state and future challenges. *IEEE Communications Magazine, 49*(11), 32–39.
15. Gao, D., et al. (2020). A framework for agricultural pest and disease monitoring based on internet-of-things and unmanned aerial vehicles. *Sensors, 20*, 1487.
16. Goldberg, Y., & Levy, O. (2014). word2vec explained: Deriving Mikolov et al.'s negative sampling word-embedding method. arXiv preprint arXiv:1402.3722.
17. Heldreth, C., Akrong, D., Holbrook, J., & Su, N. M. (2021). What does AI mean for smallholder farmers?: A proposal for farmer-centered AI research. *Interactions, 28*(4), 56–60.
18. Ibrahim, R., et al. (2018). Tools and approaches for topic detection from Twitter streams: Survey. *Knowledge and Information Systems, 54*(3), 511–539.
19. Ibtihel, B. L., Lobna, H., & Maher, B. J. (2018). A semantic approach for Tweet categorization. *Procedia Computer Science, 126*, 335–344.
20. Ingram, J. (2014). Farmer-scientist knowledge exchange. In *Encyclopedia of food and agricultural ethics* (pp. 1–8). Springer.
21. Irani, J., Pise, N., & Phatak, M. (2016). Clustering techniques and the similarity measures used in clustering: A survey. *International Journal of Computer Applications, 134*(7), 9–14. Foundation of Computer Science.
22. Jamra, H.A., Gillet, A., Savonnet, M., & Leclercq, E. (2020, June). Analyse des discours sur Twitter dans une situation de crise: Etude de l'incident a` l'usine Lubrizol de Rouen. In *INFormatique des ORganisations et Systèmes d'Information et de Décision*. Dijon (en ligne).
23. Jiménez, D., et al. (2016). From observation to information: Data-driven understanding of on farm yield variation. *PLoS One, 11*(3), e0150015.
24. Kaminska, O., Cornelis, C., & Hoste, V. (2021). Nearest neighbour approaches for emotion detection in Tweets.
25. Kenny, U., & Regan, A. (2021). Co-designing a smartphone app for and with farmers: Empathising with end-users' values and needs. *Journal of Rural Studies, 82*, 148–160.
26. Loshchilov, I., & Hutter, F. (2017). Decoupled weight decay regularization. arXiv preprint arXiv:1711.05101.

27. Klerkx, L., Jakku, E., & Labarthe, P. (2019). A review of social science on digital agriculture, smart farming and agriculture 4.0: New contributions and a future research agenda. *NJAS – Wageningen Journal of Life Sciences, 90–91*, 100315.
28. Kodinariya, T., & Makwana, P. (2013). Review on determining of cluster in k-means clustering. *International Journal of Advance Research in Computer Science and Management Studies, 1*, 90–95.
29. Kowsari, K., et al. (2019). Text classification algorithms: A survey. *Information, 10*(4), 150, arXiv: 1904.08067.
30. Li, Q., & Zhang, Q. (2021). Twitter event summarization by exploiting semantic terms and graph network. *Proceedings of the AAAI Conference on Artificial Intelligence, 35*(17), 15347–15354.
31. Liu, W., Jiang, L., Wu, Y., Tang, T., & Li, W. (2020). Topic detection and tracking based on event ontology. *IEEE Access, 8*, 98044–98056.
32. Louis, M., et al. (2020). Camembert: A tasty French language model. ArXiv abs/1911.03894.
33. Lowder, S., S'anchez, M., Bertini, R., et al. (2019). *Farms, family farms, farmland distribution and farm labour: What do we know today?* FAO Agricultural Development Economics Working Paper.
34. Mathioudakis, M., & Koudas, N. (2010). TwitterMonitor: Trend detection over the twitter stream. In *Proceedings of the 2010 ACM SIGMOD International conference on management of data* (pp. 1155–1158). ACM.
35. Mendes, J., Pinho, T. M., Neves dos Santos, F., Sousa, J. J., Peres, E., Boaventura Cunha, J., Cunha, M., & Morais, R. (2020). Smartphone applications targeting precision agriculture practices—A systematic review. *Agronomy, 10*(6), 855.
36. Messaoudi, C., Guessoum, Z., & Ben Romdhane, L. (2022). Opinion mining in online social media: A survey. *Social Network Analysis and Mining, 12*(1), 25.
37. Monteleone, M. (2018). NooJ grammars and ethical algorithms: Tackling on-line hatespeech. In *International conference on automatic processing of natural-language electronic texts with NooJ* (pp. 180–191). Springer.
38. Olatinwo, R., & Hoogenboom, G. (2014). Chapter 4 – Weather-based pest forecasting for efficient crop protection. In D. P. Abrol (Ed.), *Integrated pest management* (pp. 59–78). Academic Press.
39. Phillips, T., Klerkx, L., McEntee, M., et al. (2018). An investigation of social media's roles in knowledge exchange by farmers. In *13th European International Farming Systems Association (IFSA) Symposium, Farming systems: Facing uncertainties and enhancing opportunities* (pp. 1–20). International Farming Systems Association (IFSA) Europe.
40. Qing, Z., et al. (2021). A pest sexual attraction monitoring system based on IoT and image processing. *Journal of Physics: Conference Series, 2005*(1), 012050.
41. Rajaraman, A., & Ullman, J. D. (2011). *Data mining* (pp. 1–17). Cambridge University Press. https://doi.org/10.1017/CBO9781139058452.002
42. Roussey, C. (2021). *French crop usage*. https://doi.org/10.15454/QHFTMX
43. Sievert, C., & Shirley, K. (2014). LDAvis: A method for visualizing and interpreting topics. In *Proceedings of the workshop on interactive language learning, visualization, and interfaces* (pp. 63–70). Association for Computational Linguistics.
44. Silberztein, M. (2013). NooJ v4. In *Formalising natural languages with NooJ 2013*. Wiley.
45. Singh, V. K., Tiwari, N., & Garg, S. (2011). Document clustering using K-means, heuristic K-means and fuzzy C-means. In *2011 International conference on computational intelligence and communication networks* (pp. 297–301). IEEE.
46. Sivagnanasundaram, J., Ginige, A., & Goonetillake, J. (2019). Farmers as sensors: A crowdsensing platform to generate agricultural pest incidence reports. In *2019 International conference on internet of things research and practice (iCIOTRP)* (pp. 13–18). IEEE.
47. Steinbach, M., Karypis, G., & Kumar, V. (2000, May). *A comparison of document clustering techniques*, http://conservancy.umn.edu/handle/11299/215421. Accessed: 13 Dec 2021.

48. Thareau, B., & Daniel, K. (2019). Le numérique accompagne les mutations économiques et sociales de l'agriculture. *Sciences Eaux & Territoires, 29*(3), 44.
49. Turenne, N., Andro, M.: Maladies des cultures (2017).
50. Wang, D., Abdelzaher, T., & Kaplan, L. (2015). *Social sensing: Building reliable systems on unreliable data*. Morgan Kaufmann.
51. Xu, Z., et al. (2018). Mobile crowd sensing of human-like intelligence using social sensors: A survey. *Neurocomputing, 279*, 3–10.
52. Zahera, H. M., Elgendy, I. A., Jalota, R., & Sherif, M. A. (2019). Fine-tuned BERT Model for Multi-Label Tweets Classification. In *Proceedings of the Twenty-Eighth Text REtrieval Conference*, TREC 2019, Gaithersburg, Maryland, USA, November 13–15, 2019.

Chapter 17
A Multi-criteria Decision-Making Model for Technology Selection in Renewable-Based Residential Microgrids

Zeynep Bektas and Senay Sadic

Abstract This study proposes a multi-criteria decision-making model for technology selection for renewable-based residential microgrids, which is one of the most important decisions in the planning and installation phase of microgrids. All the three sustainability dimensions named environmental, economic, and social criteria are considered for a problem with such different aspects. Under these three dimensions, sub-criteria are determined by evaluating the structure and needs of residential microgrids. While determining the sub-criteria, the future electricity consumers of the microgrid are also considered. Therefore, a decision-making model is proposed, which considers both the power generation and load consumption sides even during the installation stage. Alternatives of the decision-making model are determined by the region of the case study to be applied. Renewable energy technologies that can be reached and installed in this region are considered. By evaluating the outputs of the model, the most appropriate technology combinations for the residential microgrid in the case study are recommended. The integrated application of AHP and TOPSIS methods is proposed for the handled problem. Quantitative data, expert opinions, and surveys are used to obtain sub-criteria values according to the type of the sub-criterion.

Keywords Renewable energy · Technology selection · Residential microgrids · Multi-criteria decision-making · Sustainability

---

Z. Bektas (✉)
Department of Industrial Engineering, Kadir Has University, Istanbul, Türkiye
e-mail: zeynep.bektas@khas.edu.tr

S. Sadic
Department of Industrial Engineering, Antalya Bilim University, Antalya, Türkiye
e-mail: senay.sadic@antalya.edu.tr

G. Kayakutlu, M. Ö. Kayalica (eds.), *Decision Making Using AI in Energy and Sustainability*, Applied Innovation and Technology Management,
https://doi.org/10.1007/978-3-031-38387-8_17

17.1 Introduction

The world energy market is shifting from conventional energy sources, especially fossil fuels, to renewable energy sources. Conventional energy sources are not only being depleted as a resource but also known to cause high levels of carbon emission, has a high price with volatile cost, and damage to the environment [7]. Solar, wind, biomass, geothermal, marine energy, and other non-fossil energy sources can be listed among renewable energy sources and are favorable due to their renewable and recyclable nature and causing much less harm to the environment [18].

While countries are generating renewable energy policies and providing incentives for country-wide energy transition at the macro-level, a group of environmentally conscious residents also take personal responsibility and choose renewable-based residential areas. Using 27% of the overall energy demand, the household sector is the third-largest energy consumer and despite the decreasing emissions, it still generates 30–40% of total greenhouse gas emissions [19]. Therefore, households need to move from fossil fuels to renewables.

United Nations have announced Sustainable Development Goals (SDG) – 17 global targets to be reached by every country, aiming to support moving toward a better planet [24]. By supporting clean energy transition and its positive effects on the environment, this study contributes to SDGs, specifically SDG 7 – Affordable Clean Energy, SDG 11 – Sustainable Cities and Communities, and SDG 13 – Climate Action.

This chapter proposes a multi-criteria model that considers all sustainability dimensions to choose the most appropriate technologies for residential microgrids (MG). In Sect. 17.2, a brief literature review on energy technology selection from a sustainability perspective has been presented. Section 17.3 contains the research methodology. Due to the literature review, three sustainability criteria and 19 related sub-criteria are listed and considered the study's geographical region and characteristics. Also, 11 technology alternatives for electricity and heating are listed. Section 17.4 covers the application of the multi-criteria decision-making (MCDM) model which includes an AHP model to find criteria weights and a TOPSIS model to rank technology alternatives and select the best option for heating and electricity. Section 17.5 mainly includes the work's analysis to see how different sustainability perspectives affect optimal decision-making within multi-criteria. Finally, in Sect. 17.6, the study is summarized, emphasizing essential aspects.

17.2 Literature Review: Renewable Energy Technology Selection from Sustainability Perspective

Energy management and planning decisions within the technology selection could be seen as complex tasks with several players and different objectives via its constraints. While each technology has advantages and disadvantages, a successful technology selection differs concerning many criteria, including user experiences, regional climatic conditions, and energy resource availability. None of the technologies available are known to be superior to the rest. Therefore, MCDM models are widely used for technology selection of renewable energy planning.

Recently, a literature review on MCDM methods used to evaluate renewable energy technologies in households has been completed. There are many criteria for selecting renewable energy technologies and numerous MCDM methods. In contrast, the leading technology alternatives are solar photovoltaic and solar thermal technologies, wind energy technology, heat pumps, biomass heating technology, etc. [19].

A comprehensive list of MCDM methods is also referred to in the chapter. Among the widely used MCDM methods, papers use one or more methods: CRITIC, AHP, TOPSIS, EDAS, WASPAS, PROMETHEE, WSM, DEA, and fuzzy applications of these methods. Rather than using a single MCDM method, integrating different methods for various stages of the process may support more reliable decision-making. Using fuzzy numbers can capture the uncertainty involved. While the authors used numerous sub-criteria in the MCDM models, the main criteria are economic, technological, social, and environmental. Risk is another criterion used to select more reliable technologies [13].

This section selects and analyzes a sample paper list from the related literature to see the technology selection applications.

A grey-based MCDM model to support the evaluation of renewable energy sources is developed based on three main MCDM methods, DEMATEL, Analytic Network Process, and VIKOR. The authors further applied that methodology to Turkey as a case study. They have emphasized the results, mentioning that they are case-specific and cannot be generalized for other technology decisions of other countries [7].

Another MCDM approach based on SWARA combined with Multi-Objective Optimization on the basis of Ratio Analysis plus the full MULTIplicative form is also developed for the renewable energy technology selection with an application on a conceptual development project in Iran [11]. Another case study from British Columbia – Canada is in the net zero community energy system where the choice of renewable energy technology is made using a fuzzy TOPSIS method [14]. Seddiki and Bennadji [17] proposed an integrated MCDM method that uses the Delphi method, questionnaire, fuzzy AHP method, and FPROMETHEE for evaluation of renewable energy alternatives for electricity generation in a residential building. Delphi methods were used at the beginning of the study to support and facilitate communication among experts.

For countries with islands or off-grid communities, renewable energy is necessary and the correct approach to provide electricity to such communities. To support the renewable energy technology selection decision in southern Bangladesh, an integrated Entropy-EDAS (Evaluation Based on Distance from Average Solution) model is designed [3]. The results suggest that the solar-wind-battery hybrid energy system technology should be selected. The authors note that the outcomes can be generalized to other isolated areas of Bangladesh. A renewable energy technology selection model combining AHP with the Combined Distance-Based Assessment method is proposed to support the off-grid residence of Rohingya refugees in Myanmar in Bangladesh. The chapter also selected solar and wind technologies as the appropriate renewable energy technologies to be implemented [4].

The dynamic nature of socio-technical and economic criteria is considered in a recent paper that proposes a model with dynamic intuitionistic fuzzy Einstein geometric averaging operator, intuitionistic fuzzy entropy, and the intuitionistic fuzzy TOPSIS [2]. Through the expertise of five energy experts, the model found biomass energy as the ideal option to implement for a case study in Lagos, Nigeria.

17.3 Methodology: AHP- and TOPSIS-Based Decision Support System for Technology Selection in Renewable-Based Residential Microgrids

The purpose of the decision-making problem discussed in our study is technology selection for residential MGs. It is necessary to determine the appropriate criteria, considering the characteristics and requirements of the problem, as well as spatial conditions. Since the focus of our study is sustainability, the problem will be handled from this perspective. Sustainability is evaluated under three dimensions which are environmental, economic, and social aspects. Making this clear distinction makes it simple to identify sub-criteria to consider in residential MGs. In the process of determining sub-criteria, the related literature was also examined [1, 6, 10, 16, 19, 20].

Environmental considerations in the technology selection of an MG at the installation stage are the environmental effects and environmental feasibility of the system. In this context, environmental sub-criteria are determined as carbon emissions, resource availability, impact on ecosystems, waste management, and noise. The second sub-criterion, resource availability, means whether the region has a resource that will be an input to the relevant technology. In addition to several cost items, incentive-like supports and system efficiencies, which will affect the amount of installed power and therefore the cost, are what should be considered economically. Thus, the economic sub-criteria are determined as initial investment cost, unit operation and maintenance cost, the energy efficiency of technology, and incentives and subsidies. The fourth sub-criterion will be evaluated in Turkey due to the location of the MG to be discussed.

The social aspect is the broadest field. Here, energy systems and in particular all social dimensions, persons, communities, or institutions that MGs may be associated with should be considered. In this context, the social criterion is expressed with ten sub-criteria. These are physical security, social acceptance, contribution to the development of the region, employment creation, impacts on human health, impacts on quality of the landscape and residential areas, national policies, international policies, energy saving habits of the consumer side, and energy usage habits of the consumer side. Physical security, which is the first social sub-criterion, is related to the risks of the relevant technology. This criterion means to the general safety concerns rather than the effects on human health and the environment. The last two social sub-criteria are related to MG's consumer side which will be discussed in the case study. Since the MG's consumer side will be known and its processed data will also be used when designing the decision-making model, it is desired to benefit from this information in the decision-making model. The sub-criterion of energy saving habits of the consumer side is based on the idea that depending on the energy source used, the behavior of the conscious consumer may vary. For example, how much consumers will implement behaviors within the scope of sustainability such as reuse-reduce-recycle in energy consumption may vary. The sub-criterion of energy usage habits of the consumer side expresses the adaptation of peak load demand to resources in terms of time slices of day. For example, a consumer group that needs more electricity consumption at night will give solar energy a low rating score.

After determining the criteria to evaluate different technology alternatives, the MCDM method that we will use according to the predicted outputs of the criteria should be selected. When the sub-criteria to be used are examined, it is seen that the evaluation results for some of them can be on different scales. In addition, it is a separate problem to determine the weights of these sub-criteria, which are very different from each other. Finally, when the structure of the decision problem and all its components are considered, it is decided that the ideal MCDM method is integrated usage of AHP and TOPSIS. First, the AHP method will be used to determine the sub-criteria weights, and then the TOPSIS method will be used to rank the alternatives using these weights. The complexity will arise since some of the sub-criteria are benefit criteria, and some are cost criteria that can be easily handled with TOPSIS.

AHP stands for Analytic Hierarchy Process and it was introduced by Saaty. It incorporates a pairwise comparison scale to analyze MCDM problems [19]. In our approach, AHP will be utilized to calculate the criteria weights. To complete our solution approach, a TOPSIS methodology was also incorporated into the AHP model. The TOPSIS method was used to select the best technology under multiple criteria. TOPSIS is based on the idea that the optimal alternative must be closest to the positive ideal solution and furthest from the negative ideal solution.

For all criteria and sub-criteria of the MCDM model, the tree structure given in Fig. 17.1 is decided. As seen in the figure, five sub-criteria are included under the environmental aspect. Four sub-criteria are also included under the economic aspect. Finally, the social aspect is divided into ten sub-criteria. As it may be understood, some of these sub-criteria are benefit criteria and some are cost criteria. The sub-criteria which are benefit criteria are as follows: ENV2, ENV4, ECO3, ECO4, SOC1, SOC2, SOC3, SOC4, SOC7, SOC8, SOC9, SOC10. The sub-criteria which are cost criteria are as follows: ENV1, ENV3, ENV5, ECO1, ECO2, SOC5, SOC6.

All possible power generation technologies are considered while determining the alternatives to be evaluated within the scope of the MCDM model. Although the aim is to develop a renewable-based MG concept, conventional energy technologies will also be considered to ensure that the proposed technology alternative combination is

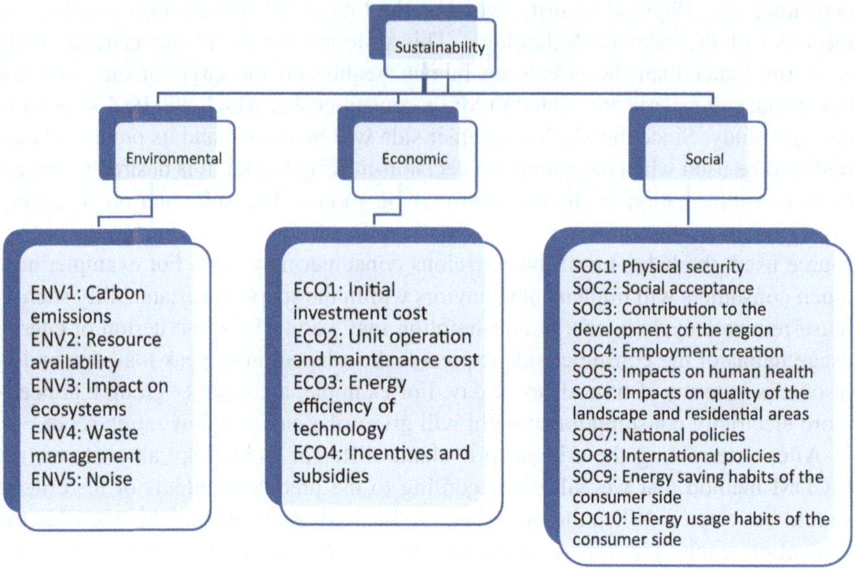

Fig. 17.1 All criteria and sub-criteria of the MCDM model

realistic. Since one of the main motivations of MGs is to meet the electricity and heat demand while doing less harm to the environment, they often include distributed renewable-based power generators. Due to the intermittent and unpredictable nature of renewable resources, they also have backup conventional resources under the aim of uninterrupted power flow. Considering the energy source choices of MGs, it is seen that they can consist of conventional or renewable energy sources, but mostly hybrid systems are preferred [9]. Therefore, for electricity generation and heating, a total of 11 alternatives to be evaluated in the MCDM model are identified. These are as follows (Here, E stands for electricity and H stands for heating.):

- Hydrokinetic energy (E)
- Coal power (E)
- Natural gas (E)
- Natural gas (H)
- Diesel generator (E)
- Diesel/fuel oil (H)
- Biofuel generator (E)
- Wind turbine (E)
- Photovoltaic panel (E)
- Solar thermal (H)
- Ground-source heat pump (H)

In the scope of the MCDM problem, these 11 alternatives are evaluated according to the given sustainability criteria tree. In this way, a ranking of energy technology alternatives will be obtained for the case to be considered. In this way, combination proposals will be made for the components of the hybrid energy system in MG, which is discussed in the case.

17.4 Application: A Renewable-Based Residential Microgrid in Antalya, Turkey

A residential MG to be established in Antalya, Turkey, is selected for a case study in the application stage. The mentioned MG will be established for employees of a multinational technology company. The company offers its employees the opportunity to work in a home office due to the COVID-19 pandemic. It also plans to build a site where they can live with below-market rent to provide the opportunities they used to provide in the office. In line with their sustainability goals, they aim to design this site as an MG and meet the electricity and heat demand with the hybrid energy system they will establish. They are aware that the system they want to be based mainly on renewable resources will also need to include the required amount of conventional capacity according to the region's characteristics. In this context, they would like to determine the most appropriate energy source combination for the relevant region, with a decision mechanism compatible with sustainability goals. Having information about the education level and sensitivity about the sustainability of the consumer side of the MG to be established also enable the decision-making model's sub-criteria about the consumption side to be used.

To apply the integrated decision-making approach to this residential MG, sub-criteria weights are first needed. For this, the steps of the AHP method to find the weight vector are used. To determine the sub-criterion weights, a pairwise comparison matrix is created in which all sub-criteria are compared with each other. This matrix is filled in by a group of experts and averages of their responses are calculated to create the final pairwise comparison matrix. Then, column normalization is done by dividing the values in each column by the total value of each column. After that, the weight of each sub-criteria is calculated by averaging the normalized matrix on a row basis. Since 19 sub-criteria are considered together, weights are obtained so that the sum of the weights of all of them is 1. The first three with the highest weight among all sub-criteria are found as physical security (SOC1) with 0.153, impacts on human health (SOC5) with 0.107, and unit operation and maintenance cost (ECO2) with 0.104. Other sub-criteria and their weights are ENV1, ENV2, ENV3, ENV4, ENV5, ECO1, ECO3, ECO4, SOC2, SOC3, SOC4, SOC6, SOC7, SOC8, SOC9, SOC10 and 0.099, 0.082, 0.050, 0.045, 0.011, 0.057, 0.050, 0.031, 0.011, 0.026, 0.022, 0.042, 0.021, 0.018, 0.029, 0.042, respectively. In addition, when the weight values obtained are considered based on criteria groups, the total weights of environmental, economic, and social criteria are 0.287, 0.242, and 0.471, respectively.

In the process of determining the ratings of all sub-criteria according to the alternatives, different sources and expert opinions are used. The ratings of sub-criteria of ENV1 [8, 10, 12], ECO1 [10, 21, 26], ECO2 [8, 10, 22], ECO3 [10, 16, 25, 27], ECO4 [23], SOC4 [5, 15], and SOC5 [10] are determined by the data collection process. To express the ratings of the sub-criterion of incentives and subsidies (ECO4), binary values are used regarding the case of Turkey. For the sub-criterion of impacts on human health (SOC5), mortality rate data is based.

For other sub-criteria, a survey is prepared to seek expert opinions. Consumer side questions (SOC9 and SOC10) were asked to several groups of company employees. Others were asked to five experts from academia and industry who have knowledge of the region and work in the energy field. In these surveys, participants are asked to make evaluations about sub-criterion and technology alternative pairs on a scale of 1–9. Because of that some sub-criteria are benefit and some are cost criteria, the meanings of scale for each sub-criterion are added to the survey file. For example, a score of 9 means the highest resource availability and a score of 1 means the lowest resource availability for ENV2. For ENV3, a cost criterion, a score of 9 means the highest bad impact and a score of 1 means the lowest bad impact. In the end, the final score of each sub-criterion and technology alternative pair is obtained by calculating the arithmetic mean of the scores given by all decision-makers.

When the TOPSIS stage is passed, first of all, normalization should be applied to sub-criteria scores on different scales. For the normalization, the vector (Euclidean) normalization technique is used. Data in each sub-criterion column is divided by the sum of the squares of all column elements. Then, a weighted standard decision matrix is constructed by multiplying the normalized values with criteria weights. The next step of the TOPSIS methodology is done by detecting the ideal points (maximum for benefit criteria and minimum for cost criteria) and computing the distance for each alternative to the ideal action via Euclidean distance. Finally, the final scores and rankings of each technology alternative are found.

Our main TOPSIS application was performed with 11 alternatives and 19 sub-criteria for a residential MG in Antalya, Turkey. Final scores of all electricity and heating alternatives are given in Fig. 17.2. Considering the ordering of final scores and differences and intervals between these scores, a hybrid electricity generation system consisting of photovoltaic panels, wind turbines, hydrokinetic energy, and a biofuel generator is proposed for the electricity demand of the MG. A hybrid heating system consisting of solar thermal and ground-source heat pumps is proposed for the heat demand of the MG.

17.5 Analysis

After our main TOPSIS application, to examine different cases, three different TOPSIS applications were also performed. In these applications, each of the three criteria groups was handled separately. Sub-criteria weights obtained from the AHP application were calculated again by normalizing for each criteria group separately. The results of these three TOPSIS applications are also given in Fig. 17.2.

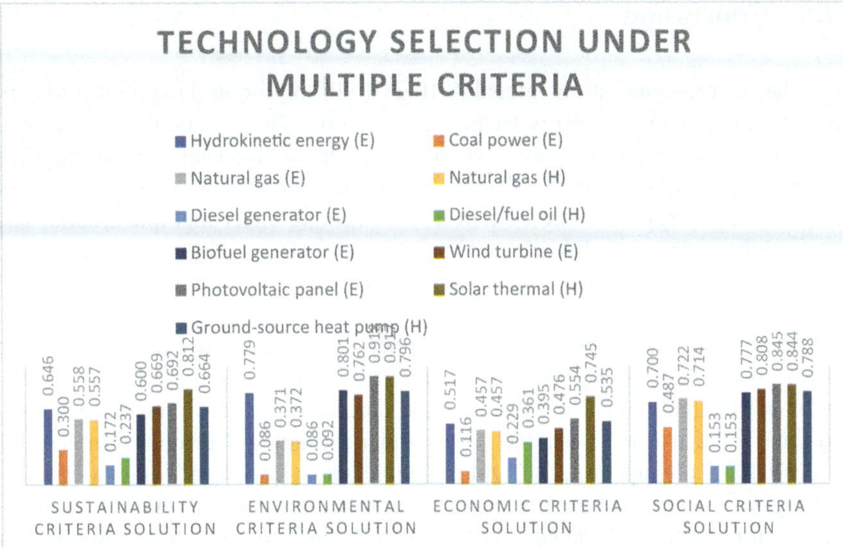

Fig. 17.2 Results of four different technology selection applications

As seen in the figure, considering only environmental criteria in the MCDM problem, a hybrid electricity generation system consisting of photovoltaic panels, biofuel generator, hydrokinetic energy, and wind turbines is proposed, and a hybrid heating system consists of solar thermal and ground-source heat pumps is also proposed. However, unlike the main application, the scores of the alternatives other than these are too low to be considered. When only economic criteria are considered in the MCDM problem, a hybrid electricity generation system consisting of photovoltaic pancls and hydrokinetic energy, and a hybrid heating system consisting of solar thermal and ground-source heat pumps would be most suitable. Besides, natural gas and wind energy alternatives are reasonable candidates. When only social criteria are considered in the MCDM problem, it is possible to propose hybrid electricity and heating systems with the most elements. In this case, a hybrid electricity generation system consisting of photovoltaic panels, wind turbines, biofuel generators, and natural gas is proposed for the electricity demand of the MG, and a hybrid heating system consisting of solar thermal, ground-source heat pumps, and natural gas is also proposed for the heat demand of the MG.

Overall, it was seen that the order of alternatives was not significantly different in any of the four applications. The components of the proposed hybrid systems are almost the same. Among the four applications, only the differences between the final scores differed in size. What they all have in common is that coal and diesel alternatives always have the lowest three scores.

17.6 Conclusion

This chapter presented an integrated AHP-TOPSIS multi-criteria approach to support decision-making in energy technology selection. The methodology developed by including sustainability dimensions and related criteria has been implemented in a new residential MG in Antalya, Turkey. This analysis has led to an awareness of solar thermal energy as the best alternative for heating under multiple criteria. The photovoltaic panel was also selected as the best electricity generation alternative, followed by other renewable energy technologies such as wind turbines and biofuel generators.

The results go along with the SDGs, highlighting those renewable resources have become sustainable energy technology alternatives along with economic criteria. The novel methodology also considers the characteristics and preferences of the consumer side. It can be applied to any other technology selection problem for residential MGs by updating MCDM problem alternatives considering the application region. Finally, it should be highlighted that the results are entirely location-dependent in terms of data and expert opinion. Future studies will lead to different results if the user consumption data is integrated into the model.

References

1. Afrane, S., Ampah, J. D., Jin, C., Liu, H., & Aboagye, E. M. (2021). Techno-economic feasibility of waste-to-energy technologies for investment in Ghana: A multicriteria assessment based on fuzzy TOPSIS approach. *Journal of Cleaner Production, 318*, 128515. https://doi.org/10.1016/j.jclepro.2021.128515
2. Aikhuele, D. O., Ighravwe, D. E., & Akinyele, D. (2019). Evaluation of renewable energy technology based on reliability attributes using hybrid fuzzy dynamic decision-making model. *Technology and Economics of Smart Grids and Sustainable Energy, 4*. https://doi.org/10.1007/s40866-019-0072-2
3. Ali, T., Ma, H., & Nahian, A. J. (2019). An analysis of the renewable energy technology selection in the southern region of Bangladesh using a hybrid multi-criteria decision making (MCDM) method. *International Journal of Renewable Energy Research, 9*, 1838–1848.
4. Ali, T., Nahian, A. J., & Ma, H. (2020). A hybrid multi-criteria decision-making approach to solve renewable energy technology selection problem for Rohingya refugees in Bangladesh. *Journal of Cleaner Production, 273*, 122967. https://doi.org/10.1016/j.jclepro.2020.122967
5. Bahaj, B., & Batten, W. (2021). Job creation in Europe from ocean energy. Available at: http://www.spok.dk/seminar/marine_job%20creation_20.pdf. Accessed: 15 Nov 2021.
6. Campos-Guzmán, V., García-Cáscales, M. S., Espinosa, N., & Urbina, A. (2019). Life cycle analysis with multi-criteria decision making: A review of approaches for the sustainability evaluation of renewable energy technologies. *Renewable and Sustainable Energy Reviews, 104*, 343–366. https://doi.org/10.1016/j.rser.2019.01.031
7. Çelikbilek, Y., & Tüysüz, F. (2016). An integrated grey based multi-criteria decision making approach for the evaluation of renewable energy sources. *Energy, 115*, 1246–1258. https://doi.org/10.1016/j.energy.2016.09.091
8. EIA. (2021). *U.S. energy information administration.* Available at: https://www.eia.gov/. Accessed: 11 Nov 2021.

9. Elsied, M., Oukaour, A., Gualous, H., & Lo Brutto, O. A. (2016). Optimal economic and environment operation of micro-grid power systems. *Energy Conversion and Management, 122*, 182–194. https://doi.org/10.1016/j.enconman.2016.05.074
10. Feitosa, F. E. B., & Costa, A. L. (2021). Application of a multicriteria methodology for evaluation of energy alternatives for hydrogen production for the automotive sector – Case study. *International Journal of Hydrogen Energy, 46*, 20799–20814. https://doi.org/10.1016/j.ijhydene.2021.03.209
11. Ijadi Maghsoodi, A., Ijadi Maghsoodi, A., Mosavi, A., Rabczuk, T., & Zavadskas, E. (2018). Renewable energy technology selection problem using integrated H-SWARA-MULTIMOORA approach. *Sustainability, 10*, 4481. https://doi.org/10.3390/su10124481
12. Kaldellis, J. K., & Apostolou, D. (2017). Life cycle energy and carbon footprint of offshore wind energy. Comparison with onshore counterpart. *Renewable Energy, 108*, 72–84. https://doi.org/10.1016/j.renene.2017.02.039
13. Kamari, M. L., Isvand, H., & Nazari, M. A. (2020). Applications of multi-criteria decision-making (MCDM) methods in renewable energy development: A review, *1*, 47–54. https://doi.org/10.22044/RERA.2020.8541.1006
14. Karunathilake, H., Hewage, K., Mérida, W., & Sadiq, R. (2019). Renewable energy selection for net-zero energy communities: Life cycle based decision making under uncertainty. *Renewable Energy, 130*, 558–573. https://doi.org/10.1016/j.renene.2018.06.086
15. Ram, M., Aghahosseini, A., & Breyer, C. (2020). Job creation during the global energy transition towards 100% renewable power system by 2050. *Technological Forecasting and Social Change, 151*, 119682. https://doi.org/10.1016/j.techfore.2019.06.008
16. Roth, S., Hirschberg, S., Bauer, C., Burgherr, P., Dones, R., Heck, T., & Schenler, W. (2009). Sustainability of electricity supply technology portfolio. *Annals of Nuclear Energy, 36*, 409–416. https://doi.org/10.1016/j.anucene.2008.11.029
17. Seddiki, M., & Bennadji, A. (2019). Multi-criteria evaluation of renewable energy alternatives for electricity generation in a residential building. *Renewable and Sustainable Energy Reviews, 110*, 101–117. https://doi.org/10.1016/j.rser.2019.04.046
18. Shao, M., Han, Z., Sun, J., Xiao, C., Zhang, S., & Zhao, Y. (2020). A review of multi-criteria decision making applications for renewable energy site selection. *Renewable Energy, 157*, 377–403. https://doi.org/10.1016/j.renene.2020.04.137
19. Siksnelyte-Butkiene, I., Zavadskas, E. K., & Streimikiene, D. (2020). Multi-criteria decision-making (MCDM) for the assessment of renewable energy technologies in a household: A review. *Energies, 13*. https://doi.org/10.3390/en13051164
20. Sitorus, F., & Brito-Parada, P. R. (2020). A multiple criteria decision making method to weight the sustainability criteria of renewable energy technologies under uncertainty. *Renewable and Sustainable Energy Reviews, 127*, 109891. https://doi.org/10.1016/j.rser.2020.109891
21. Stefánsson, V. (2002). Investment cost for geothermal power plants. *Geothermics, 31*, 263–272. https://doi.org/10.1016/S0375-6505(01)00018-9
22. SunShot-2030. (2021). *New solar opportunities for a new decade*. Available at: https://www.energy.gov/eere/solar/sunshot-2030. Accessed: 13 Dec 2021.
23. The Ministry of Development. (2018). *On Birinci Kalkınma Planı (in Turkish language)*. Available at: https://www.sbb.gov.tr/kalkinma-planlari/. Accessed: 8 Nov 2021.
24. United Nations. (2021). *Sustainable development goals*. Available at: https://sdgs.un.org/goals. Accessed: 28 Oct 2021.
25. University of Michigan. (2021). *Wind energy factsheet*. Available at: https://css.umich.edu/factsheets/wind-energy-factsheet. Accessed: 12 Nov 2021.
26. Vazquez, A., & Iglesias, G. (2016). Capital costs in tidal stream energy projects: A site-specific approach. In *Progress in renewable energies offshore – Proceedings of 2nd international conference on renewable energies offshore, RENEW 2016* (pp. 883–890). https://doi.org/10.1201/9781315229256-104.
27. Zarrouk, S. J., & Moon, H. (2014). Efficiency of geothermal power plants: A worldwide review. *Geothermics, 51*, 142–153. https://doi.org/10.1016/j.geothermics.2013.11.001

Chapter 18
Energy Management in Power-Split Hybrid Electric Vehicles Using Fuzzy Logic Controller

Anılcan Özkan, S. Davoud S. Garmroodi, and Senem Kurşun

Abstract Electric motor torque and battery operation strategy are two critical points for optimal utilization of hybrid electric vehicle (HEV). Using the optimal engine torque control scheme dramatically reduces the fuel consumption of internal combustion engine (ICE) as well as appropriate battery charge/discharge strategy helps to operate the battery at the proper defined state of charge (SOC), thus increasing the state of health (SOH) and the battery lifetime. Both of these factors contribute to reducing environmental pollution. Due to the complexity of the hybrid drivetrain and the synergic operation of the components in a power-split hybrid electric vehicle, it is difficult to construct an accurate mathematical model. In this chapter, we propose a fuzzy logic-based energy management strategy in which electric motor torque and SOC of the storage unit are considered the inputs and the speed of ICE is considered the output. Fuzzy inference rules are determined to cover all operating conditions of HEV. MATLAB® fuzzy logic toolbox is used to design a fuzzy logic controller and define the rules, and HEV is modeled using AVL CRUISE™ vehicle driveline simulation software. Finally, a worldwide harmonized light vehicle test procedure (WLTP) drive cycle is used to complete the simulation process. Simulation results show that all of the design objectives of the controller are satisfied. Fuel consumption is reduced by 27%. In addition, according to SOC results, the battery is prevented from overcharging/discharging, so SOH is improved.

Keywords Fuzzy logic controller · Energy management · Hybrid electric vehicles · Fuel consumption · Battery state of health (SOH)

A. Özkan
Istanbul Technical University, Laboratory of Automotive, Istanbul, Türkiye
e-mail: ozkana17@itu.edu.tr

S. D. S. Garmroodi
Istanbul Technical University, Energy Institute, Istanbul, Türkiye
e-mail: seyedgarmroodi19@itu.edu.tr

S. Kurşun (✉)
Istanbul Technical University, Faculty of Mechanical Engineering, Istanbul, Türkiye
e-mail: kursuns@itu.edu.tr

G. Kayakutlu, M. Ö. Kayalica (eds.), *Decision Making Using AI in Energy and Sustainability*, Applied Innovation and Technology Management,
https://doi.org/10.1007/978-3-031-38387-8_18

18.1 Introduction

Nowadays, the pace of development of communities and the dependent sectors are very rapid while traditional energy resources are limited and, in the long run, cannot meet all demands. Also, the exploitation of these resources causes environmental pollution. The US Department of Energy [1] report shows that about 26% of total US energy in the year 2020 is consumed by the transportation system; it has a significant share in the production of greenhouse gas (GHG) emissions. Reducing fuel consumption and the GHG are the most critical reasons for the automotive industry to pay special attention to HEVs. Another reason is, unlike EV, HEVs don't have limitations over long distances [2]. Energy management system (EMS) in hybrid electric vehicles pursues several goals. The main objective of an EMS is reducing fuel consumption, which also makes HEV more attractive in the market compared to conventional vehicles. To reduce fuel consumption, ICE should operate in the most efficient region. Another main concern of the EMS is the battery unit's lifetime and state of health (SOH). The battery must operate properly by avoiding the deep charge/discharge to address this issue.

With the advent of the HEV, these cars' energy management and control issue has received much attention from researchers, and new strategies are presented every day by introducing new technologies. For instance, in Ref. [3], four different control strategies are analyzed: namely, Thermostat, Fuzzy logic, Adaptive Equivalent Consumption Minimization Strategy (A-ECMS), and Q-learning considering battery aging. A parallel hybrid electric vehicle is compared concerning battery aging and fuel economy. While dynamic programming offers the optimum global performance in fuel economy and emissions, it is computationally expensive and prior knowledge requirement of the driving cycle, making them less suitable for online implementation in the vehicle. For online implementation, strategies such as rule-based control, thermostat, equivalent consumption minimization strategy (ECMS) are more ideal as they are computationally cost-effective and do not require future information of the driving cycles. The ECMS, which is a real-time realization of the Pontryagin's Minimum Principle (PMP), involves the calibration of the equivalence factor (an approximation of the co-state in PMP) in the form of piecewise constant function [4, 5]. Traditional optimal control techniques can be used only with mathematical models of the system, assuming perfect knowledge of the entire optimization horizon (time frame over which the optimization is defined); since both these conditions are usually not respected by real systems. Optimal control implementation in a dynamic physical system, whose future is unknown, is necessarily *suboptimal* [6].

HEVs are classified into three categories based on powertrain architecture: series, parallel, and series-parallel. In this work, we will concentrate on power-split HEV. Power-split is a type of series-parallel powertrain architecture that includes generator, motor, and engine, connected by planetary gear, as shown in Fig. 18.1. Power-split has great potential in improving fuel economy by taking the advantage of both series and parallel hybrid configuration. According to the driving condition, the

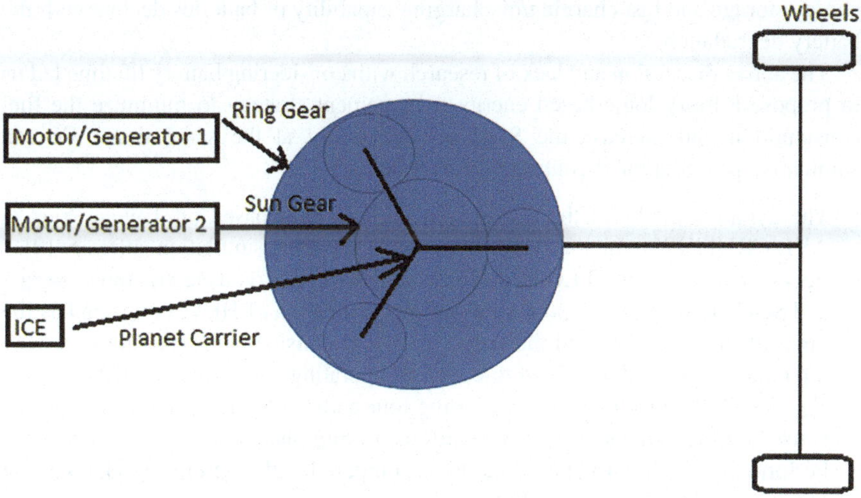

Fig. 18.1 Toyota Prius simulation in AVL cruise which generator, motor, and engine connected with planetary gear

mechanism simultaneously switches between two architectures that provides more freedom and enables the components to allow the components to operate in the most efficient region to attain the required torque.

Staying in this region requires an EMS, which controls the power flow between components and SOC of the battery unit by tracking engine speed, wheel-torques according to different driving situations. EMS keeps the SOC of the battery in the predefined safe operating interval, prevents the battery from overcharging/ discharging, and improves SOH and battery lifetime. In addition, proper managing SOC creates coordination between motor and generator and makes fuel consumption more balanced [3–5]. The EMS's controlling strategy (CS) is divided into two categories: optimization-based and rule-based control. Optimization-based controlling process is generally characterized by an objective function to achieve specific purposes such as maximizing HEV performance in various driving cycles [6] or minimizing fuel consumption and exhaust emission [7], and the dynamic programming (DP) method is used to find the global solution. But DP is not practical because it needs prior knowledge of driving conditions while the whole route is unknown before starting. Also, due to the complexity of hybrid drivetrain and power-split configuration and the synergic operation of the components in a power-split HEV, it is difficult to construct an accurate mathematical model [8]. Rule-based strategy is more intuitionistic than mathematical model and has received more attention due to the simple structure, and naïve control algorithm. So, a rule-based control strategy based on fuzzy logic is used in this study. Since the most critical factors of EMS for HEVs are fuel consumption and battery life, the main objectives are determined to reduce fuel consumption and battery degradation. However, despite in simplicity, inputs and outputs and control parameters should be appropriately selected [2]. The energy storage and fast charging/discharging capability of batteries decline with the battery degradation.

The above discussion and lack of research with considering battery lifetime led us to propose a fuzzy logic-based energy management strategy to minimize the fuel consumption and increase the SOH of the battery at the same time. We can summarize procedure and contribution as follows:

- The dynamic model of the power-split HEV is built based on Toyota Prius in AVL CRUISE™, which is a flexible vehicle driveline simulation software [9].
- Fuzzy logic controller (FLC) is designed in MATLAB® and electric motor torque and SOC are two critical points for optimal utilization of HEV, is selected as the input of the controller and the speed of ICE is considered as the output. Fuzzy inference rules are determined to cover all operating conditions of HEV.
- The WLTP [10] driving cycle, including four parts with different average speeds, is used during simulation to cover various driving phases.
- Performance of the FLC-based model is compared with reference model without fuzzy controller for different SOCs.

The rest of the chapter is organized as follow: Sect. 18.2 explains the power-split hybrid configuration and its energy management, Sect. 18.3 describes FLC and propose structure in detail; Sect. 18.4 describes the FLC implementation in AVL CRUISE™; Sect. 18.5 represents the simulation results and comparisons; Sect. 18.6 gives conclusion and makes remarks (Fig. 18.2).

18.2 Energy Management and Control Strategy in Power-Split HEV Configuration

The drivetrain of power-split HEV (Fig. 18.1) is a combination of series and parallel HEV configurations, and it has advantages of both the series and parallel drivetrains. Power-split transmissions provide higher flexibility to operate the electric motor and the ICE to fulfill the request torque for vehicle propulsion. This flexibility allows opportunity to operate the ICE in high-efficient region [11]. In this structure, Motor/Generator 1 is the primary source for traction and is connected to the ring gear; Motor/Generator 2 (Generator) is connected to sun gear, and ICE is tied to the planet carrier. There is no mechanical coupling between ICE and driven wheels. ICE is mainly used to provide electrical charge via Motor/Generator 2 [7]. For this study, we assumed that Motor/Generator 1 is connected to driven wheels, and the electric motor mainly provides torque demand. ICE is operated according to the SOC of the battery unit to charge the batteries and assist to the propulsion of the vehicle. In the following section structure of the FLC, fuzzification, and the defuzzification of the inputs and outputs will be explained.

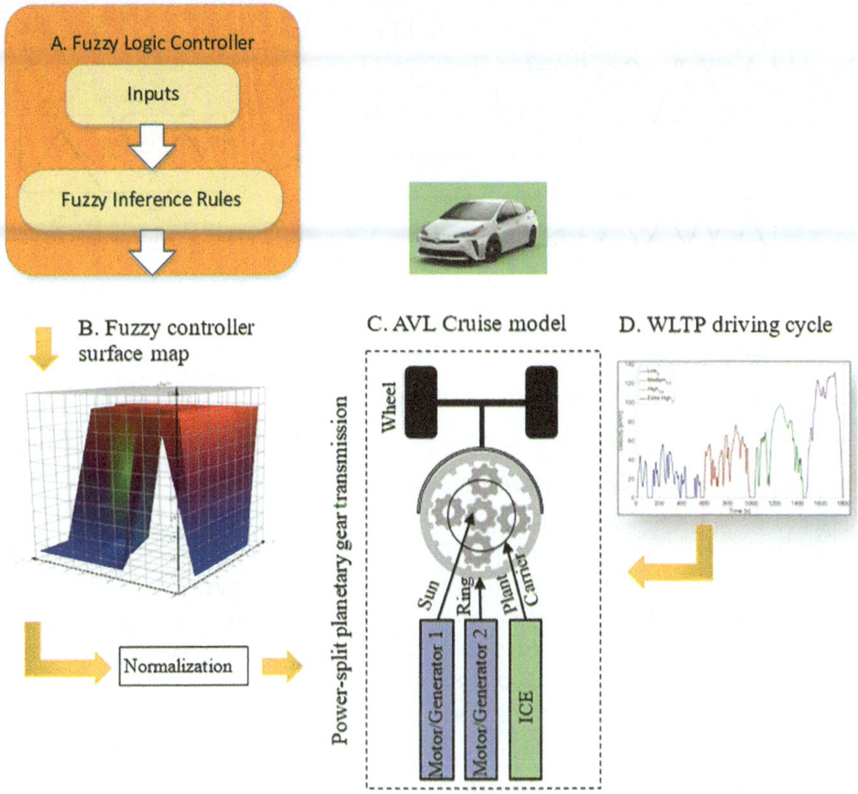

Fig. 18.2 Toyota Prius simulation in AVL cruise, WLTP driving cycle, and FLC which includes the inputs, fuzzy inference rules

18.3 Fuzzy Controller Design for Energy Management

Power flow between different components of the HEV makes it challenging to establish an accurate model as well as it complicates the energy management of a power-split HEV. On the other hand, the Fuzzy controller does not require an accurate mathematical model of the system that makes it a suitable approach for proper energy management of HEV. However, to achieve the desired goal, it is necessary to know the system performance, and inputs and outputs of the system must be selected precisely. Therefore, the Fuzzy controller is designed with two inputs and one output, considering that ICE has low fuel efficiency at the low-speed region. Thus, ICE is primarily operated between medium and high-speed regions to avoid high fuel consumption, and fuzzy rules are constructed in a way that covers all possible driving scenarios. So, we used MATLAB fuzzy logic toolbox to design a fuzzy logic controller that consists of 22 inference rules. Mamdani is chosen as an inference system.

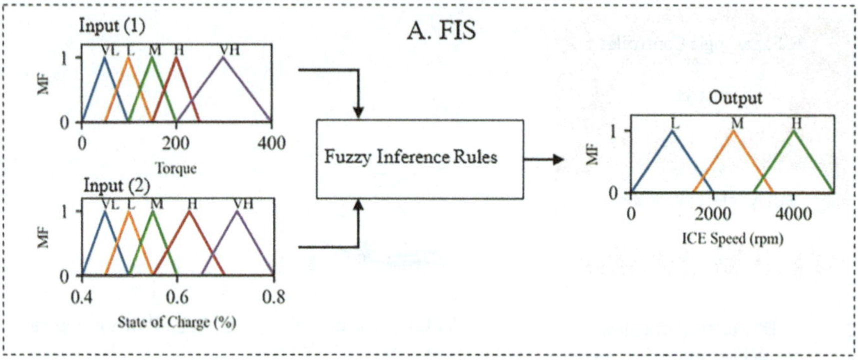

Fig. 18.3 Triangular membership function of the inputs and outputs

Table 18.1 Inputs and output membership functions

Input 1		Input 2		Output	
Torque (Nm)	MF	SOC (%)	MF	ICE (rpm)	MF
0–50–100	VL	40–45–50	VL	0–1000–2000	L
50–100–150	L	45–50–55	L	1500–2500–3500	M
100–150–200	M	50–55–60	M	3000–4000–5000	H
150–200–250	H	55–62.5–70	H		
200–300–400	VH	65–72.5–80	VH		

18.3.1 *Fuzzification of Inputs*

For fuzzification of the inputs, triangular membership functions (Fig. 18.3) are selected. Range of membership functions are defined by considering operating conditions of electric motor and battery. Electric motor torque is regarded as 0–400 Nm. For the SOC, the operation range is limited to 40–80% considering battery state-of-health since SOC over 80% results in high temperature in batteries and affects SOH negatively; SOC less than 40% also has adverse effects and causes high charging and discharging rates. These two inputs are divided into five fuzzy sets, namely, very low (VL), low (L), medium (M), high (H), and very high (VH). Membership function values of torque and SOC as Input 1 and Input 2 are given in Table 18.1.

18.3.2 *Fuzzy Inference System*

Mamdani inference system is used in the fuzzy controller. Twenty-two inference rules are determined based on the different operating conditions and design objectives. Fuzzy rules are given in Table 18.2.

Table 18.2 Fuzzy inference rules

#	Input1	Input2	Output	#	Input1	Input2	Output	#	Input1	Input2	Output
1	VL	VL	H	9	H	H	M	17	VH	VH	H
2	VL	L	H	10	H	VL	H	18	VL	M	M
3	L	VL	H	11	H	L	H	19	L	M	M
4	L	L	H	12	H	VH	M	20	L	H	M
5	M	L	H	13	VH	VL	H	21	VL	H	L
6	M	M	H	14	VH	M	H	22	VL	VH	L
7	M	H	M	15	VH	H	H				
8	H	M	M	16	VH	L	H				

18.3.3 Defuzzification of Output

The centroid method is used for defuzzification of output. Output of the fuzzy controller is determined as ICE speed. Membership functions of the output are determined by considering fuel consumption map of the ICE. The output ICE are given in Table 18.2 and presented in Fig. 18.1 as an output.

Defuzzified output is normalized between 0 and 1 by using Eq. 18.1 and forwarded to AVL Cruise. The output surface of the fuzzy inference system is depicted by the surface plot diagram as shown in Fig. 18.2.

18.4 Implementation of Fuzzy Controller in HEV Model Using AVL CRUISE™

Power-split drive train is modeled in AVL Cruise™, and output surface map of Fuzzy Logic Toolbox is implemented to this model. The AVL Cruise model connects ICE, generator, and electric motor to planetary gear. Part of ICE power is used to charge the battery pack, and the rest of ICE power is used for propulsion. AVL Cruise model is given in Fig. 18.2.

$$\text{Input}_{\text{ICE}} = \frac{n_{\text{ICE}} - n_{\min}}{n_{\max} - n_{\min}} \tag{18.1}$$

In the following section, the simulation results with different driving cycles and different initial state of charge will be presented and discussed.

18.5 Simulation Results and Discussion

After the fuzzy logic controller is connected to the AVL Cruise model, the simulation runs by WLTP. Since the battery is not constantly fully charged, two different scenarios are defined; first, we assumed that the battery is almost fully charged by 80% of its nominal capacity and starts the duty; then we changed the initial SOC to 45% and run the simulation. This make the simulation more realistic and the results more reliable because the controller performance measured under different initial conditions. In addition, the same scenarios defined for the typical case in which model does not have the controller and the case that FLC is added to the structure. So, we can compare the two models and present how much FLC improves the performance of the model and how much fuel consumption is reduced for the different initial conditions in driving cycle. We can summarize the scenarios as follows:

18.5.1 Eighty Percent Initial SOC Without Fuzzy Logic Controller (Case A)

The initial SOC is set to 80% in this scenario, and FLC is not considered. The results show that fuel consumption is higher in this case. As Fig. 18.4b shows, the engine operated with more frequent start-stop because the engine is in low-frequency region and engine fuel efficiency is low in this region; as a result, fuel consumption increased. SOC graph (Fig. 18.4c) also shows higher charge/discharge fluctuation; hence more power of the vehicle is generated by battery unit; thus, the depth of discharge is increasing, which leads to battery degradation and reduce battery life. Cumulative mass and cumulative volumetric fuel consumptions are 1.3 kg and 1.6 L, respectively, and SOC at the end of the simulation cycle is about 48%, as presented in Table 18.3.

18.5.2 Eighty Percent Initial SOC with Fuzzy Logic Controller (Case B)

Same as Sect. 18.5.1. the SOC level is set to a maximum permissible value, and the vehicle is simulated in the WLTP drive cycle considering the FLC is connected. As shown in Fig. 18.4b, the engine works more steadily, and the engine start-stop is lower than the case without a controller. Also, the speed curve (Fig. 18.4b) shows lower fluctuations which certifies the engine operates in a more steady state. In Fig. 18.4c, except the first part of the cycle that sharply decreases in the SOC level, there is no apparent deep charge/discharge and fluctuations are within a specific range and depth of discharges are about 1%. The reason for this sharp discharging is

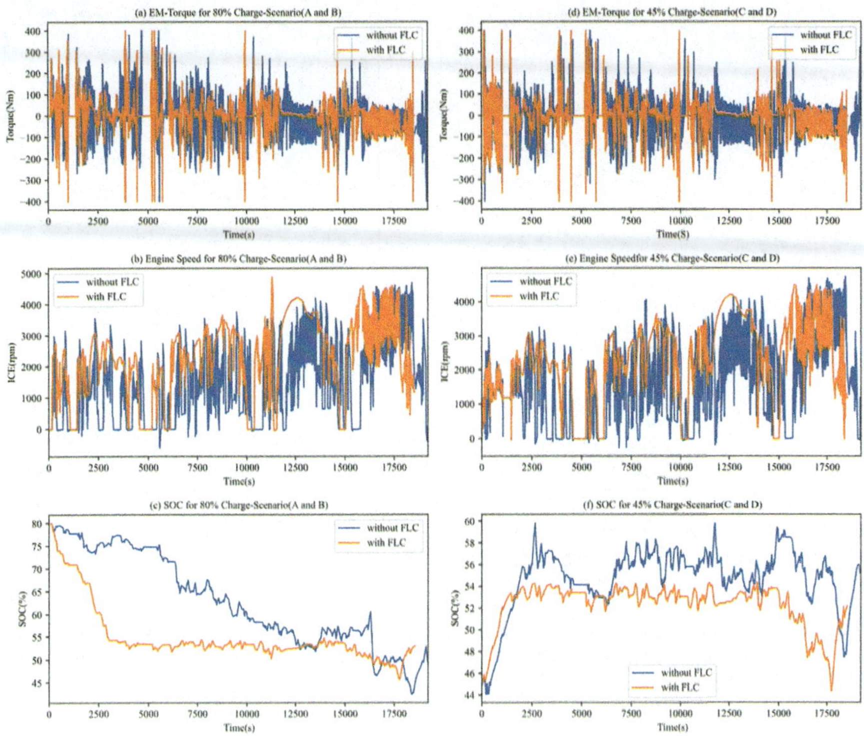

Fig. 18.4 Simulation result at the end of the simulation; left shows the result of torque, ICE speed, and SOC for 80% initial SOC, respectively; right shows the same for 45% initial SOC

Table 18.3 Simulation results at the end of the driving cycle

Case	Mass fuel consumption (Kg)	Volumetric fuel consumption (Liter)	SOC at the end of the cycle (%)
Case A (without FLC)	1.3	1.6	48
Case B (with FLC)	1	1.3	53
Case C (without FLC)	1.4	1.8	56
Case D (with FLC)	1.1	1.5	52

because of the design and the priority that is intended to reduce fuel consumption over battery lifetime.

As Fig. 18.5 shows, at the end of the simulation, fuel consumption, both in terms of mass and volume, reduced significantly, and it is much less than the case without a controller. According to the Table 18.3., cumulative fuel consumption in terms of

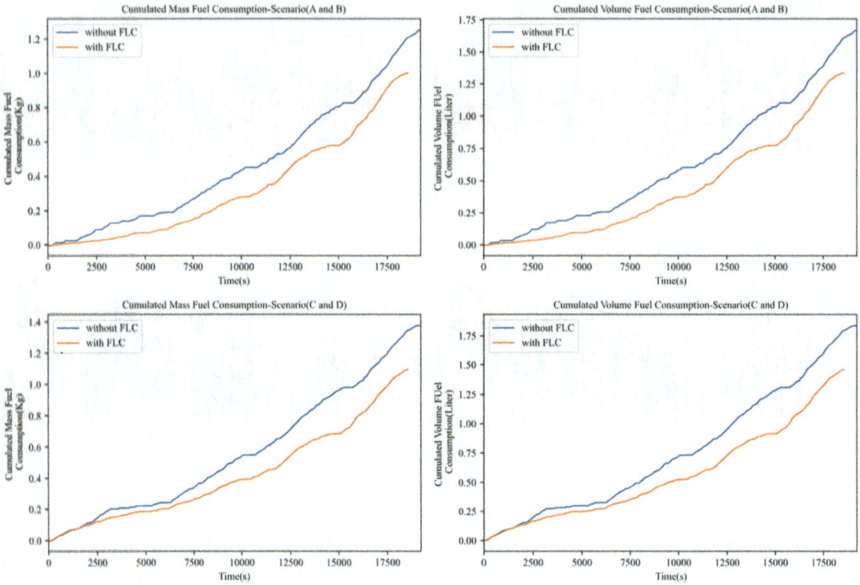

Fig. 18.5 Simulation result for fuel at the end of the simulation; left shows the result of cumulative mass fuel consumption for 80% and 45% initial SOC, respectively; right shows the cumulative volumetric fuel consumption for 80% and 45% initial SOC, respectively

mass reduced from 1.3 kg to 1 kg and in terms of volume decreased from 1.6 l to 1.3 l which shows decrease about 23% and 19%, respectively.

Case A and B can be considered best cases where the battery is almost fully charged, so the energy required to produce high torque is available, in which case the actual performance may not always be maintained and the amount of fuel consumption. As a result, the simulation results will also not be valid for real cases. Therefore, in the following, Case C and D are defined for the situation where the battery is not fully charged.

18.5.3 Forty-Five Percent Initial SOC Without Fuzzy Logic Controller (Case C)

The initial SOC is set to 45% for case C, and FLC is not considered in the structure. Figure 18.4d, e shows that engine operates at high speed with more frequent start-stops, which results in increased fuel consumption; also, generated torque shows rapid fluctuations. In general, the intensity and fluctuation pattern of case C is similar to Case A. Depth of charge/discharge is high because, at the first cycle, most of the power generated by the engine is used to charge the battery; also the depth of charge of the battery is about 36%, which is a risk for battery lifetime. During the other parts of the cycle, the depth of charge/discharges is significantly high, which shows high and frequent ICE operation. This transient behavior is also not desired and reduces the battery lifetime.

18.5.4 Forty-Five Percent Initial SOC With Fuzzy Logic Controller (Case D)

For this case, the initial SOC is set to 45%, and FLC is considered in the structure. The results are similar to case B, fuel consumption is reduced; for the first cycle, depth of charge is about 17% that is lower than case C (Fig. 18.4 (f)); in the following, the battery is charged/discharged more steadily without any noticeable deep charge/discharge. Eventually, because of the FLC, fuel consumption in terms of mass and volume are decreased from 1.4 kg to 1.1 kg and 1.8 l to 1.5 l, respectively, which is equivalent to 21% and 17% reduction at the end of the driving cycle.

18.6 Conclusions

In this chapter, a rule-based energy management control strategy based on FLC is developed for power-split HEV. Both ICE and battery SOC are considered multi-controlling factors that affect fuel consumption and thus greatly contribute to increase environmental pollution. Rules are defined to cover all operating conditions to improve the efficiency of ICE, fuel economy, and the lifetime of the battery by controlling the depth of charge/discharge. The proposed rule-based strategy is compared with the case without FLC to clarify the model's effectiveness. Also, different initial SOC of batteries are added to the simulation process to investigate the model's adaptability with the real driving condition. Results indicate that the proposed strategy successfully reduces the fuel consumption by about 10–20% with respect to the initial SOC compared to the model without a controller. Energy transfer is also done in such a way that charge/discharge is more stable which reduces the depth of charge/discharge so helps to increase battery lifetime. Therefore, the improvement of the energy management strategy will lead to optimal and efficient use of HEV on the market, which will prevent the waste of resources. In addition, fuel consumption will be reduced, which improves the communities' economy and creates a healthy environment for inhabitants by reducing environmental pollution.

References

1. Homepage – U.S. Energy Information Administration (EIA).
2. Wu, J., Ruan, J., Zhang, N., & Walker, P. D. (May 2019). An optimized real-time energy management strategy for the power-split hybrid electric vehicles. *IEEE Transactions on Control Systems Technology, 27*(3), 1194–1202. https://doi.org/10.1109/TCST.2018.2796551
3. Syed, F. U., Ying, H., Kuang, M., Okubo, S., & Smith, M. (2006). Rule-based fuzzy gain-scheduling PI controller to improve engine speed and power behavior in a power-split hybrid electric vehicle. *Annual Conference of the North American Fuzzy Information Processing Society – NAFIPS*, 284–289. https://doi.org/10.1109/NAFIPS.2006.365423
4. Zhang, J., Shen, T., & Kako, J. (2020). Short-term optimal energy management of power-split hybrid electric vehicles under velocity tracking control. *IEEE Transactions on Vehicular Technology, 69*(1), 182–193. https://doi.org/10.1109/TVT.2019.2950042
5. Montazeri-Gh, M., & Mahmoodi-k, M. (2015). Development a new power management strategy for power split hybrid electric vehicles. *Transportation Research Part D: Transport and Environment, 37*, 79–96. https://doi.org/10.1016/J.TRD.2015.04.024
6. Wirasingha, S. G., & Emadi, A. (2011). Classification and review of control strategies for plug-in hybrid electric vehicles. *IEEE Transactions on Vehicular Technology, 60*(1), 111–122. https://doi.org/10.1109/TVT.2010.2090178
7. Gökce, K., & Ozdemir, A. (May 2014). An instantaneous optimization strategy based on efficiency maps for internal combustion engine/battery hybrid vehicles. *Energy Conversion and Management, 81*, 255–269. https://doi.org/10.1016/J.ENCONMAN.2014.02.034
8. Li, P., Jiao, X., & Li, Y. (Feb. 2021). Adaptive real-time energy management control strategy based on fuzzy inference system for plug-in hybrid electric vehicles. *Control Engineering Practice, 107*, 104703. https://doi.org/10.1016/J.CONENGPRAC.2020.104703
9. AVL CRUISE™ – avl.com
10. WLTPfacts.eu. Worldwide harmonised light vehicle test procedure.
11. Navale, V., & Havens, T. C. (2014). Fuzzy logic controller for energy management of power split hybrid electrical vehicle transmission. *IEEE International Conference on Fuzzy Systems*, 940–947. https://doi.org/10.1109/FUZZ-IEEE.2014.6891785

Correction to: The Power of Combination Models in Energy Demand Forecasting

Erkan Isikli and Seyda Serdarasan

Correction to:
Chapter 9 in: G. Kayakutlu, M. Ö. Kayalica (eds.), *Decision Making Using AI in Energy and Sustainability,* Applied Innovation and Technology Management,
https://doi.org/10.1007/978-3-031-38387-8_9

The book was published with one of the author's name as Seyda Serdar Asan. It should be Seyda Serdarasan.

The author's name has now been corrected as below:

Given name: Seyda
Family name: Serdarasan

The updated original version of this chapter can be found at
https://doi.org/10.1007/978-3-031-38387-8_9

G. Kayakutlu, M. Ö. Kayalica (eds.), *Decision Making Using AI in Energy and Sustainability*, Applied Innovation and Technology Management,
https://doi.org/10.1007/978-3-031-38387-8_19

Correction to: Single-Valued Neutrosophic CRITIC-Based ARAS Method for the Assessment of Sustainable Circular Supplier Selection

Arunodaya Raj Mishra, Raghunathan Krishankumar, Fausto Cavallaro, Ram Kishun Lodhi, and K. S. Ravichandran

Correction to:
Chapter 3 in: G. Kayakutlu, M. Ö. Kayalica (eds.), *Decision Making Using AI in Energy and Sustainability*, Applied Innovation and Technology Management,
https://doi.org/10.1007/978-3-031-38387-8_3

The original version of this chapter was published with incorrect affiliation. The actual affiliation of the author 'Raghunathan Krishankumar' was erroneously added to the affiliation of the author 'K.S. Ravichandran' (Amrita Vishwa Vidyapeetham). The chapter has now been corrected.

The correct affiliation is:

Raghunathan Krishankumar
Information Technology Systems and Analytics Area,
Indian Institute of Management Bodh Gaya
Bodh Gaya, Bihar, India

K. S. Ravichandran
Department of Mathematics, Amrita School of Physical Sciences, Amrita Vishwa Vidyapeetham, Coimbatore, Tamil Nadu, India

---

The updated version of this chapter can be found at
https://doi.org/10.1007/978-3-031-38387-8_3

G. Kayakutlu, M. Ö. Kayalica (eds.), *Decision Making Using AI in Energy and Sustainability*, Applied Innovation and Technology Management,
https://doi.org/10.1007/978-3-031-38387-8_20

The manufacturer's authorised representative in the EU is Springer Nature Customer Service Centre GmbH, Europaplatz 3, 69115 Heidelberg, Germany. If you have any concerns regarding our products, please contact ProductSafety@springernature.com

Printed and bound by CPI Group (UK) Ltd, Croydon, CR0 4YY
28/09/2026
02224075-0013